G. Drysdale (George Drysdale) Dempsey, George Drysdale Dempsey

Rudimentary Treatise on the Drainage of Bowns and Buildings

G. Drysdale (George Drysdale) Dempsey, George Drysdale Dempsey

Rudimentary Treatise on the Drainage of Bowns and Buildings

ISBN/EAN: 9783744675437

Printed in Europe, USA, Canada, Australia, Japan

Cover: Foto ©berggeist007 / pixelio.de

More available books at **www.hansebooks.com**

RUDIMENTARY TREATISE

ON THE

DRAINAGE

OF

TOWNS AND BUILDINGS:

SUGGESTIVE OF

SANATORY REGULATIONS CONDUCIVE TO THE HEALTH
OF AN INCREASING POPULATION.

BY G. DRYSDALE DEMPSEY, C.E.,

Author of " The Practical Railway Engineer," and of the " Rudimentary Treatise
on the Drainage of Districts and Lands."

REVISED AND GREATLY EXTENDED: WITH NOTICES OF THE METROPOLITAN
DRAINAGE, THAMES EMBANKMENT, AND WATER SUPPLY SCHEMES.

THIRD  EDITION.

LONDON :
VIRTUE BROTHERS & CO., 1, AMEN CORNER,
PATERNOSTER ROW,
1865.

PREFACE.

Two volumes of the *Rudimentary Series*—" The Art of Draining Districts and Lands," and the work now submitted to the reader—relate to the subject of drainage generally, with water supply as an auxiliary or necessary contingent : the removal of surplus waters and refuse on the one hand, and the supply of pure water on the other. The one volume applies to *Districts* and *Lands ;* the other to *Towns* and *Buildings.*

It becomes necessary to remark in this (the *Third*) Edition of the present work, that when the late Mr. Dempsey prepared the *First* and *Second* Editions, the whole subject of the *Drainage of the Metropolis* was in confusion. The most eminent engineers were in conflict as to the best mode of attaining the desired end. The plan eventually adopted, and now (1865) in progress, differs from that which Mr. Dempsey and many other engineers recommended. This, of course, is not conclusive as to the relative merits of the different schemes. Who were right and who were wrong, in these speculations, we shall not know for many years to come ; until the Intercepting Main Drainage plan sha'l have had a fair trial by long-continued working. On this account it seems desirable to leave Mr. Dempsey's calculations and deductions in their original form ; because they embody, or rather illustrate, one particular principle of Drainage, which may probably apply to many other large towns, irrespective of its adoption or rejection in the metropolis. By greatly enlarging the APPENDIX, space has been found for a succinct account of all that has been done in relation to the Main Drainage Scheme since the publication of the Second Edition of this volume, in 1854.

Plain facts are stated, without any prediction concerning the degree of success that may attend the operations now in progress. In a later portion of the APPENDIX, relating to the *Utilisation of Sewage*, it will be seen that this important question remains nearly in the same unsettled state as when Mr. Dempsey prepared the Second Edition. Much has been said, and much written; but the world has yet to learn whether the sewage of the great metropolis is to be rendered available as an agricultural fertiliser. We may here refer, for fuller information on this important subject, to another volume of this Series (Vol. 146), Mr. Robert Scott Burns' Rudimentary Treatise for Students of Agriculture, called, "Outlines of Modern Farming;" the fifth volume of those Outlines relates to Utilisation of Town Sewage, Irrigation, and the Reclamation of Waste Lands. The *Embankment of the Thames* being now recognised as an important feature in the Main Drainage Scheme, we have deemed it useful to devote one portion of the APPENDIX to this subject.

Another volume of the *Rudimentary Series* * treats generally of Water Works, and the Supply of Water to Towns. It has been considered only necessary here, therefore, to notice briefly in the APPENDIX one or two advances which have been made between 1854 and 1865, in improving the systems sketched in the text by Mr. Dempsey, especially in regard to London and Glasgow.

LONDON, 1865.

* "A Treatise on Waterworks for the Supply of Cities and Towns; with a Description of the principal Geological Formations of England, as influencing the Supply of Water; details of Engines and Pumping Machines for Raising Water; and description of Works which have been executed for Procuring Water from Wells, Springs, Rivers, and Drainage Areas."—By Samuel Hughes, F.G.S., Civil Engineer. Vol. 82***

CONTENTS.

DIVISION III.

DRAINAGE OF BUILDINGS.

APPENDICES.

DRAINAGE.

DIVISION II.

SECTION I.

Classification of Towns according to Position and Extent.—Varieties of Surface, Levels and Inclinations.—Application of Sewage Manure.—Metropolitan Sewage Manure Company.—Methods of treating Sewage.—Magnitude of London Sewers.—The Fleet Sewer.—Metropolitan Commission of Sewers.—The Tunnel Scheme.—Great London Drainage Bill.—Messrs. Stephenson and Cubitt's Evidence.—General Board of Health.

194. ACCORDING with our definitions (Part I. p. 1), we propose to treat of the *supply* of *water* to *towns* and *buildings* as a branch of the general subject of *Drainage*, since the purposes of the art cannot be effected without an adequate and regulated supply of water by a combination of natural and artificial agencies, the extended control over which constitutes the purpose of water-supply for all highway, manufacturing, and domestic uses.

195. The means of obtaining water for towns, and of conducting the drainage matters from them vary, mainly, according to their position with reference to the sources of water; and, in a subordinate degree, according to their superficial extent. The sources being those already enumerated in our First Part, viz. rivers, rains, and springs, the command of one or more of these will be presented as the most economical means of deriving the necessary supply for each town under consideration. Towns situated on the banks of tidal rivers, or in near proximity to them, may be

B

usually sufficiently supplied from these sources, unless some
parts of the district extend upward to such elevation above
the river-level that the raising of this supply requires expensive artificial power; in which case springs at higher
levels may be advisably resorted to, or the drainage waters
from superior lands may be so conducted as to assist the
supply.   Towns which are far distant from rivers are commonly entirely dependent upon springs or drainage waters
for their artificial supply.

196. The refuse matters to be discharged from towns and
buildings,—consisting of the disintegrated materials of
street paving and roads; of superfluous rain water; of excrementitious matters, solid and liquid; of the waste products of combustion; of the refuse of animal and vegetable
substances; besides the various waste matters used in manufactures,—require arrangements of different kinds to be
provided with regard to the purposes to which these matters may be usefully applied.   For such discharges of these
matters as are to take place through subterranean channels,
one principle is, however, common to all, viz. that the receptacle to which they are conducted must be situated at a
level somewhat lower than that from which they are forwarded.   The arrangements for this purpose will, therefore,
be varied according to the nature of the site of the town.
If this be low in relation to the surrounding country, and
level, the refuse may be indifferently collected within or
without the town, with, however, the advantage in the latter
plan of avoiding such exposure of the decomposed matters
as tends to pollute the atmosphere, and at the same time
saving distance in the transfer of such portions of those
matters as are destined for agricultural uses.   If the site of
the town be a valley with lower ground in the midst of it
than is found anywhere without its limits, the readiest point
of collection will be the lowest level in the town itself at
which the drainage can be united, and artificial power will
be required to distribute such matters as are intended for
agricultural purposes around the higher ground outside.

From towns which occupy elevated sites, having lower lands around them, the refuse matters and drainage waters should be conducted away at once; or, if found necessary to collect them, a point or points should be selected for this purpose altogether beyond the limits of the town itself.

197. In the several cases here supposed, the question of disposing of the refuse matters should be treated without any reference whatever to the presence of a river through or contiguous to the town, except upon the single consideration that such river, being in all probability situated at the lowest level of the site, may afford facilities, after the refuse has been collected in reservoirs near its banks, for its conveyance in suitable barges or vessels towards the higher lands for which some portion of this refuse is ultimately destined. Former practice in the art of town-draining has indeed regarded the one question of river or no river, as the grand determinal one for the disposal of drainage and refuse matters. How to get rid of the animal ordure created within the walls of a town, was formerly deemed to be satisfactorily answered provided a river flowed beneath, and offered a tide to wash away in boundless wastefulness those matters which, properly applied, will endow barren lands with the richest fertility.

198. Although reluctant to dwell upon the trite subject of the *importance* of draining, we claim attention to this great leading principle in the drainage of towns and buildings, viz. that the ultimate economy of the art comprehends two distinct purposes, whereof the second—the disposal and utility of the refuse matters—is little less in importance than the first—the discharge of these matters from the dwellings and highways of men. And the accomplishment of this second purpose involves the beneficial appropriation of refuse matters so as to make them actually productive, and avoid interference with those healthy uses of inland waters for which they are properly adapted. In illustration of this principle we will endeavour to estimate the value for agricultural purposes of the excrementitious matters

flowing from a town, from which estimate the pernicious effects of discharging those matters into the courses whence the supply of water is derived for the several uses of the population may be readily inferred.

199. The value of manures as promoters of vegetation is known to result from their possession of the essential element, nitrogen, in the form of ammonia, with the subordinate properties of alkalies, phosphates, and sulphates. Now, the experiments of Boussingault and Liebig have furnished us with the means of estimating the quantity of nitrogen contained in the excrements of a man during one year, at 16·41 lbs., upon probable data, and also that this quantity is sufficient for the supply of 800 lbs. of wheat, rye, or oats, or of 900 lbs. of barley. "This is much more than it is necessary to add to an acre of land, in order to obtain, with the assistance of the nitrogen absorbed from the atmosphere, the richest crops every year. By adopting a system of rotation of crops, every town and farm might thus supply itself with the manure which, besides containing the most nitrogen, contains also the most phosphates. By using, at the same time, bones and the lixiviated ashes of wood, animal excrements might be completely dispensed with on many kinds of soil. When human excrements are treated in a proper manner, so as to remove this moisture, without permitting the escape of ammonia, they may be put into such a form as will allow them to be transported even to great distances."* Making reasonable allowance for the reduced quantity produced by children, we shall be safe in assuming the nitrogen thus resulting from any amount of population to be equal to the supply required for affording 2 lbs. of bread per diem for every one of its members! Or assuming an average of 600 lbs. of wheat to be manured by each individual of the population of London; and taking this at two millions, for a rough calculation, the manure thus produced is sufficient to supply the growth of wheat of a total weight of 1200 mil-

* Liebig.

lions of pounds, or 535,714 tons. The total manuring matters, solid and liquid, produced in a town, allowing for those which are produced in manufactories and sewage water, are probably equal in weight to one ton annually for each member of the population, or two millions of tons produced in the metropolis.

200. That this vast quantity of manure should be made available for agricultural production is a principle which cannot be denied, and which is properly limitable only by the consideration of expense as weighed against the value of the results. The expense will be made up mainly of three items, viz. of the *collection*, of the *raising*, and of the *distribution* of the refuse matters. The collection being an item common to all methods of disposal, will not be chargeable entire in any comparative estimate, but as modified by the peculiarities in the collection of which the plan is susceptible. The cost of raising is of course wholly chargeable to a system of artificial dispersion, as distinguished from the prevailing modes of self-discharge into low channels, but the former system will be debited only with the excess of expense (if any), beyond that incurred by the present methods of distributing the manuring matters for use upon the land. The cost of each of these works, however, may be reduced to a minimum by skilful arrangements, and our experience is yet insufficient to enable us to determine these with that precision which further practice will secure, or to estimate their total with the exactness necessary for forming a just comparison between the present and the proposed methods.

201. In a subsequent part of our work we propose to consider the items of cost in carrying out an efficient system of town-drainage ; being satisfied, at this stage of our subject, in declaring the fundamental principle that the refuse of a town, including not only excrementitious, but all other waste matters and sewage, is far too valuable to be thrown away ; and that the question of its appropriation should be made dependent only upon rules of a liberal

economy, which ought, moreover, to be severely criticised before admitted to practical consideration.

202. The palpable inference from this principle is, as already stated, that the contiguity and position of a river, with reference to a town, have no necessary connection with the arrangement of its drainage beyond the facilities which may be thus afforded for the passage or subsequent convey-ance of the sewage matters for their ultimate disposal. For it is quite certain that no correct general views of town-drainage can prevail while we continue to regard a river as the natural and suitable trunk sewer into which all colla-teral and main courses of brickwork are to discharge their fœtid contents, which, according to the state of the river, are either immediately spread upon its banks to contami-nate the air of the town, or duly infused in its waters, to be afterwards exposed with the same vicious effects.

203. From the principles here laid down, it will be un-derstood that in the twofold purposes of the drainage of towns, viz. the supply of water, and the discharge and dis-posal of the refuse matters, the relative levels of the town, with the adjacent districts, and of the several portions of the town with each other, are the main considerations upon which the peculiar methods to be adopted in each case are determinable; but it will also be evident that, generally, those surfaces which are the most favourable for an econo-mical water supply are the least so for the ready disposal of refuse matters, and the converse is equally true; those sur-faces which present facilities for dispersing drainage-matters being commonly the least accessible to water.

204. Thus the flat districts on the margins of rivers and inland streams of adequate capacity are the most favourable sites for towns for the supply of water, but for drainage they are the least so; since the main channels or sewers are required to be laid at low levels, and the raising of their contents for use upon the neighbouring lands, which are probably much higher, becomes a very expensive process. On the other hand, a town on a hill-top is the most readily

and cheaply drained ; but its supply with water, whether from springs, rivers, or surface drainage—all at lower levels —is a work of great and constant cost.

Let us consider the several kinds of site which towns may occupy.

205. *First.*—A plain or flat surface, with surrounding country of similar character. *Water* from rivers, springs, or from the surface of lands in the neighbourhood. Artificial power will be probably required to raise the water, however derived. The *drainage matters* must be conducted into one or more main sewers, and raised by artificial power for dispersion upon the land.

206. *Second.*—A plain or flat surface, with surrounding country rising from the town. Unless well situated with regard to a river, the supply of *water* will probably be the most economically obtained from springs on the hills, or from the collection of the waters which accumulate upon their surface. The *drainage matters*, if destined for the higher lands, should be generally conducted by mains towards the outskirts of the town, and the question of levels will evidently derive additional importance from the necessity of raising the sewage to levels naturally above that of the town itself.

207. *Third.*—A plain or flat surface, with surrounding country falling from the town. The supply of *water* beyond that derived from wells and springs will require artificial power, while the *drainage matters* may be collected in main sewers, and, in all probability, dispersed without any application of power, by the force of their own gravity.

208. *Fourth.*—An inclined surface on the side of a hill. *Water* will be derivable, probably, from several sources. If a river flow at the base of the site, the lower parts of the town will be most economically supplied from it, while for the higher the surface water from lands above or springs will be the most readily available. The general system of collecting and distributing the *drainage matters* will be chiefly dependent upon the localities where they are in-

tended to be ultimately disposed of. If these be on the lower part of the hill, the method will be very simple, requiring only that main sewers be laid at the base of the town, from which the sewage may be distributed without any application of artificial power. But if the disposal of the sewage be inevitably desired on the lands above the town, the site constitutes one of the least favourable for economical drainage, which will require a constant expenditure of artificial power.

209. A river-valley and a hill-top will evidently present a repetition or duplication of similar features to those here described, the only limitation in the resemblance being that, in the case of a town on the summit of a hill, the water supply will, most probably, be derivable only from lower sources by artificial power.

In these sketches the general superficial features of the site are of course only referred to. Intermediate undulations which may exist will affect the determination of the details of any arrangement of channels designed for serving the drainage of the town.

210. With reference to the artificial power which may be required for the supply of water, or the discharge of the drainage matters, if a tidal river can be commanded, it becomes a question of the highest importance whether this cannot be, and, if so, in what way, made available as a source of the power required. And another question deserving the most attentive consideration is, whether the ebb tide may not be rendered efficient in aiding the discharge of the sewage where the fall is inadequate to insure its self-discharge.

211. As a general principle in town drainage, however, it should be so arranged and conducted as to require no artificial supply of water. The surface water should always be made sufficient to carry away all refuse matters, solid as well as liquid. Two reasons exist for this : first, the economy of the water, which in many cases is a paramount, and in all should be a leading, consideration; and, secondly,

the dilution of the sewage with any unnecessary liquid involves more capacious arrangements for its diffusion, and in most instances an extravagant amount of power to raise it.

212. The utmost economy of water for draining purposes can be secured only when a sufficient inclination in the sewers can be obtained. The methods of making a fall the most effectual are, therefore, deserving of the most careful attention in every scheme for town-drainage. The application of the tidal waters for assisting the discharge of the sewage can consequently be entertained only with reference to the principal main sewers at the lowest level, and with such adaptation, if practicable, as will admit the subsequent separation of the proper sewage matters from the water thus introduced to aid their progress and discharge.

213. Although the rules here suggested should be kept in view as leading objects in all arrangements of town-drainage, they will in many cases be admissible in part only, owing to the reference to existing works which is imposed upon us. Thus, in all towns for the attempted drainage of which some arrangements or other have already been executed, our practical operations become doubly difficult, since we are constrained to endeavour to reconcile these with the improved details which correct principles induce us to prefer. By way of illustration, which will be found fully instructive, let us turn our attention to the works now in action upon and beneath the surface of our own metropolis, and consider how the principles here stated can be the best applied to improve the means of its drainage.

214. LONDON, standing upon a bed of clay, the substrata to which, successively, are plastic clay, chalk, and gault, occupies a part of the valley through which the river Thames has its course. The site of the town in some places rises gently from the river, and at others is below the level of high water, extending in dead flat districts.

B 3

For the principal part, however, the surface rises above the river, which therefore came to be regarded as the natural and proper channel for all the drainage of the town, the main sewers having been arranged to discharge their contents into it. Indeed, so thoroughly was this purpose of the river formerly recognised, that the Thames was familiarly termed the "*Grand Sewer of London.*"

215. Now, in order to make this method effective of only one of the true purposes of drainage, viz. the mere getting rid of the sewage matters,—it is evident that the arrangement must be such, that the whole of these matters are duly collected in the buildings and streets, and delivered into the sewers; also, that these are so constructed and situated that the matters they receive shall pass as rapidly as possible, and certainly without any interruption that would amount to stagnation, into the main sewers, and that these again faithfully and promptly convey the sewage into the final receiver, the river. So far, however, from being fulfilled in their entirety, no one of these conditions is fully and satisfactorily discharged. Thus, in many parts of the town, the refuse matters are collected in holes beneath the houses, and removed only when these holes become filled, and the surrounding soil permeated to supersaturation. Some districts have no sewers or drains of any description; and again, of the sewers which are constructed, very—very few, are formed with a rate of declivity sufficient for the self-discharge of the sewage, while many of them are laid *perfectly level.* Attempts are made in some districts to obviate the evils of insufficient declivity, by a flushing of water through the sewers, the water being, for this purpose, accumulated for a time, and then suddenly released, so as to produce the effects of a powerful current. Of these methods some details will be found in a subsequent part of this treatise; but they can be regarded only as palliatives, and expensive ones, applying, moreover, to one only of the many imperfections of the present system. The crowning defect, however, exists at the last stage of this machinery,

where the outfalls of the sewers into the river are so low, that their contents are delivered at, or a little above, low-water level. The decomposing matters are consequently delivered upon the banks of the river, and left there to stagnate and poison the atmosphere, and to be brought up with the next tide for the thorough pollution of the waters. This is an irremediable evil of the present arrangement, by which no adequate fall can be obtained for the sewers, consistent with their discharge into the river near the high-water level, the only position in which the sewage could be effectually conveyed away from the higher towards the lower districts. Into some of these sewers the water of the tide is permitted to enter, the immediate consequence of which necessarily is, that the discharge of the sewage is suspended, and the gases engendered by the decomposing matter within the sewers are driven back towards the town. The return of the tide of course assists the outflowing of the contents of the sewers to some small extent; but, notwithstanding this expedient for assisting the discharge, the sewers are found to require periodical cleansing by hand, the foul matters being raised to the surface in buckets, and conveyed away in carts.

216. Of the many details of imperfection which mark the existing combination of arrangements constituting the sewerage of the metropolis, all who have studied the subject are, to some extent, cognizant, and all are equally prepared to admit the magnitude of the several improvements which have been made within the last twenty years, and which tend to alleviate some of the most palpable evils of the prevailing system; but no thorough rectification can be effected until the correct principles of town-drainage are recurred to, and applied with such modifications as may enable us to make the best use of existing arrangements, without sacrificing objects of greater magnitude and importance.

217. All the real difficulties of the drainage of London have their origin in the great error of attempting to convert

the river Thames into the common receiving sewer. In the attempt to accomplish this improper object, the sewage is brought down from the high lands, distances of miles, and heights of many feet, away from the very points where it should have been collected, and would be at once available for agricultural purposes. In the same attempt it has been found necessary to construct the lower ends of these sewers of immense size, in order to contain the accumulated sewage with which they are thus loaded. As the buildings have been extended, or a necessity has arisen for accommodating lower levels, the main sewers have been industriously removed, and rebuilt below their former position, and their capacity enlarged, to provide for the increased quantity poured into them. In the same obstinate attempt to pollute the waters of the river, miles of sewers have been constructed without any declivity whatever, on an absolute level, in which, as a necessary consequence, the refuse matters accumulate and solidify, until some happy rush of surface-waters sends them onward to the common receptacle, from which the population has the privilege of afterwards supplying their personal and household wants. Now, let us banish the notion of turning our rivers into sewers, and consider how the sewerage could be best arranged, supposing it had to be done as an entirely new work, and that we were unfettered by any consideration of making present subterranean structures available for the purpose.

218. Without seeking records of the actual levels of any one of our river-watered towns, we may assume as a feature common to many of them the existence of several ranges of elevation, running parallel, or nearly so, to the direction of the river, which we will suppose to be generally east and west. These ranges of elevation will be interrupted at intervals by ancient water-courses, and also by small ridges running north and south. These several features of the natural surface will determine the most economical courses for the general system of drainage. Thus the highest of the ranges (which we will call A) should have a course of

sewers for the especial service of the districts above it; the
next range (B) should have another course of sewers to serve

Fig. 67.

SOUTH   NORTH

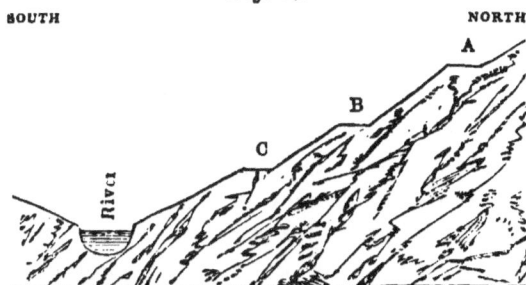

the district between A and B; the following range (c) should
similarly be provided with its course of sewers to drain the
district between B and c, and so on, as in the sketch, fig. 67.
The general inclination, east and west, of each of these
courses of sewers would be determined by the position
of the ridges and hollows running north and south, as
shown in fig. 68, where the highest points of the several

Fig. 68.

WEST   EAST

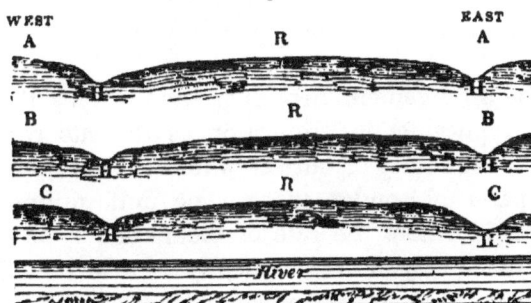

courses would be at R, and the lowest at H successively.
By this arrangement, means would be obtained of collecting
the sewage at each level or range of elevation, and disposing of it with the minimum power to be expended in
raising it for manuring purposes.

219. The next great question to be determined would

be, the most economical power that could be obtained at
each of these points, H, for the purpose of raising the
sewage for dispersion upon the land. For the lower level
or range, the pumps could be worked by wheels driven by
the tide of the river-water; and, in all probability, those
for the upper levels could be worked, at any rate partially,
by streams of water conducted, in suitable channels, from
the uplands. At least, while we can command the immense
water-power of rivers, and of the accumulated surface
drainage from the large districts above, the best means of
making this available for our purpose deserve all consideration, before we resort to the expensive power of steam.
For the extended flat districts of towns which bound tidal
rivers, the tides of the river could also be made available, to
a great extent, in doing the work required. The completion of the scheme would then want the details of the
arrangements to be made suitable to the superficial features
and relative levels of each part of the site for the construction, &c., of the sewers, and the machinery to be
applied for raising and distributing the sewage; and we
should finally be prepared to arrange the minor channels
or drains so as to subserve the efficient cleansing of every
inch of the surface, and of every individual tenement in
the town.

220. The arrangement here suggested will be the more
peculiarly applicable, in proportion as the site resembles
the general regularity of surface which we have supposed.
In many parts of London there is no uniformity of inclination, and in others the rate of inclination is so trifling,
that the surface may be treated nearly as a level. But the
illustration we have referred to shows the general principle
of the arrangement which would promote the greatest
economy in the drainage of a town, the site of which
resembles, in its principal features, the section given in
fig. 67.   In the application of this general system to
London, *as it is*, these departures of superficial character
from the theoretical regularity must of course receive due

attention; and, beyond this, the existing arrangement of sewers must be carefully noted and consulted, with the view of rendering them, as far as possible, available as parts of the general plan. Until this existing arrangement is presentable to the eye, upon plans and sections of the metropolis that shall exhibit every peculiarity of surface and of subterranean structure, by which the details of the plan to be adopted would be affected, no correct estimate can be formed of the extent of modifications that would be required, nor how weighty may be the reasons for relinquishing, in some parts of the town, the scheme of a succession of levels or ranges of elevation. An acquaintance with these details will probably show the expediency of making use of some of the existing main sewers through out the principal part of their length, but INTERCEPTING their contents before they reach the river, and forming tanks or other receptacles in which these matters should be collected.

221. The two objects—the public health, and economy—being kept distinctly in view in the design and execution of these arrangements, it becomes necessary to show that the sewage can be collected; treated, raised, and dispersed, without any detriment to the first of these objects; and that these purposes can be effected at such cost as will be at least balanced by the advantage of applying the sewage as manure, or a material for irrigation.

222. The contents of the sewers, consisting of human and animal excrements, earthy matters carried down by the surface water from the roads and streets, with some portion of decayed vegetable and animal substances, &c., although at first partly solid, afterwards become reduced to a thick liquid state, of tolerably uniform quality. During the putrefaction of these matters, the ammonia they contain (and which is one of their useful constituents) is disengaged: and if this process take place in the open air, it is of course mingled with the atmosphere in the form of carbonate of ammonia, and leaves the sewage in a less valuable condition. Now this volatile carbonate of ammonia may be

fixed in many ways.   Thus says Liebig:—" Gypsum, chloride of calcium, sulphuric or muriatic acid, and superphosphate of lime, are substances of a very low price; and if they were added to urine until the latter lost its alkalinity, the ammonia would be converted into salts, which would have no further tendency to volatilize.   When a basin, filled with concentrated muriatic acid, is placed in a common necessary, so that its surface is in free communication with the vapours issuing from below, it becomes filled, after a few days, with crystals of muriate of ammonia.   The ammonia, the presence of which the organs of smell amply testify, combines with the muriatic acid, and loses entirely its volatility, and thick clouds or fumes of the salt, newly formed, hang over the basin.   In stables the same may be seen.   The ammonia escaping in this manner is not only lost, as far as our vegetation is concerned, but it works also a slow, though not less certain, destruction of the walls of the building.   For, when in contact with the lime of the mortar, it is converted into nitric acid, which dissolves gradually the lime.   The injury thus done to a building by the formation of soluble nitrates has received (in Germány) a special name—*salpeterfrass* (production of soluble nitrate of lime).   The ammonia emitted from stables and necessaries is always in combination with carbonic acid.   Carbonate of ammonia and sulphate of lime (gypsum) cannot be brought together at common temperatures, without mutual decomposition.   The ammonia enters into combination with the sulphuric acid, and the carbonic acid with the lime, forming compounds destitute of volatility, and consequently of smell.   Now, if we strew the floors of our stables, from time to time, with common gypsum, they will lose all their offensive smell, and none of the ammonia can be lost, but will be retained in a condition serviceable as manure.   (Mohr.)" *

223. Chemistry thus supplies us with the means by which

* " Chemistry in its Application to Agriculture and Physiology," by Justus Liebig.   Edited by Drs. Playfair and Gregory.   Fourth Edition, 1847, p. 189.

all the offensive and detrimental properties of the sewage
may be suppressed, and all the useful properties safely
retained. There is evidently no necessary reason why a
tank or receptacle in which the sewage is collected and
stored should be, in any respect, more disgusting to the
senses, or injurious to the health of human beings, than a
reservoir of the purest water.

224. Can the remaining question of *cost* be disposed of
with equal satisfaction, so as to show that the application
of the sewage to manuring purposes may be effected with
due economy? Will the agricultural value of the sewage
pay the expenses of applying it? We believe it may.
These expenses will embrace the construction of the tanks
or stores for the sewage, of the pumps and raising machinery, and the means of treating the sewage with gypsum
or other agent, and of distributing it upon the lands to be
served; but against the total, in a comparative estimate,
would have to be placed the cost of the present partial
removal of night-soil from cesspools, the immense additional cost incurred by the necessity of having immense
sewers, the cost of outfalls into the river, and the expense
of cleansing the present sewers by hand. Now, in order
to form a rough estimate of these several expenses with
which the present system is to be debited, we will assume
the population of the metropolis to be two millions,* and
the number of houses 200,000, that is, ten persons to each
house on an average; that half of these houses are still
drained into cesspools, and that the cleansing of each of
these costs one pound annually. We shall then have
100,000*l.* as the annual cost of removing the contents of
the cesspools of the metropolis. The number of miles
of sewers constructed during the ten years from 1833 to
1843, throughout the metropolis, was about 120, or, annually, twelve miles on an average; and the excess of
capacity in these sewers, made necessary by the deficiency
of declivity, and the great length to which they are extended, probably involved a cost, in construction, equal at

* See Appendix No. 3, p. 194, for details applicable to later years.

least to 2000*l.* per mile.   This item would thus amount to
24,000*l.* annually, in which the expense of outfalls to the
river may be included.   To these are to be added the expenses of cleansing the sewers by hand, which may be
moderately computed at 10,000*l.* annually throughout the
metropolis.   We shall thus have an amount of 134,000*l.*,
which would be annually saved in these items by the proposed system.

225. A very reduced estimate of the value, for manure,
of the excreta of human beings (reduced avowedly for the
sake of gaining public belief), represents it at 5*s.* for each
person annually.   The value of the produce of the population of London would thus be 500,000*l.* per annum.   Admitting one-half of this to be now made available, we shall
have the other half, amounting to 250,000*l.*, gained by
the proposed mode of collection, and adding this to the
134,000*l.* estimated saving (224), we have a total of 384,000*l.*
annually available for the expenses of construction and repair of apparatus, and current cost of collecting, raising, and
treating the sewage of the metropolis.   This sum will endow
thirty-eight stations with an annual income each exceeding
10,000*l.* for interest of capital in first construction and current expenses of working and treating.   And this number
of stations appears fully adequate to realize all the economy
of power which can be attained by judiciously providing for
several levels in each district of the metropolis.

226. In order to show that this rough estimate as to the
value of the sewage, and the cost of applying it, is not
formed upon fallacious data, calculated to induce an un-
founded preference for the method recommended, we may
refer to the authority of the Superintending Inspectors
under the General Board of Health, Messrs. Cresy and
Ranger.   Mr. Cresy, in reporting upon the present sanitary
condition of the borough of New Windsor, and offering his
official recommendations for its improvement, estimated the
population at 10,200, and the annual value of the sewage
manure at from 1000*l.* to 1500*l.*   And for the first cost of

the apparatus for distributing this manure, he allowed an expenditure of 4000*l.*, being 3000*l.* for 10 miles of pipe, and 1000*l.* for pumping engine, &c. Mr. Ranger estimated the value of the sewage matters of the town of Uxbridge to be at least equal to 1700*l.* per annum, the population being 3219 (in 1841), or about 10*s.* for each person; and for the first cost of the distributing apparatus, he allowed 3500*l.* In reporting upon the town of Eton, of which the population was 3526, in 1841, Mr. Cresy estimated the value of the excreta at 500*l.* annually, and the cost of main sewers and tanks for the sewage at 1000*l.*

227. These estimates of the value of sewage vary from 2*s.* to 10*s.*, 6*d.* per individual, between which our average of 5*s.* is certainly a safe medium. And the allowance for first cost of apparatus for pumping, &c., varies from about 6*s.* to 1*l.* 2*s.* per individual. If we assume 1*l.* as a safe average, the annual interest, at five per cent., upon two millions of pounds, being 100,000*l.*, we shall have 284,000*l.* left for the current expenses of our thirty-eight stations, or about 7470*l.* each annually, which must be admitted to be a very liberal estimate of the cost of pumping and treating the sewage.

228. Of the current expenses of distributing the liquid sewage upon the land, and of first conveying it from the stations to the districts to be supplied, whether by a system of piping, or by vessels, or carts, it will not be necessary to offer any estimate here. These duties will probably involve an expenditure which would have the appearance of being heavy, if not fairly compared with the cost now incurred in imperfect manuring on the one hand, and on the other, with the vastly-increased value given by the application of the liquid sewage to the products of arable and pasture lands. When the costs and the results of the two methods can be, from actual and extended experience, placed thus in juxtaposition, we are justified in anticipating that a large balance of advantage and economy will appear incontesta-

bly due to the system of applying and distributing the liquid sewage as here described.

229. The multiplicity of stations necessary to carry out the scheme of ranges of elevation may appear to involve practical difficulties, and objections of a serious character, that should be adverted to, and their real value shown in contrast to the advantages which this arrangement offers. The primary consideration which must be satisfactorily fulfilled is, the practicability of accomplishing this method without any sacrifice of health, by the raising and distribution of the sewage at and from the several stations. The treatment with gypsum, already alluded to, may, it is presumed, be carried on at such a cost as shall not impair the ultimate economy of the process, and in such a manner that no offence shall be committed against the most fastidious delicacy of sense. Indeed, the completion of the process would perhaps require that the gypsum should be administered either constantly, or at stated intervals, within the collateral or the main sewers, by which means all disengagement of foul gases would be prevented, and the sewage would arrive at the receiving tanks in an innocuous and fixed condition. But until this method is carried out, the tanks may be so covered in, and their contents excluded from association with the external air, as that the latter shall not suffer any contamination; and similarly the sewage, pumped up, may be distributed in closed pipes, or loaded into river or canal boats, or into wheeled vessels, so constructed of iron that the sewage, already purified in the tanks, shall be at once conveyed away, without exposure in the slightest degree. The great advantage, however, of effecting this purification at the earliest possible stage (and the only *perfect* system is that which shall provide for doing it in the drainage of each individual house) is too apparent to be disregarded or lightly estimated. It has been well remarked, that so long as our covered sewers are permitted to emit the noisome gases engendered by the putrefying

matters within them from the air grates and gully holes into the streets, they remain, to all intents and purposes, *nuisances*, and are nearly as dangerous and offensive as if they were open sewers. And when it is remembered that these very gases, which bring pestilence and death to our poor population, would, if retained within the sewage, constitute its most valuable properties, we surely find abundant reason for seeking the best practicable method of applying the purifying process before these dangerous properties are developed, and of transmitting the fructifying matters to our fields before their value has been thus grievously impaired.

230. Another objection which the system of ranges of elevation will probably meet with is, that the sewage matters cannot be made available at any great number of points within the town and its suburbs. This may be admitted. It is, indeed, very likely that no profitable use can be made of the sewage within some few miles of the metropolis; but it does not therefore follow that it can be more advantageously conveyed in sewers to the distant station, where it may, perhaps, be applied at once. If we had the materials for instituting a fair and exact comparison between the first cost of constructing and current expenses of maintaining, in a healthy condition, the deep and large sewers required for any system of sewage which is not mainly determined by the minor varieties of surface, and the expenses of transporting the sewage from many points by pipes or by river, railway or canal, with all the facilities we can now command for any general and extended system, the latter would be found to be recommended as strongly by its economy, as it undoubtedly is by its superior efficiency in subserving the health and well-being of the population. The economy of transmitting liquid matters in pipes is well known, and the cost has been estimated, by engineers, at $2\frac{1}{2}d.$ per ton for a distance of five miles, and to a height of 200 feet. This includes interest of capital and all current expenses. Railway tolls and canal dues

may indeed be regulated so as to afford but little facility for the transmission of manure at present; but wonderful modifications will be volunteered in these matters, so soon as the system becomes general, the demand for accommodation be increased, and the mechanical and chemical appliances for purifying and transporting the sewage rendered more ready and effective.

231. Any system of collection which attempts to concentrate the sewage at a few points must be attended with extravagance in one or more of three different ways. First, and constantly, in the enlarged capacity of sewers required, for two reasons, viz. the great quantity to be conveyed, and the extra size needed to compensate for the deficiency of declivity. Secondly, in the greater distance over which the sewage has to be transmitted. Or, thirdly, in the greater cost of raising it from the level of the sewers to the high surface above it. Thus arise objections to the propositions which have been made for conducting the whole of the sewage of London to one point, situated at a distance of several miles eastward of the town, in which case retransmission would become necessary, in order to supply districts in all other directions. And, on the other hand, if it be collected at any point towards the north or north-west, where probably the greatest demand for it would be found, the reservoir would be necessarily constructed at a depth from the surface which would entail immense and needless expense in raising the sewage for application on the land.

232. Another important point in which the many-station system offers advantages, which the single-station system does not, is the facility for storing the sewage for any required time, until it may be available for agricultural use. It is well known that the manuring matters are required to be applied at certain seasons and intervals, according to the nature of the vegetation. If the sewage be collected in one reservoir, the size of it, in order to serve the true purpose of a reservoir or store-place, must be immense, almost to impracticability; but if divided, as proposed, between many

points, tanks of moderate size would be amply sufficient to contain the accumulation of long intervening periods, while the facility of distribution would, moreover, probably induce the use of the sewage for varieties of culture, that would tend to equalize the demand for it.

233. Sewers are, properly, *mere passages* for the sewage, but they are so only while the matters sent into them are induced to continue moving by the declivity of the sewer, or by the artificial force of water, or other agent, to drive or carry them forward. Without one or other of these aids the large sewers we have been constructing are known to become reservoirs of sewage, or cesspools, in which, during dry seasons, the refuse matters remain decomposing for days and weeks, sending up the most pernicious gases into the drains and water-closets of the houses, and through the air-grates and gully-holes into the streets of the town. The system which adopts ranges of elevation and varieties of surface, as indications by which to lay small sewers with rapid declivities, and leading into tanks situated at short distances from each other, obviates these evils, by constructing the sewers so that they maintain their proper character of mere passages, and the tanks are made to perform the double purpose of storing and purifying the sewage. The chamber of the tank into which the sewers discharge may be trapped, by turning down into water, so that no impure exhalations can possibly return along the sewer towards the houses and streets.

234. In adapting this arrangement to London, or any other town already provided with sewers, those in which sufficient fall exists may be retained and made to deliver into tanks at the lowest point of each range or district. But all sewers without fall, or with less than is sufficient for the rapid self-discharge of their contents, should be at once abandoned, and a separate system of sewers constructed, dipping into one or more tanks at the centre and other points of the flat district. As a general principle, to be very reluctantly departed from, the surface-water and

waste household-waters should be admitted as the only dilu-
tants of the solid sewage. Artificial scouring or flushing
of the sewers may be regarded as an expensive and troublesome correction of some of the évils occasioned by deficient
declivity, but one sometimes attended with a most mischievous consequence, viz. the forcing up of the sewage into the
streets from some of the lower sewers, which become surcharged with the flushing water during the process. Another
expedient which is occasionally recommended, and in some
cases practised, the scouring by admitting the tidal water
of the river, is inapplicable to any method which purposes
to preserve the sewage for agricultural uses ; and must be
admitted to have the effect of thoroughly defiling the lower
banks of the river at which the discharge takes place.

235. Having thus shown that the true principle upon
which towns should be classified in order to consider the
best system of arranging their general drainage, is that of
ranges of elevation and varieties of surface without any reference to contiguous rivers or water-courses, and having,
in support of this principle, stated that the system of conducting the sewage to the river, besides wasting that which
is really of great value and poisoning the water, necessitates
immense mis-expenditure in large and deep sewers, it will
be desirable, before closing this section, to cite some further authority as to the value of sewage manure, and notice
the plans which have been proposed for applying that produced in London ; and also to quote some few instances of
the works which have been executed to provide for delivering the sewage of London into the river Thames.

236. The sewage of the city of Milan is collected in two
concentric canals, the inner one of which is called the Sevese,
and the outer the Naviglio, and delivered into one called the
Vettabbia, which flows from the southern part of the city
and, after a course of about ten miles, discharges into the
river Lambro. Throughout its course this stream of sewage is made to flow over a large extent of meadow-land, and
is found to possess such valuable fertilising properties that

the deposit, which has to be periodically removed in order to preserve the level of irrigation, is bought by the neighbouring agriculturists, and esteemed an excellent manure. Some of the meadows, which are thus irrigated by the sewage water of Milan, yield a net rent of 8*l.* per acre, besides paying taxes, &c. These meadows are mowed in January, March, April, and November for stable-feeding. They besides yield three crops of hay, viz. in June, July, and August; and in September they furnish ample pasture for the cattle till the irrigation in the winter.

237. The late Mr. Smith, of Deanston, in reporting to the " Commissioners for Inquiring into the State of Large Towns and Populous Districts," upon the " application of sewer-water to the purposes of agriculture," gave an interesting description of the method which has been adopted for upwards of 30 years in applying the sewage-water of part of Edinburgh to the irrigation of grass-land. " The sewer-water, coming from a section of the Old Town, is discharged into a natural channel or brook, at the base of the sloping site of the town, at sufficient height above a large tract of ground extending towards the sea, to admit of its being flowed by gravitation over a surface of several hundred acres. The water, as it comes from the sewers, is received into ponds, where it is allowed to settle and deposit the gross and less buoyant matter which is carried along by the water, whilst it flows on a steep descent. From these tanks or settling-ponds the sewer-water flows off at the surface, at the opposite end to its entrance. The water so flowing off still holds in suspension a large quantity of light flocculent matter, together with the more minute *débris* of the various matters falling into the sewers, and chiefly of vegetable and animal origin. The water is made to flow over plats or plateaus of ground, formed of even surface, so that the water shall flow as equally as possible over the whole, with various declinations, according to circumstances; and it is found, in practice, that the flow of water can easily be adjusted to suit the declination." " The practical result

c

of this application of sewer-water is, that land, which let formerly at from 40s. to 6l. per Scotch acre, is now let annually at from 30l. to 40l., and that poor sandy land on the sea-shore, which might be worth 2s. 6d. per acre, lets at an annual rent of from 15l. to 20l. That which is nearest the city brings the higher rent chiefly because it is near, and more accessible to the points where the grass is consumed, but also, partly, from the better natural quality of the land. The average value of the land, irrespective of the sewer-water application, may be taken at 3l. per imperial acre, and the average rent of the irrigated land at 30l., making a difference of 27l., but 2l. may be deducted as the cost of management, leaving 25l. per acre of clear annual income due to the sewer-water." Mr. Smith calculated that 17,920 gallons of sewage-water, containing 5 cwt. of dissolved and suspended matter, are equal in fertilising power to 2½ cwt. of guano, or 15 tons of farm-yard manure; and he estimated the expense of the material and process, as applied to one acre of land, as follows :—

	£	s.	d.
Cost of manuring one acre of land with			
17,920 gallons of sewer-water . .	0	12	9
2½ cwt. of guano at 8s.	1	0	0
15 tons of farm-yard manure at 4s. . .	3	0	0

He further calculated that the comparative economy of the sewer-water manuring will increase with the greater quantity of each kind of manure applied ; thus :—

Cost of manuring one acre of land with			
35,840 gallons of sewer-water . .	0	16	6
5 cwt. of guano at 8s. . . .	2	0	0
30 tons of farm-yard manure at 4s. . .	6	0	0

238. Mr. G. Stephen in his " Essay on Irrigated Meadows," published in 1826, had previously described the system of sewer-manuring with commendation. Mr. Stephen says, " Edinburgh has many advantages over many of her sister cities, and the large supply of excellent spring-water

is one of the greatest blessings to her inhabitants, both in respect to household purposes, and in keeping the streets clean; and, lastly, in irrigating the extensive meadows selected below the town by the rich stuff which it carries along in a state of semi-solution; where the art of man with the common shore-water has made sand-hillocks produce riches far superior to anything of the kind in the kingdom, or in any country. By this water about 150 acres of grass-land, laid into catch-work beds, is irrigated, whereof upwards of 100 belong to W. H. Miller, Esq., of Craigintinny, and the remainder to the Earls of Haddington and Moray, the heirs of the late Sir James Montgomery, and some small patches to other proprietors. The meadows belonging to the last-mentioned noblemen, and part of the Craigintinny meadows, or what are called the old meadows, containing about 50 acres, having been irrigated for nearly a century, they are by far the most valuable, on account of the long and continual accumulation of the rich sediment left by the water; indeed, the water is so very rich, that the proprietors of the meadows lying nearest the town have found it advisable to carry the common shore through deep ponds, where the water deposits part of the superfluous manure before it is carried over the ground. Although the formation is irregular, and the management very imperfect, the effect of the water is astonishing; they produce crops of grass not to be equalled, being cut from four to six times a year, and given green to milk cows. The grass is let every year by public auction, in small patches, from a quarter of an acre and upwards, which generally brings from 24l. to 30l. per acre. This year (1826) part of the Earl of Moray's meadow gave as high as 57l. per acre."

239. The results of experiments tried at Clitheroe, in Lancashire, showed that the fertilising properties of sewage-water were nearly four times as great as those of common farm-yard manure. Mr. Thompson applied 8 tons of the sewage-water to one acre, and 15 tons of the ordinary manure to another, and the produce of the former was as

c 2

1·875 to 1 of the latter. Comparing the produce with the weight of manure, the proportion of the one to the other will therefore be as 1·875 to ·532, or nearly fourfold.

240. The sewage-water of Mansfield, in Nottinghamshire, has been so applied to lands by the Duke of Portland, that, with a preliminary expenditure of 30*l.* per acre, to put the land in a condition fit for irrigation, its annual value has been raised from 4*s.* 6*d.* to 14*l.* per acre. Mr. Dickinson treated some land at Willesden, in Middlesex, with liquid manure, derived from horses, and obtained fine crops of Italian rye grass, although the land had previously been deemed unworthy of cultivation. Mr. Dickinson obtained ten crops in twelve months. In the year 1846, the first crop (cut in January) yielded more than 4 tons per acre ; the second gave nearly 8, and the fourth, cut in June, produced 12 tons per acre.

241. At Ashburton, where they have applied liquid manure for 50 years, and at other towns in Devonshire, the land thus treated produces grass at least a month earlier than lands not so treated, and is valued at 8*l.* to 12*l.* per acre, while land not so improved is considered worth only from 30*s.* to 40*s.*

242. One of the earliest, if not *the* earliest, of the suggestions for saving and applying the sewage of London appears to have emanated from the late Mr. John Martin, the artist, in the year 1828. Mr. Ainger, in 1830, published a plan for " preserving the purity of the water of the Thames," by constructing covered drains along the sides of the river to receive the minor drainage, and Mr. Martin in July, 1834, presented to the select committee of the House of Commons, then inquiring into the state of the law respecting sewers in and near the metropolis, with a view to suggest amendments, a " Plan for improving the air and water of the metropolis by preventing the sewage being conveyed into the Thames, thereby preserving not only the purity of the air, but the purity of the water; and likewise for manure and agricultural purposes." The objects of this

plan were described to be—"first, to materially improve the drainage of the metropolis ; secondly, to prevent the sewage being thrown into the river, and to preserve in its pure state the water which the inhabitants are necessitated to use; thirdly, to prevent the pollution of the atmosphere by the exhalations from the river and the open mouths of the drains; and, fourthly, to save and apply to a useful purpose the valuable manure which is at present wasted by being conveyed into the river."

243. The details of the plan embraced the formation of a receptacle at Bayswater, on the north side of the Uxbridge Road, for the drainage of Kilburn, part of Paddington, Bayswater, &c., &c., and of another receptacle above Vauxhall Bridge to receive the contents of the present King's Scholars' Pond Sewer. For the body of the city, Mr. Martin proposed a grand sewer to commence at College Street, Westminster, and run parallel with the river, and to be extended to a convenient point near the Regent's Canal at the east end of London. And for the south side of the river, a similar plan was recommended, the sewer in this case to commence near Vauxhall Bridge, and pass along the bank of the river to Pickle-Herring Stairs, thence, branching off through Rotherhithe, to a convenient spot adjoining the Grand Surrey Canal, where a grand receptacle should be constructed similar to that by the Regent's Canal on the north side of the river.

244. These grand sewers were proposed to be constructed of iron, the bed of them being on the same level with the shore, and following the inclination of the river, about 7 in. per mile. The top of them to form quays, at least 2 ft. above the highest possible tide. To prevent the possibility of those sewers being burst by the accumulations of floods, they were to be provided with flood-gates; and to afford facility for inspecting, and, if necessary, cleansing them, light iron galleries were designed to be suspended from the roof. The sewers were to be built up of iron, the bottom paved with brick, and the top arched with sheet iron, with

wrought-iron ribs; the size of the sewers being on the average 20 ft. in depth and width, and the estimated cost of
their construction 60,000*l*. per mile, including sewer, pier
or quay, strong quay-wall of cast-iron, towards the river,
&c., &c.   The whole length of the two sewers proposed was
about 7½ miles.   The description of the receptacles in
which the sewage was to be collected will be best quoted
from the proposer's " Plan " submitted to the committee.

215. " The grand receptacle at the end of this great
covered sewer should be 20 yards deep, and 100 yards
square, with a division down the centre, separating it into
two compartments, each 50 yards in width, with a flood-gate
at the inner angle of each compartment for the sewage to
run in at; and at the opposite extremity, within about
13 ft. of the top, there should be an iron grating, 5 ft. wide
by 50 yards long, through which the lighter and thinner
parts of the sewage would rise; the heavier and grosser
parts would sink to the bottom, and gradually fill up the
base of the drain, when the gate should be closed, and the
one leading into the second division of the receptacle
opened.   At the extremity of the receptacle, between the
two compartments, there should be an engine to raise the
manure into barges, and also to pump the water in case of
extraordinary tide; in this way the expense of an extra receptacle for the water accumulating whilst the tide is up
would be saved; this, however, would only be required in
spring tide.   The receptacle would be so firmly built, and
covered with a roof of wrought iron, supported by cast-iron
pillars, that a road could be made over it; or it might be
built upon, and thus no room would be lost; and, that a
particle of smell might not be allowed to escape, there
should be a communication for the foul air to pass from the
receptacle to the fire of the engine, which would then completely consume it."   It is of course unnecessary to remark
that these estimates and particulars are quoted for their
historical interest, and not for any practical value that
attaches to them.

246. About the year 1845, a company was organised, we believe under the original auspices of Mr. Martin, and promoted by Mr. Smith of Deanston, which proposed to carry into effect, for the general benefit of the metropolis, a plan for collecting the sewage by means of a receiving sewer which should cut the existing sewers at a mean distance of 620 yards from the river. Mr. Martin, the founder of the Company, however, objected to this, preferring to receive the contents of the sewers near their outfall into the river. The Company referred to—the " Metropolitan Sewage Manure Company,"—contemplated the "conveying of the sewage-water of London, by means of a system of pumping engines and pipes, analogous to that of the great water companies, and thus distributing the fertilising fluid over the land in such manner and proportions as may be best adapted to the various kinds of field and garden cultivation." The proposals of the Company were eventually limited to an operation upon the sewage belonging to the King's Scholars' Pond Sewer, one of the principal sewers in Westminster, and with this reduced purpose an Act of Parliament was obtained for prosecuting the scheme in 1846.

247. The actual proceedings of the Company, their results and defects, were described officially in a Report (dated 25th December, 1851) to the General Board of Health, by one of their superintending inspectors, W. Lee, Esq., and from this Report we extract the following :—" I thought it my duty to visit the works of the Metropolitan Sewage Manure Company, at Fulham, to examine the market-gardens and other land to which the sewage *has been applied* there, and to ascertain from some of the consumers their opinions as to the result upon various crops. The Company has a station on the west bank of the Kensington canal, at Stanleybridge, where they have erected a steam-engine to pump the sewage water over the top of a stand-pipe 75 feet high. This altitude gives a sufficient pressure for the whole district, and the fertilising fluid is conveyed from the standpipe and distributed, by about 15 miles of mains and services, varying from 14 inches diameter at the works

down to 2 inches diameter in the fields and gardens where
it is used. The consumers have plugs or hydrants fixed at
convenient distances in their lands, and with their own servants, and hose and jet pipe, apply it when they please,
paying to the Company the sum of three pounds ten shillings per acre per annum.

" Hitherto the Company has only taken the Walham Green
sewer-water, which contains, probably, less fertilising matter than any other in the metropolis, in consequence of the
less density of the population within its area, and the great
volume of the upland rain water. It is, however, the most
conveniently accessible to the works. When the arrangements for using the Ranelagh sewer-water are completed, it
is expected that more beneficial results will be experienced.

" The first place I examined was a field from which two
crops of grass have been taken during the year, and it is
now a most excellent pasture. I examined the large market-garden occupied by Mr. G. Bagley, who said, in answer
to my inquiries :—

" ' I have found that all crops grow faster and are healthier
from the application of sewer-water. The size of vegetables
is increased in a fine season; in some descriptions the quality is also better, and consequently they would obtain better prices. I may particularly mention vegetable marrow,
a plant that requires much nutriment; it has produced a
large crop in warm dry weather. In dry weather, the sewage is of great assistance to scarlet runners and French
beans, but if rain were to come after its application, the
consequence would be injurious. I have found it of great
service to cucumbers, especially from the convenience of its
application. I think, however, it is almost better for cabbage and brocoli, than anything else. I have not perceived
any offensive effluvia at the time of its application or afterwards—not enough: if there had been a little smell, the
manure would perhaps have been better than it was.'

" Mr. E. Bransgrove, manager of Mrs. Harwood's extensive
market-gardens, said :—

" ' We have used the sewer-water ever since the Company

commenced. The improvement in the plants is very ma-
nifest; I think it helps to lessen the blight. The growth
of vegetable marrow is much improved, also of scarlet runners, and the same of lettuce and celery. This plot was
celery, and the liquid manure was applied to it. After the
crop was got off, it was planted with cabbage, without any
further application of the sewage, but the plants are still
benefited by the former application. The next patch of
savoys had it applied to them about three months since,
and are a very good crop. There was very little smell
either at the time of its application or afterwards.'

'I likewise visited the market-gardens of Mr. W. Bagley,
Mr. R. Steele, Mr. J. Crouch, and Mr. J. Bagley, to all of
which, as well as many others in the district, the sewage
manure has been applied. In one place, a field of cabbage
plants was pointed out, and I was informed that the sewage
manure had been applied to about one half and not to the
other. Those to which it had been applied were darker
in colour, larger, and much more vigorous than the remainder.

"I have, in my former report, already considered the objection urged in the provinces, *that liquid manures could only be
beneficially applied in* WET WEATHER, and have shown that
the solution used was too strong. It is a remarkable fact,
that the tenants of the Metropolitan Sewage Manure Company all concur in the statement, that *the greatest advantages
are produced by its application in* DRY WEATHER, when manuring and watering are combined in the same operation.
These facts, and my own experience, lead to the conclusion that *the town sewerage water should be collected, and raised
to the required altitude in as concentrated a condition as possible ; but that it should be distributed and applied to the land, in
such a state of dilution with water, as may be required by the
season of the year, the state of the weather and the quantity of
moisture in the soil.* This may be easily accomplished, and
almost without expense, by a pipe from the water-works,
discharging into the downward arm of the sewage stand-

pipe, at the command of the person in charge of the station. By these means, the sewerage water may be applied with advantage at all times. In dry weather, and upon thirsty soils, a weak solution would be used; and in the winter season—in wet weather, on uncropped land—or with gross-feeding plants, as strong a solution as might be requisite, to produce the greatest possible fertility.

"Having said this much of the Company's operations, I feel bound to add that the failure of a sewer manure company in the metropolis, as a commercial speculation, was predicted by Mr. Smith, of Deanston and other competent agriculturists, who withdrew from it. A disproportionately large outlay of capital and machinery was made from a wrong sewer, that is to say, from a sewer which, in fact, was an old watercourse, and chiefly adapted to the removal of storm water; and the supply of manure from this was carried to a wrong district, or one of the least suitable districts, namely, to a district already supplied at the cheapest rate with powerful manure: of that district only a small proportion received the Company's supply. The whole of the available capital was expended before the pipes could reach the ordinary farming district, where there appeared to be a demand for it. Nevertheless, the fertilising power of the extremely-diluted manure, when applied to the land, already highly charged with manure, was extraordinary. On this land, where town manure had been applied at an expense of 20l. per acre, and labour in working the land to about the same extravagant amount, an augmentation of nearly one-third of the ordinary large crops was obtained. The range of the distribution by the jet was visible in the extra growth of the vegetation, although much of this success was no doubt attributable to the water itself, irrespective of the matter in solution."

248. Every sanitary reformer must have regretted that the Company has not been more successful; but when its operations are pointed to as a reason why works for the application of the sewage of towns should not be undertaken,

it becomes necessary to indicate the causes of its success
being partial only. The errors committed are in matters
of detail; the principle—that town sewage water is a valuable fertiliser—is fully confirmed, even by the Company's
operations.

249. During the inquiries upon this subject which were
instituted by the Sewage Company, some proposals were eli-
cited which differed considerably from those upon which
the Company is now acting. Two of these may be noticed,
which were examined by the select committee, appointed
in 1846, to consider plans for the application of the sewage
of the metropolis to agricultural purposes. It should be
mentioned, that the original plan of the Metropolitan Sewage Manure Company embraced the formation of reservoirs,
in which the sewage was to be collected from the sewers,
but these reservoirs were relinquished by the promoters of
the bill, in consequence of the objection made against
them by the owners and occupiers of land.

250. One of these proposals was to carry away the entire
sewage of the metropolis, in a tunnel of from 8 ft. to 12 ft.
in diameter, and at a depth of from 40 ft. to 80 ft. beneath
the surface. This sewer was to commence at the Grosve-
nor Road, and pass through Westminster to the Strand,
and on the south side of St. Paul's Churchyard, Cannon
Street, &c., to the Commercial Road, thence under the
River Lee, and in a straight line through the West Ham
marshes, to a large reservoir and works, to be constructed
in an angle between the western banks of Barking Creek
and the northern banks of the Thames. The sewer-water
was to be raised by steam-engines from the receiving reservoir into other reservoirs, sufficiently elevated to permit the
solid matters being deposited at a level above the Trinity
high-water mark. From these reservoirs the solid matter
would be periodically removed, dried by artificial means,
and then compressed and packed for transmission by land
or water. The liquid matter was to be discharged as worth-
less, at all states of the tide. Mr. Wicksteed, the proposer,

calculated that engines would be required of an aggregate power equal to that of 1060 horses, and capable of raising, when worked at full power, 18,000,000 cubic feet to a height of 50 feet, in twenty-four hours, being considered equal to two and a half times the present ordinary quantity of sewer-water. The waste of the liquid sewage contemplated in this scheme destroys the purpose of what was intended to be a simple method of collecting the contents of the existing sewers.

251. The other proposal which received the attention of the select committee, and to which we will allude, was suggested by Mr. Higgs, the patentee of a method of treating sewage matters. Mr. Higgs's patent is dated 28th of April, 1846, and the object is entitled, " The means of collecting the contents of sewers and drains in cities, towns, and villages, and of treating chemically the same; and applying such contents, when so treated, to agricultural and other useful purposes." The scheme comprises the construction of tanks or reservoirs for the sewage, with suitable buildings over them in which the gases evolved are to be collected, condensed, and combined with chemical agents; and also an arrangement of spars or bars, on which the salts formed by this combination may crystallize. Apparatus is devised also for distributing the chemical agents over the mass of sewage, and the claims of the patentee extend to the use and application of them for the purpose of precipitating the solid animal and vegetable matters contained in the sewage, and of absorbing and combining with the gases evolved from it. " Hydrate of lime," or " slaked lime," and " chlorine gas," were the agents proposed to be employed for these purposes, and the solid matters were to be cut into suitable shapes, and dried ready for use. The committee, however, did not feel justified in recommending these elaborate processes, in the then immature state of the public mind upon the subject. *

252. Prior to the month of December, 1847, the sewers of London were controlled by seven separate boards of ma-

* See further in Appendix No. 6, pp. 225—228.

nagement or district Commissions, six of which belonged
to London north of the Thames, and one to London south
of the Thames. The six former districts were: 1, Westminster and part of Middlesex; 2, The Regent's Park;
3, Holborn and Finsbury; 4, The City of London; 5, The
Tower Hamlets; and 6, Poplar Marsh or Blackwall. The
one southern district included certain portions of the counties of Surrey and Kent. The several boundaries and extent of these districts are not worth definition, because they
are now all (except the City division) included in one Metropolitan district, which is roughly stated to comprise a circular area of about twelve miles diameter, or about 114 square
miles,* and to contain about 2000 miles of streets, and
1000 miles of sewers, consisting of old and new sewers,
and open ditches.† The City Sewers District is about one
square mile in extent, and comprises about 50 miles of
sewers.‡ The progress of sewer-building in this City District may be imagined from the following facts, viz. that the
commission began building sewers in 1756, and up to the
year 1832 had completed 9 miles and 1035 yards. In the
year 1843, Mr. Kelsey, their surveyor at that time, stated,
above 35 miles of sewers were completed. The length of
50 miles now constructed is said to have completed the
sewers as far as the *streets* of the district are concerned,
leaving only *courts* and *alleys* to be provided for.

253. Let us briefly glance at the magnitude of some of
the sewer-works which have been constructed, in the extravagant attempt to convey the entire sewage of London into
the river Thames, and to maintain the subterranean channels in a clean and healthy condition. The Irongate sewer,
which was formerly the city ditch, varies in height from
6 ft. 6 in. to 11 ft., and in width, from 3 ft. to 4 ft. The
Moorfields sewer is 8 ft. 6 in. by 7 ft., and at the mouth,

* Evidence of R. Stephenson, Esq., before Select Committee on the Great
London Drainage Bill, 13 June, 1853.
† Evidence of Mr. J. W. Bazalgette, before the same, 10 June, 1853.
‡ Evidence of Mr. W. Haywood, before the same, 13 June, 1853.

10 ft. by 8 ft.; and at the north end of the Pavement this
sewer is 27 ft. below the surface. The Fleet Ditch, which
drains the land from Highgate southward, is partly formed
in two distinct sewers, which run on each side of Farringdon Street; they are from 12 ft. to 14 ft. high, and each is
6 ft. 6 in. wide, yet they are liable to be flooded by the immense rush of waters from the northward, and a single storm
will raise the water 5 ft. in height in both of them almost
instantaneously. The culvert constructed at the mouth of
this sewer was severely injured, in 1842, by the flood consequent upon a thunderstorm. The Bishopsgate Street
sewer, which receives the drainage of Shoreditch and adjacent places, is 5 ft. by 3 ft., is sometimes overcharged, and
returns the waters from the high ground.

254. By way of conveying a ready idea of the vast size

Fig. 69.

of the Fleet sewer, the great length and extended functions
of which have just been noticed, the two figures 69 and 70

Fig. 70.

are introduced, the one showing the section of the sewer at
the city boundary, where its dimensions are 12 ft. 3 in. by
11 ft. 7½ in., and the other showing the size of the mouth
of the sewer, which is 18 ft. 6 in. by 12 ft., and of ample
capacity to *admit one of the largest locomotive engines*, the

gigantic dimensions of which are sufficiently familiar to all
who have found it necessary to cross the platform of a railway station. Yet this sewer has often been surcharged;
and only within the last year (1842) the culvert, so ably
constructed at its mouth by Mr. James Walker, was severely
injured by the flood consequent upon a thunderstorm.*

255. The large sewers constructed in the Tower Hamlets division are 4 ft. 6 in. by 3 ft., and the cost per foot
varied from 15s. to 1l. 5s., *according to the depth.* One of the
sewers, from Hoxton New Town, is laid, for a length of
3000 ft., on a dead level, and discharges into another, which
is also on nearly a dead level, because a fall could not be
obtained. From the 45 miles of sewers in this division,
about 2000 loads of sewage have to be annually removed by
hand. All the outlets are into the Thames, below London
Bridge, and are provided with valves, which sometimes fail,
owing to some matter issuing from the sewers that prevents
their closing, and of course the tide rushes in. Up to the
year 1843, 13,000 ft. of sewers were rebuilt, at a lower level
than formerly, in order to accommodate levels not then provided for, and at the same time maintain the communication
with the river.

256. In the Surrey and Kent division, all the main arched
sewers are necessarily provided with flaps and penstocks,
nearly all the district being below the level of high water in
the river. Some of the sewers in this division have only
2 ft. of fall per mile.

257. In the Holborn and Finsbury division, 38,000l.
were expended upon the Fleet sewer between the years
1826 and 1843, and the surveyors were enabled to reduce
the size of a part of it from 12 ft. by 12 ft. to 10 ft. by 9 ft.,
"from an advantage in the difference in the fall, there being
more fall in this situation, which rendered the proportion
of the current through the smaller sewer equal to the larger

* Evidence of the Surveyor to the City Commission of Sewers, before
the " Commissioners of Inquiry into the state of large towns and populous
districts." (1843.)

one." The extra expense incurred by the increased depth, which sometimes occasioned a passing through sand and treacherous materials, may be inferred from one item in the cost of construction, viz. that in such cases it was found necessary to use close timbering to strut the sides of the excavation ; the sand also, sometimes, despite all precautions, rose 6 inches in one night ; and the building of some of the sewers in Pentonville " had the effect of loosening the ground, or causing the ground to slip, the whole way up the hill," and thus seriously damaging the walls of the gardens above. The Fleet sewer, as already described, takes the water from Highgate and Hampstead, conducting it towards the river Thames, and it has been known " to rise six, eight, and ten feet in a night ; there have been instances of persons being carried away by it." The fall in that part of it which is 12 ft. by 12 ft. is at the rate of a quarter of an inch in 12 feet ; and in the 10 ft. by 9 ft., the fall is three-quarters of an inch in 10 feet. The effect of this difference of declivity is, that " if the 12 by 12 were completely full, the other would not be full by 2 ft."

258. The following details respecting the Fleet sewer are quoted from a report dated 25th October, 1853, made by Mr. Haywood, the Surveyor to the Commissioners of City Sewers. The Fleet sewer is the largest in the metropolis. An area of land six or seven times the size of the whole City of London is drained by it, the waters of that area being carried into it by many hundred collateral sewers. Near the outfall—that is, at the Crescent, Bridge Street, Blackfriars—the flow of these accumulated waters is probably from 18,000 to 20,000 gallons per minute during the day in dry weather ; with but slight rain, the flow is so much increased that no man can stand against it ; with a very moderate quantity of rain, falling uniformly over the drainage area, it is questionable if a horse could maintain its footing against the current in the sewer ; with heavy rain the sewer is half filled. The difficulties attending repairs to the Fleet and similar sewers are very great. The

City Sewers Surveyor reported that a portion of the Fleet
sewer, between Fleet Street and Blackfriars Bridge, being in
such a condition as to require extensive and immediate repair, it was determined to put in a new invert, and under-
pin the side walls, and for this purpose to construct a dam
across the sewer, so as to lay the bed of it dry.   In order
to avoid complaints of street interruption or other public
nuisance, the works were commenced and carried on without opening any portion of the sewer itself, the only means
of ingress and egress being one small hole 3 feet wide and
3 feet long in a branch sewer in Crescent Place.   Down
this small opening the workmen descended and conveyed
all materials required, at great labour and time, and, necessarily, cost.   Moreover, mechanical appliances, which are
used with so much advantage in many similar works, and
by the aid of which, with sufficient space, an amount of
work five hundred times that done in the Fleet sewer
could have been performed in the same space of time
with ease, could not be used, owing to the insufficiency of
space, and, therefore, manual labour, applied at immense
disadvantage, was all that could be adopted.   The men
were frequently working with two thirds of their bodies immersed in water, amid an uncertain light, and a deafening
roar of water, and at times, with but little chance of their
lives, if they had once lost their footing.   Half of the
24 hours the tidal water was in the sewer; when it was out,
the men could with difficulty work three hours each morn-
ing, especially on Saturdays; and at other times, if any
rain came down, the work had to be suspended at once.
The sewer was never entirely free for 24 hours either from
rain actually falling, or the drainage waters from the upland
districts after the rain; and actual progress was made only
in one tide out of every three or four.   Two or three times,
when dams were all but complete, heavy floods of rain
washed the larger portion of them away.

259. The following extracts from the evidence given by
Mr. Roe, surveyor to the late Holborn and Finsbury Com

mission of Sewers, before the Commissioners of Inquiry into the state of large towns and populous districts, in 1843, will give some notion of the state of the sewerage of that division. " The Holborn and Finsbury divisions are peculiarly situated, having no immediate communication with the river Thames. The waters from these divisions have to pass through one or other of the adjoining districts, namely, the city of London, the Tower Hamlets, or the city of Westminster, before reaching the river. The sewers of the Holborn and Finsbury divisions have, therefore, of necessity, been adapted to such outlets as the other districts respectively afforded; and these having formerly been put in without due regard to an extended drainage, the sewers of these divisions have not had the benefit of the best fall that could have been obtained. Of late years many of the outlets have been lowered in the adjoining districts; but to alter the existing sewers of these divisions to the amended levels would require the rebuilding of about 323,766 ft. of sewer, at an expense of nearly a quarter of a million sterling." In this state of things, Mr. Roe conceived that the ordinary current of water which passes along the covered sewers (in some constant, in others periodical), in these divisions, which, in numerous cases, does not prevent deposit from accumulating, might yet be made available for that purpose; and accordingly "a series of experiments were commenced, in order to ascertain what velocity could be obtained in the sewers; and it appeared that deposit might be removed by the means of dams placed in certain situations to collect heads of water, at less expense, than by the usual method. Another series of experiments were made for the purpose of endeavouring to ascertain the proportion of decomposed animal and vegetable matter, and detritus from the roads, carried through the sewers to the river Thames by the common run of water. Several square boxes were constructed, to hold 1 cubic foot of water each. These were filled with water from different sewers. After allowing the

turbid water to clear itself by precipitation, I ascertained
the relative amount of the precipitate. The following were
some of the results :—From the river Fleet sewer, near the
outlet, the proportion of decomposed animal and vegetable
matter, and detritus from streets and roads, held in mechanical suspension, was 1 in 96. The run of water was
10 in. in depth, and 10 ft. in width, having an average
velocity of 83·47 ft. per minute, passing 692·8 cubic feet of
water per minute ; the matter conveyed being 7·21 cubic
feet per minute, or 103·060 cubic yards per annum. The
river Fleet sewer conveys the drainage of about 4444 acres
of surface, or about four-sevenths of the surface of these
divisions. That great quantities, in addition to the above,
are carried away by the force of water in rainy weather is
certain ; allowing this source, and the remaining three-
sevenths of the district to only equal the discharge by the
river Fleet sewer, there appears to be a quantity of upwards
of 200,000 cubic yards of matter carried to the Thames per
annum from these divisions in mechanical suspension, and
by the force of velocity, weight, and volume of water." In
one part of this division (at Canonbury) the sewer is *sixty-
eight feet below the surface*, and *the drainage of the houses is
provided for by a subsidiary sewer.*

260. In the Westminster division the outfalls to the river
vary from 10 ft. to 15 ft. below the level of high-water mark,
that is about 5 ft. above low-water mark, and some of them
are provided with flaps. The cost for cleansing sewers by
hand, made necessary by deficient fall, amounted, in the
year 1842, to 1850*l.* ; the average for seven years, being
about 1,550*l.*, and the deposit is so hard that it is sometimes found necessary to use the pick-axe to dislodge it.
Some of the main sewers have a fall of only ½ inch in 100
feet. During the ten years ending 1843, 27,056 yards of
sewers were constructed by the Commissioners for this
division, and these were principally old sewers *built at a
lower level*, or diverted along the public way. The size of
the main sewers is 3 ft. in width, and 5 ft. 6 in. in height,

and of this size 32,000 yards were constructed from 1833 to 1843 jointly by the Commission and by individuals. In 1843 one of the sewers in this division, the King's Scholars' Pond Sewer, was described to have from its commencement at Shepherd's Well to the flood gates at the Thames, a total fall of 285 ft. 4 in., yet the fall is in some parts very deficient. In the Pimlico district adjoining the palace, the fall is for the last 5500 ft. only 5 ft.;—less than a foot in each thousand feet, and the outlet was still so low that flood gates were necessary at its outlet. During the rising of the tide, therefore, these gates were closed for six hours, and the sewage of course remained pent up or thrown back towards the houses.

261. In the Kent and Surrey divisions, in which there are several open sewers, the drainage is so conducted that " during the time the tide is up in the river, the sewers have to receive all the water, making its way into them, and must be sufficiently capacious to hold both that quantity, and all rain that may fall, until the fall of the tide allows a discharge." The covered main sewers are 5 ft. 3 in., by 4 ft. 9 in. The whole of the district is many feet under high-water mark.

262. The facts here quoted are certainly sufficient to show that the system of draining into the Thames-is attended with vast extra expense in the depth and size of sewers, and with great inconvenience and inefficiency in the want of declivity thereby laboured under, which are powerful considerations against the propriety of this method, and the weight of which has to be added to that of the loss occasioned by the utter waste of most valuable manuring matters, and the injurious effects of saturating these matters in the water from which the daily supplies for the metropolis are mainly derived.

263. Having thus briefly noticed some of the details of the sewerage arrangements of the metropolis, as carried on under separate commissions prior to the year 1848, it is desirable to notice some of the proceedings of the United

or " Metropolitan Commission," since that time, including
their proposals, estimates, and actual works.   There is
good reason for extending this notice, sufficiently, moreover, to embrace a general review of these proposals, since
they present to us the labours of the highest engineering
talent yet brought to bear upon the great question, How
can London be properly drained ?   The new commission,
appointed in December, 1847, consisted of the following
members :—Lord Ebrington, Lord Ashley, Dr. Buckland,
Mr. Hume, M.P., Hon. F. Byng, Dr. Arnott, Dr. Southwood Smith, Sir J. Clark, Rev. W. Stone, Professor Owen,
Sir H. De La Beche, Messrs. R. A. Slaney, M.P., J. Bid-
well, J. Bullar, W. J. Broderip, R. L. Jones, J. Leslie, and
E. Chadwick.   In the Session of Parliament 1848, an Act
(11 & 12 Vict. c. 112), was passed, entitled, " *An Act to
consolidate and continue in force for Two Years, and to the end
of the then next Session of Parliament, the Metropolitan Commissions of Sewers.*"   This Act of 1848 was amended by
another (12 & 13 Vict. c. 93), passed August 1st, 1849,
entitled, " *An Act to Amend the Metropolitan Sewers Act.*"
By these Acts powers were given for making district rates,
and levying contributions for building sewers, for making
sewer rates partly prospectively and partly retrospectively,
for levying and recovering " improvement rates," and borrowing money on mortgages and annuities " *to pay off any
debt not actually due.*"

264. On July 23rd, 1849, at a Special Court of the
Sewers' Commission, a report was read from the surveyor
of the Metropolitan Commission at that time (Mr. Phillips),
embracing plans and estimates he had prepared for the
drainage of the metropolis.   In this report, the principles
of town-drainage were laid down as follows :—

" 1. That two outfalls, independent of such other, should
be provided, one for the discharge of natural, or land and
surface waters, and the other for the discharge of artificial,
or house and soil drainage

" 2. That in order to perfectly drain the subsoil of the

town so as to free it from damp, and to carry off, as quickly as possible, the natural waters, a system of permeable land drains and sewers should be provided to discharge into the natural watercourses and rivers.

" 3. That as outfalls are already provided by streams and rivers for the discharge of the natural waters, it is only necessary to provide separate and proper outfalls for the discharge of the artificial, or house and soil drainage, which outfalls should convey the sewage, as fast as it is produced, to a depôt at a convenient and unobjectionable place, quite clear of and below the town.

" 4. That in order to carry off the house and soil drainage without contaminating the atmosphere of the town by the escape of effluvia through the numerous inlets, as is at present the case, a system of impermeable drains should be provided, distinct and separate from the permeable land drains and sewers, to discharge, without intermission, into the said artificial outfall independently of the river.

" 5. That at the main outlet a depôt should be formed, and works established for raising the sewage, and for converting and distributing the same for agricultural and horticultural purposes."

265. For the discharge of artificial, or house and soil drainage, the surveyor proposed a tunnel extending from the Plumstead marshes to Twickenham, and crossing under the River Thames eleven times. The proposed course and construction were as follows :—From the depôt at Plumstead to pass under the river to East Ham and Plaistow marshes, recrossing the river from below Bow Creek to the Greenwich marshes, where a shaft would be provided, and a branch drain to serve Charlton, Greenwich, &c. Then to cross the river again to Blackwall, where another shaft and main branch drains for Poplar, &c., would be constructed. Continuing between the export and import West India Docks, the tunnel would reach the margin of the river, and a shaft and branch drain for Limehouse be provided.

Crossing to the Surrey side, the tunnel would run through Rotherhithe, and have two shafts and branch drains for Rotherhithe, Deptford, and Hatcham, westward, and following the bank of the river eastward.

The tunnel would leave Rotherhithe just below the entrance of the Grand Surrey Canal, and continue under the river to the Middlesex side between the entrance to Shadwell Dock and the Rotherhithe tunnel, and run south of the London Docks through Wapping to the margin of the river above the entrance of Hermitage Dock. Two shafts would be built along this part of the line, and connected with branch drains, one westward to the Tower and along Thames Street to Blackfriars, intercepting all the sewage running into the Thames as far as the Temple Gardens. From the east shaft a branch would be carried for Wapping. Leaving the entrance of Hermitage Dock, the tunnel would cross the river to the Surrey side near Shad Thames, and continue through Bermondsey, Southwark, and Lambeth, below Westminster Bridge. Four shafts and deep main branch drains would be constructed on this length, to serve the house drainage of Bermondsey, Southwark, Lambeth, Newington, Kennington, Walworth, Peckham, Camberwell, Dulwich, Norwood, Streatham, Stockwell, Clapham, Brixton, Battersea, Wandsworth, Tooting, Mitcham, Carshalton, Merton, &c.

From the Surrey side the tunnel would pass under the river below Westminster Bridge to Westminster, and continue through that district to the King's Scholars' Pond sewer at Pimlico; thence to Chelsea, and between Fulham and Hammersmith, crossing the river to Barnes, recrossing to Mortlake, and finally crossing above Richmond Bridge to take up the sewage of Twickenham and the neighbourhood. The line here described, although thus appearing circuitous, may be seen on a map to thread the windings of the river in a uniform curve.

266. The tunnel here proposed was to have a fall of

49 ft. 3 in. throughout its length (19½ miles), a diameter
of 8 ft. at its outlet, and 6 ft. at the upper end. The following was the estimate for the work :—

Constructing 19½ miles of tunnel sewer, at an
  average of £23,709 per mile, including shafts   £512,323
Constructing the shaft at the terminus   4,099
Steam engine, pumps, and boilers, complete   . 95,000
Engine house, &c.   .   ,   .   33,577

Total estimated cost of 19½ miles of tunnel
  sewer, with machinery, &c. £644,999

The annual cost of maintenance, consumption of coals,
and superintendence, was set down at 15,000l., and it was
calculated that the rated value of the property draining
into this tunnel being 10,000,000l., an annual rate of 1
penny in the pound for 22 years would suffice to pay the
principal and interest. The depôt (at Plumstead marshes)
would be 10 ft. deep, and elevated above the surface 20 ft. ;
an average of 10,000,000 cubic feet of sewage were expected to be discharged into it during the day, and this
would be precipitated and filtered until the liquid parts
were reduced as nearly as possible to pure water, while the
solid matter remained. When there was no demand for
liquid manure, it might be directed into the Thames or the
sea, or otherwise disposed of as circumstances might render
advisable. This gigantic conception appears to have served
only for the Commissioners to talk about, and to elicit the
geological objections of Sir H. De la Beche and Dr. Buckland, upon the ground that the tunnel could not pass
through a continuous bed of clay, but must, for a great
part of its length, be bored through beds of sand, gravel,
and chalk, all of which are percolated by immense quantities of water, and would prove extremely unfavourable for
such operations as those contemplated.

267. At the end of 1849 or beginning of 1850, the Commissioners appear to have become thoroughly puzzled by
the conflicting advice of their own surveyors, and affrighted

at the task before them; and they desperately solicited
anybody and everybody to tell them what they would recommend for the drainage of London.   The invitation
thus given for the sending in of "plans," was responded
to by some 150 or 160 amateur drainers, of whom the only
fact known is, that each of them became entitled, some time
after they had tendered their recommendations, to receive
a printed map from the Commissioners, showing what the
sewer system of London at that time *was*, a kind of information which one might have supposed should have *preceded*
their suggestions for its amendment.

268. In October, 1851, when a new Commission was
formed, of which Mr. R. Stephenson and Sir W. Cubitt, the
eminent engineers, were members, these 150 or 160 plans,
which had lain till then unexamined, were exhumed, and
carefully considered and reported on by a sub-committee
appointed for that purpose. "Upon the report of that Committee," says Mr. Stephenson,* "Mr. Forster, the then engineer of the Commission, proceeded to design a plan for
the general improvements of the sewage of London, and
more especially with a view to the interception of it, to
remove it from the Thames as far as possible in the vicinity
of London.   I was connected with Mr. Forster, at least I
took a great interest in the question, and every step he took
I was cognizant of.   I examined the greater part of London, and more especially directed my attention to the lower
districts on the south side of the Thames, which are so
badly drained at present, and which are, in fact, the seats
of diseases of all kinds. We proceeded so far as to complete our plans for the south side, and were about to let the
contracts; but a question was raised as to the powers of
borrowing money, and pledging the rates of the districts.
It was found we had no such power; therefore we had no
means of proceeding with our contracts.   We had then an
interview with Sir George Grey on the subject, with a view

---

* Evidence before Select Committee on the Great London Drainage Bill,
13 June, 1853.

of bringing a Bill into Parliament, to authorise us to raise money for those purposes. Very shortly after that interview, the Government was changed. I do not know what steps were taken by the next Government, but I believe none, and I believe none are now being taken by the Government to relieve the Commissioners of Sewers from their embarrassments."

269. The plan designed by Mr. Forster, and here referred to, comprised two sewers, a high-level sewer, and a low-level sewer, the former to receive the natural drainage or rainfall and surface-water, and conduct it to a low part of the district, as Bow Creek, by gravitation simply ; the low sewer to collect all the house drainage, and convey it also by gravitation to Bow Creek, where it would be pumped up and descend with the surface-water to the outlet at Barking Creek. By this plan, the districts that were proposed to be drained by *gravitation* comprised a very large proportion of the whole area of the north side of the Thames, admitting of a sewer falling 4 ft. a mile, commencing at the extreme west of London—Bayswater—and continuing to the River Lea, where Mr. Forster proposed that there should be reservoirs, if necessary, for the purification of the sewage; but that would be conveyed into the Thames conveniently at all times. With respect to the *low-level* tract, it was proposed (commencing, as in the high-level sewer, at the lifting point on the eastern bank of the River Lea) to construct a sewer at a depth of 47 feet below the invert of the high-level sewer, proceeding beneath the River Lea near to Four Mills Distillery, taking the north-western bank of the Limehouse Cut, at which point it receives a branch intended to intercept the sewage of the Isle of Dogs ; thence continuing along the bank of Limehouse Cut, through a portion of the Commercial Road, Brook Street, and beneath the Sun Tavern Fields, into High Street, or Upper Shadwell, at a point opposite to the church ; thence along Ratcliffe Highway, and Upper East Smithfield, across Tower Hill, through Little and Great Tower Street, Eastcheap,

D 2

Cannon Street, Little and Great St. Thomas Apostle,
Trinity Lane, Old Fish Street, and Little Knight Rider
Street; thence beneath houses in Wardrobe Terrace, and
on the eastern side of St. Andrew's Hill, along Earl Street
and Blackfriar's Bridge. From Blackfriar's Bridge it was
proposed to construct the sewer along the river shore to the
junction of the Victoria Street sewer at Percy Wharf, which
sewer, between Percy Wharf and Shaftesbury Terrace,
Pimlico, would thus become an integral portion of the intercepting line. The whole length of this main sewer, from
the River Lea to Bayswater, was about 8 miles, the section
of sewer 8 ft. by 5 ft. or 5 ft. 6 in., and the estimated cost
1,200,000*l*.

270. The want of powers to raise money, under which
the Commission laboured, as already explained in quoting
Mr. Stephenson's Evidence, necessitated the suspension of
proceedings, and shortly afterwards (in 1851) the death of
Mr. Forster deprived the Commission of their professional
adviser, and opened the door for a careful reconsideration
of the plans he had proposed. In October, 1851, Mr. R.
Stephenson and Sir W. Cubitt accepted the appointments
of Consulting Engineers to the Commission, and various
minor works or those of "drainage," merely, as distinguished from the ulterior purpose of "intercepting" the
passage of refuse matters into the Thames, were, during the
following two years, sanctioned and executed.*

* The following record of estimates for works ordered by the Metropolitan
Commission may be useful :—
Ordered 6th September, 1850.

					£	s.
340 ft. of	12 in.	pipe sewer,	Shepherd's Bush		70	0
120 „	12	„	Kent Road		22	0
550 „	15	„	Hanway Street			
175 „	12	„	John's Court	Oxford Street	320	0
220 „	12	„	Petty's Court			
133 „	9	„	Lisson Grove		25	0
50 „	9	„	Portman Market . . .		20	0
40 „	6	„				
500 „	9	„	Portland Street, Walworth . .		50	0

271. From a statement of the affairs of the Metropolitan Sewers Commission, read at a General Court, on the 6th September, 1853, it appeared that when, previous to July, 1852, the power of the previous Commission to call for a rate of 6*d*. in the pound was reduced to 3*d*., 200,000*l*. was the sum required for their ordinary annual expenditure; but that after this, they were empowered to raise not more than 100,000*l*., although they had entered into contracts which could not be withdrawn; that when the New Commission was issued in July, 1852, after giving credit for all the rates, there was still a balance of 36,000*l*. against the Commissioners; that in the last 12 months, the present Commission had executed, at the public expense, brick and pipe sewers to the extent of 28½ miles in length, con-

450 ft. of	2 ft. brick culvert, in line of Falcon Brook	.	. 120	0
360 „	12 in. pipe sewer, Brick Street, Park Lane .	.	. 90	0
350 „ 12	„	Crown Court, Temple Bar	.	. 87	10
145 „ 15	„	} Moor Street, Marylebone	.	. 95	14
200 „ 12	„				
75 „ 12	„	Lamb Court, Clerkenwell	.	. 19	10
626 „ 12	„	Fulham Road . .	.	. 106	0
50 „ 9	„	Ebury Square, Pimlico .	.	. 10	0
800 „ 18	„				
940 „ 15	„	{ Camberwell Grove .	.	. 450	0
1140 „ 12	„				
420 „ 9	„				

Ordered 21st July, 1852.

170 ft. of 12 in. pipe sewer, in Chester Mews, Pimlico. Gradient 1 in 60 . . . . 32 10

660 ft. brick sewer, 3 ft. 9 in. × 2 ft. 6 in., in New Street Mews. Gradients 1 in 43, and 1 in 200. Materials—stock bricks, Portland cement, and blue lias mortar 350 0

100 ft. 12 in. pipe sewer, in Woburn Mews, West. Gradient 1 in 100. (Contribution £2 per house) 18 0

260 ft. brick sewer, 4 ft. × 2 ft. 8 in., in Highgate Road. Gradient 1 in. 122. Materials—stock bricks, Portland cement, and blue lias mortar 225 0

120 ft. of half-brick barrel sewer, 2 ft. in diameter, in Wandsworth Road 30 0

(*See page* 72 *for prices in* 1854.)

tracted for or put in hand 12½ miles, making a total of 41 miles in length, at the cost of 142,853*l.*; but there were many other works considered as public, though constructed at the expense of proprietors abutting on public roads; there had been 11½ miles of these public sewers constructed in that period, at a cost of 36,413*l.*; these sewers were built at private expense, but afterwards became the property of the public, and were cleansed and repaired at the public expense, the original proprietor receiving a contribu tion towards the expenses from all parties who drained into them; that besides these works and those executed at the public expense, there was a large amount of pipe sewers that could hardly be separated from public works. The length of these pipe sewers was 70 miles, executed at a cost of 44,108*l.* The works superintended by the Commissioners cost 80,521*l.*, and the new sewers, executed by themselves, 142,853*l.*, making a total for works superintended during the year of 223,374*l.* The works of the Commissioners, however, did not end there. There had been repairs, and the cleansing of open and covered sewers, and on these three objects they had expended a sum of 36,880*l.*

272. The state of funds in the several districts under the jurisdiction of the Commissioners, on the 30th June, 1853, was reported as follows:— Credit, 35,754*l.*; debt, 29,407*l.* 12*s.* 7*d.*; available balance; after taking credit for outstanding rates, 81,418*l.* 14*s.* 9*d.*; current ordinary expenses per annum, 59,177*l.* 5*s.*; estimated amount of extraordinary works ordered, 66,429*l.* 10*s.* 10*d.* Total prospective expenses, 125,606*l.* 15*s.* 10*d.* Available balance for further extraordinary works:—Credit, 8,723*l.* 16*s.* 9*d.*; debt, 52,911*l.* 17*s.* 10*d.*; amount to be derived from new rates of 6*d.* in the pound, 216,370*l.* Of this latter item an amount of 140,000*l.* might be collected within the year. Amount unprovided for, to be met by new rates, 44,188*l.* 1*s.* 1*d.* This statement does not embrace the following liabilities:— Permanent loans, 99,492*l.* 13*s.* 1*d.*;

claim for Ordnance survey, 24,212*l.* 17*s.* 5*d* ; total, 123,705*l.* 10*s.* 6*d.* Special contracts to the amount of 46,508*l.* have been entered into, and duly executed, during July and August, in respect of the most important of the works ordered, but not commenced, or only partly executed, up to 30th June, 1853. The following are the amounts produced by a 6*d.* rate on the larger districts :—Ranelagh, 24,000*l.* ; western division of Westminster, 37,000*l.* ; eastern ditto, 21,000*l.* ; Holborn, 23,000*l.* ; Finsbury, 24,000*l.* ; Spitalfields, 17,500*l.* ; Surrey and Kent, 39,000*l.* ; Counter's Creek, 5,200*l.* ; Greenwich, 4,000*l.* ; Fulham and Hammersmith, 2,500*l.* ; Ravensbourne, 2,000*l.* ; Richmond, 1,000*l.*

273. In October, 1853, the surveyor and consulting engineers of the Commissioners appear to have completed their plans for the main drainage of the metropolis, north and south of the Thames, which, after being received and adopted in Committee, were adopted at a Special Court of the Commissioners, held October 20th, 1853.

The report by the surveyor (Mr. J. W. Bazalgette) was mainly as follows :—

NORTH SIDE OF THE THAMES.

" The ' Hackney Brook, or Northern High Level Intercepting Sewer,' will divert the whole of the sewage and flood-waters of 14 square miles of the upper districts from the low districts, and from the river Thames, thereby reducing the size and cost of construction of the sewers in the low districts, and preventing the destruction of property by floods, as at present.

"For instance, the Fleet sewer is now overcharged by the amount of drainage falling into it; it is uncovered for a great part of its length, through a dense population; and in other places it has fallen in, and has been temporarily repaired ; but the Commissioners have never been able to apply an effectual remedy to it, on account of its present defective route, and the enormous cost of rebuilding so large a sewer in a more direct line through a crowded population; and, as it has always been intended ultimately to divert its contents from the Thames, such a work could only have been of a temporary character; whereas now, by the diversion of the upland drainage, these difficulties will all be removed, and it and many similar sewers reduced to economical and permanent works.

"The amount of drainage received by this upper line will govern the sizes of the lower lines of intercepting sewers; and, as Parliament has now sanctioned the construction of a tunnel railway along the New Road, which will intercept our present main line of sewers, and necessitate the immediate construction of an intercepting sewer along its route, this upper line is rendered the more immediately imperative.

"This upper sewer will be in itself a complete and perfect work, and independent of any other intercepting sewer; it will afford an advantageous means of flushing and cleansing the sewers in all the low districts above which it passes, and thus improve their drainage; and it will relieve the neighbourhoods of Stoke Newington, Kingsland, Clapton, and Hackney, from the nuisance of an offensive open sewer, known as the ' Hackney Brook,' and the latter places from frequent inundations.

"The route of this work commences with a sewer 4 ft. 6 in. in diameter, by a junction with the 4 ft. circular sewer now constructing as the outfall for the Hampstead drainage, in the line of the open Fleet sewer, its inclination being 1 in 72. It passes along Gordon-house Lane, crosses the Highgate Road at an inclination of 1 in 134, and intercepts the second branch of the Fleet sewer, being at this point 6 ft 9 in. in diameter. It then passes across the fields into the Tufnell Park Road, turns down the Holloway Road to its junction with the Tollington Road; at this point its size is increased to 10 ft. 6 in. in diameter, and it is laid at such a level as to enable it to receive a branch from the Edgware Road and Regent's Park, diverting the waters of the upper portion of the Ranelagh and King's Scholars' Pond Sewers, and again intercepting the Fleet sewer at Kentish Town. From the Holloway Road it passes under the Great Northern Railway and New River Cut, taking nearly the same direction as the present open Hackney Brook sewer, for which it will act as a substitute, and falling at an inclination of 1 in 476 to the High Street, Stoke Newington, near to Abney Park Cemetery, where its size is increased to 11 ft. 6 in. in diameter. It then passes along the Rectory and Amherst roads to Church Street, Hackney; and from this point it is increased to 12 ft. 6 in. in diameter, and its inclination is 1 in 1320. Thence it crosses under the East and West India Dock Railway, along Albion Road, West-street, and through the Victoria Park, and under Sir George Duckett's canal, up to which point it is kept up to such a level that it may be carried over or under the River Lea Navigation; or the sewage may be diverted from the point at Sir George Duckett's canal down below the Lea Navigation, at Four Mills, retaining the present as a flood outlet; or the point selected would afford an advantageous position for the erection of sewage manure works. With all these alternatives held in view, it is proposed for the present to terminate this work with two parallel culverts, each 7 ft. 6 in. in height, by 8 ft. in width, discharging into the River Lea at the present Hackney Brook outfall. This sewer is kept at such a level (for the purpose of being carried over or under the River Lea, and discharging at high water by gravitation), that a portion of the drainage of the Hackney Marshes cannot be turned into it; this must either drain into the Lea, as proposed by Mr. Forster, or eventually be carried into the low level sewer, to be raised by pumping.

" It is further proposed to fix tide-gates at the mouth of the present Hackney Marsh open sewer, to prevent the ingress of the tide, and by this means, and the diversion of the upland sewage from it, to prevent the floods to which the Hackney Marshes are now subjected

" The estimated cost of the whole of this work, from Kilburn and Hampstead to the River Lea, and a branch to
Four Mills, will be about 271,290*l*., and the probable cost
of that part of it which it is now proposed to execute will
be about 181,290*l*.

" The areas which would drain into this sewer amount to
about 8913 acres, or 14 square miles."

Sir William Cubitt, one of the consulting engineers,
having approved the plan, referred to the cost in the following words :—

" I have most carefully gone over the estimates, the
prices of which are guided, in a great degree, by works of
the same kind recently executed, and at this time in progress for the Commissioners, to which heavy percentages
are added for contingencies and compensations, for which
no exact estimate can be made. The total amount, as
shown by the estimate, is 271,290*l*.; but, reasoning from
analogy in similar cases, I strongly recommend that the
probable total cost of this measure should not be stated,
or, at least, I do not think myself safe in stating it at less
than 300,000*l*."

SOUTH SIDE OF THE THAMES
SURREY AND KENT DRAINAGE.

" In preparing a design for a high-level or catchwater
sewer, which, in conjunction with a low-level sewer, will
complete the main features of a design for the effectual
drainage of the metropolitan districts on the south side of
the Thames, and the diversion of the sewage to a lower
point in that river, it has been sought, as far as practicable,
to comply with the resolutions of the Commissioners in
1850, communicated to their engineer, the late Mr. Forster,
on the subject of the main drainage; and, in describing
the proposed scheme, it will be necessary, first, to call
particular attention to the natural features of the district
to be drained, which covers an area of about 24 square
miles.

" The urban districts and more closely populated suburbs
are bounded on the north by the river Thames, on the
east by the river Ravensbourne, and on the west by
the river Wandle; the southern boundary being a summit,
or water-shed level, about Mitcham, Streatham, Norwood,
and Sydenham. The river Thames now receives the drainage of this district from the Heathwall, the Effra, the Bat-
tle Bridge, the Great St. John's, the Duffield, the Earl, and
several minor sewers along its course, for a length of
11 miles, between the outlets of the rivers Wandle and
Ravensbourne. It is not proposed to intercept the rainfall
upon the lands adjoining these two last-mentioned rivers;
they will continue to receive it as heretofore. But it is
proposed to divert from those lands, as far as practicable,
the drainage of such roads and buildings as may be built
upon them.

" The northern half of the entire area is mostly urban;
and its general surface varies from the level of Trinity high
water to 6 ft. below it. This tract has, doubtless, at one
period, been a marsh covered by the waters of the Thames,
although many of the houses upon it have now basements
and cellars below the ground.

" The southern half is suburban, and its surface falls
rapidly towards its northern or lower margin, from an elevation of 350 ft. higher; so that during heavy rains the
floods from the whole of this upper district descend with
great rapidity into the sewers of the two lower districts,
and, the mouths of those sewers being closed by the waters
in the Thames for about eight hours each tide, the sewers
themselves, moreover, being insufficient to store the storm
waters, the result is, that during that period the lower districts are flooded.

" It will be evident that it is impossible to maintain a
continual and unintermitting flow in the sewers of the two
lower districts, or to drain the cellars and subsoil, and make
them dry and healthy, without the aid of pumping.

" The first object, therefore, has been so to intercept the

waters from the suburban, or high district, as to carry off the largest possible amount by gravitation, and, having thus relieved the low districts from periodical inundations, whereby their size is reduced to a manageable limit, to deal with them as hereinafter described.

"The high-level intercepting sewer commences at the north-west corner of Clapham Common, being of the ordinary form (3 ft. 9 in. by 2 ft. 6 in.), and falling at the rate of 1 in 341 to its junction with its northern branch, its route to this point being through Clapham, Park Road, Acre Lane, and Cold Harbour Lane, where it intercepts the Effra sewer near the Brixton Road, its size increasing in proportion to the amount of drainage to be received by it, up to 9 ft. 6 in. in diameter at the point where it receives the drainage from 2200 acres. From this point the main line falls at the rate of 1 in 1760, or 3 ft. per mile, to its outlet. And from the junction with the northern branch to the junction with the southern branch, it is 12 ft. in diameter, and drains 3500 acres, its direction being through Cold Harbour Lane, Love Walk, and fields to the Lyndhurst Road, where it receives the southern branch. From this point to the outlet it is increased to two parallel culverts, each 11 ft. in diameter, passing mainly through fields and garden ground, and under the Brighton and North Kent Railways, near New Cross stations, to the mouth of the Ravensbourne, at Deptford Creek, into which it will discharge the drainage from 7850 acres by gravitation at high water of the highest tides.

"The northern branch commences with a sewer 4 ft. in diameter, by a junction with an existing sewer in Rectory Grove, Clapham; and it falls at an inclination of 1 in 175 through Larkhall Lane to Union Road. Here it increases to 6 ft. in diameter, and falls to its junction with the main line in Cold Harbour Lane, at the rate of 1 in 1472, again intercepting the Effra sewer at a lower point in the Brixton Road, and draining 650 acres through a 7 ft. sewer.

"The southern branch, or Effra diversion, commences

by a junction with the Effra sewer at Croxted Lane, near Dulwich. It falls, at inclinations of from 1 in 273 to 1 in 100, in a direct line to its junction with the main sewer in the Lyndhurst Road, being 9 ft. in diameter, and receiving the drainage from 3100 acres.

" These branches are important parts of the high-level sewer, and without them it will not answer. In particular, the southern branch will save the covering of an equal length (about two miles) of the present open Effra sewer, the waters of which, instead of descending by a circuitous route, and with a flat gradient into the lower districts, will be diverted at a high level to a point nearer the outfall; and thus the size and cost of the upper portion of the main line will be reduced to one-half of what would otherwise be required, while a deep outfall for a new district now subjected to floods, namely, Dulwich, will be provided.

" Taking the more comprehensive view of the drainage of the whole area, it is proposed to extend and slightly alter the design for the low-level sewer laid before the Committee on the 20th ult., so as to include the drainage of Battersea and Battersea Fields. The present scheme commences with a circular sewer, 5 ft. in diameter, at the level of the bed of the Falcon Brook, Battersea. It passes along Battersea Road, under the South-Western Railway, along Priory Road, Lansdowne Road, and the Clapham Road, to St. Mark's Church, Kennington Common, up to which point its inclination will be 1 in 1760, or 3 ft. per mile, and its size will then have been increased to 8 ft. in diameter. From this point it proceeds in the direction of Mr. Forster's line, by the side of the Surrey Canal, to the Old Kent Road, at an inclination of 1 in 2112, or 2 ft. 6 in. per mile, being at this point 9 ft. in diameter. It then proceeds along the Old Kent Road, in Coldblow Lane, where it leaves Mr. Forster's line, taking a direct course across the fields and market gardens, under the Brighton and the North Kent Railways, along New Douglas Street

and Griffin Street, to High Street, Deptford, its inclination
for this length being 2 ft. per mile, or 1 in 2640, and its
size 10 ft. in diameter. Up to this point it receives the
drainage from 4450 acres. At High Street, Deptford, its
size is increased to 12 ft. in diameter, being intended
eventually to receive here an important branch sewer from
the Lower Deptford Road, which will divert from the
Thames the drainage of about 3300 acres. The main line
then passes between the Gas Works and the Vitriol Works,
near the Greenwich Railway, and under Deptford Creek, to
a vacant piece of ground, well adapted for a pumping station, or for sewage manure works, or for continuing the
line at some future time further eastward.

" The lift from the bottom of the sewer to Trinity high-
water mark would here be 30 ft. ; and from 1500 to 2000-
horse power would have to be provided at the pumping
station.

" The lengths of the proposed sewers would be as follows :—

High-Level Sewer.	Miles.	Fur.	Miles.	Fur.
Main line	6	2		
North branch	2	0		
South branch	2	1		
Total			10	3
Low-Level.				
Main line	7	4		
North branch	2	3		
Total			9	7
Grand Total			20	2

" The sections of these sewers are of a favourable character; their inverts generally average from 20 to 30 ft. below
the surface ; and a very small portion of the work will have
to be tunnelled

" Careful attention has been paid to the character of the
surface under which the sewers are to be constructed; pri-

vate property and houses have been avoided as much as possible; but, owing to the streets being irregularly laid out, it has not been practicable to effect this altogether, and in such cases the comparative value of the property has been duly considered.

"It will be observed that, to save the permanent cost of pumping, the gradients of the low-level sewer have been reduced; but the increase in the size of a sewer compensates for a loss of fall; and the sewers will still maintain, through their whole length, a velocity of current varying from $1\frac{1}{2}$ to 2 miles per hour, with the ordinary flow of sewage, and during rains that velocity will be increased. In this sewer there will be a constant current of a large body of water, so that the average velocity will be sufficient to keep it clear from ordinary deposit; but there will also be a good opportunity of flushing it, together with the other sewers in this district, from the catchwater sewer and the Thames at high water.

"The *minimum* velocity of current in the high-level sewer will be about $2\frac{1}{2}$ miles per hour, while at the upper end it will be considerably more.

"The drainage of those portions of the metropolis at present requiring a deeper outfall must not be made dependent upon any scheme of interception, unless such scheme is adequate to carry off the rainfall, as well as the sewage, from them. It therefore follows that, whatever scheme of interception is eventually adopted, if it be short of such capacity, the sewerage and drainage of the metropolis must be perfected and completed independently of it.

"If this reasoning be sound, it is manifestly more economical to construct one work which will answer every purpose, than to complicate it by the execution of duplicate sewers, in order to gain the same object which may be obtained by one; and it remains only to show that these sewers are sufficient to carry off the rainfall, as well as the sewage, from the districts for which they will be the outfall.

"An average of an annual rainfall amounting to 30 inches, added to the *maximum* daily flow of sewage, would give only 3000 cubic feet per minute upon the upper district, and 4000 cubic feet upon the lower district, while extraordinary thunderstorms, amounting to 1 or 2 inches in an hour, are recorded. These, however, generally occur during the hot seasons, when the ground absorbs and the air evaporates a large portion of the rain. Such storms are generally partial, covering but a small district at one time, and are never of long duration, so that the whole of the waters that have fallen upon the district do not reach the outfall sewer until after the storm has abated; and, a small portion only of the sewer being pre-occupied at the commencement of the storm, the remaining portion for some time stores the flood waters. A heavy and uniform rain of long duration is probably the most severe ordeal to which sewers are subjected, when, the ground and the air having become charged to their full extent, absorption and evaporation abstract but little from the sewers, and the feeder sewers, having also become charged, would be delivering to their full extent.

"The largest quantity of rain which fell in London in one day during the wet season of 1852 measured 1¼ inch; but storms, amounting to 2½ inches in a day, although of rare occurrence, have been recorded, and these would deliver into the high-level sewer 45,000 cubic feet per minute, while it is capable of discharging upwards of 46,000. The low-level sewer will discharge 21,400 cubic feet per minute; and it is here proposed that the existing main sewers in the low districts should act as reservoirs during such extraordinary storms. From their construction, and the remarkable flatness of their level, they are peculiarly adapted to this object; and it is estimated that on these occasions they would store the surplus floods without the rising of the latter to an inconvenient height, and would again deliver their contents into the Thames at low water, or into the intercepting sewer, as it became free to receive them Thus the provision of engine power for such rare

occasions would be saved, and the existing sewers made
available.

" The estimated cost of this work will be as follows :—

High-Level Sewer.	£	£
Main line	195,600	
North branch	20,400	
South branch . .	42,000	
		258,000
Low-Level Sewer.		
Main line . . .	140,000	
North branch	32,000	
Engines, pumps, &c. . .	100,000	
Land, buildings, &c., required for the permanent establishment	100,000	
		379,000
Total		£637,000"

*Extracts from the Report of the Consulting Engineers on the
Surrey and Kent Drainage, or the Drainage of that portion
of the Metropolis which lies South of the River Thames.*

" The object of the measure is to drain, take off, or intercept all the rain-water and sewage which now fall into the
Thames from the district which lies between the rivers
Wandle and Ravensbourne, and does not at present fall
into those rivers, the whole of which comprises an area of
at least 24½ square miles, or 15,680 acres, of one-half of
which the rainfall can be intercepted and taken off by gravitation; but the other, or lowest half, the greater part of
which lies below high-water level, would have to be pumped
up and delivered into the river at some point below London
—say Deptford Creek—in the first instance, and subsequently at some point lower down the river, should such an
extension be deemed advisable.

" It may be here observed that the pumping-station at

Deptford Creek would afford a most excellent situation for
the manufacture of 'sewage manure,' by which means the
sewage-water would be deodorised and delivered into the
Thames, divested of its most offensive and deleterious
matter, which by that process would (as is affirmed) be
separated and converted into a most valuable product for
agricultural purposes.

" To prevent mistakes and misapprehensions on the vast
subject of draining this great metropolis, and particularly
as regards its cost, we deem it proper to state in this place
that the special object of this report is not the 'sewerage,'
properly so called, of the metropolis south of the Thames,
but, in reality, the construction of a work, by means of
which the sewerage of the southern portion of London
may be freed from flooding in the first place, and well
drained in the next; in short, to produce the same effect
that would be produced by raising the land on the south
side of the metropolis 20 ft.; in which case, every main
sewer could discharge itself into the river, either deodorised
or not, as might be deemed advisable, without pumping,
and the cost of sewerage would be the expense of the
sewers only to effect the drainage of the basements of all
the houses ; and the whole expense of what is herein recommended is the construction of a set of underground
main intercepting drains or 'arteries,' sufficiently low to do
that by the artificial means of pumping which the Thames
cannot do of itself by natural means ; and it follows, of
course, that the cost of putting things in this position, and
constructing this 'arterial drainage,' is altogether over and
above the usual cost and expenses of carrying into effect
the sewerage of a town, a circumstance which cannot, in
our judgment, be too strongly impressed upon the minds
of those who are crying out (and certainly not without reason) for improved sewerage ; for, supposing that the 'arterial
drainage,' with its plant of steam-engines and pumps, costs,
in the whole, three-quarters of a million sterling, it will
probably cost another quarter to complete the 'sewerage'

of the district even as it now exists, leaving alone further extensions of the southern portions of the metropolis.

"Now, looking to the fact that no real benefit can be derived till the plan be fully carried out to the extent only as herein contemplated, and that the sooner it can be done the better, the question is, Can 750,000*l.* be raised forthwith to complete this 'arterial drainage,' and secured by special rates upon the whole district to which it equally applies, leaving the raising and expending of money for the common sewerage to be managed as at present in the different districts; the first would be a uniform and standing rate for a given long period, till the sum borrowed was paid off, with interest, which would be followed by a small uniform rate, to cover the current expenses of pumping, &c. ; and the second, a rate varying both in district and amount, according to the wants and current expenditure of the district in which the new sewers or other works are situated.

"The raising of the funds necessary to carry any measure of this kind into effect, and securing the repayment, is of equal, or, perhaps, even greater importance, than the measure itself, and cannot, in our opinion, be too strongly impressed upon the Commissioners, and by them upon that department of the Government with which such matters rest ; but we again repeat our opinion that the capital to be raised for the arterial drains, such as the Hackney Brook sewer, on which we have already reported, on the north of the metropolis, and the high and low level arterial drains on the south, should be considered as separate and distinct, both in the mode of raising and rating, and also of repaying the money, from that for the common sewers, which are made to discharge into the arterial drains.

"We have, &c.,
"W. Cubitt,
"R. Stephenson."

274. On the 27th of February, 1854, a Special Court of

Sewers, composed of members both of the city and metropolitan Commissions, was held, for the purpose of receiving, considering, and determining upon, the report of the surveyors to the commissioners, "upon the sewage interception and main drainage north of the Thames, a portion of the works described in such report having relation to sewers made or to be made within the City of London and liberties thereof." Appended to the report, which described minutely the localities through which the main sewers were proposed to be carried, was an estimate of the probable expense of the works requisite on the north side of the Thames, amounting to 1,378,190*l*. This is exclusive of the proposed "Hackney Brook" northern high-level intercepting sewer, estimated at 271,290*l*., including the cost of the work from Kilburn and Hampstead to the river Lea, and a branch to Four Mills. Sir W. Cubitt, one of the consulting engineers of the Commission, added a revised estimate of his own as to the probable actual cost of the works described in the report, and also of those proposed for the south side of the Thames, including the extension to the Thames in Plumstead Marshes. As to the north of Thames intercepting system and conveyance down to Barking Creek, Sir William estimated 1,750,000*l*. instead of 1,378,190*l*. And for the south side he allowed 750,000*l*. for the works as terminating at Deptford Creek, and previously estimated by Mr. Stephenson and himself, besides 500,000*l*. for the extension of such works through Greenwich and Woolwich to an outlet in Plumstead Marshes, making a grand total of 3,000,000*l*. for completing the arterial and intercepting drainage of the entire districts north and south of the Thames.

275. It has appeared desirable to quote these reports at considerable length, in order to record the latest proposals emanating from the highest engineering authorities yet consulted on metropolitan drainage. The Commission, under whose authority these reports were prepared, however, on the occasion of the meeting last referred to, held on

February 27, 1854, adjourned *sine die*, in consequence
of the receipt of a letter forwarded by order of Lord
Palmerston, the Secretary of State for the Home Department, avowing his opinion that the system of drain
age advocated by the Board of Health, (as distinguished
from that adopted by the Commissioners) was " that which
ought to be adopted, as combining the greatest degree of
efficiency with the greatest degree of economy." Now, the
essential difference between the " systems " advocated by
the " Commission" on the one hand, and the " Board" on
the other, will be apparent from the following extract from
a letter addressed to Lord Palmerston by Mr. F. O. Ward,
and referred to in his lordship's communication to the
Commissioners just quoted :—

" The Commissioners of Sewers and the Board of Health
are at issue as to the cheapest and best way of draining
houses.

" The Board of Health advocate the drainage of each
house block by tubular submain running behind the houses,
and receiving the sewage of each by a short tubular branch.
They recommend a large reduction of the sizes of drains
hitherto employed ; for the single house drain they recommend a 4-in. pipe ; for the submain receiving several of
these a 6, gradually expanding to 9, 12, and so on up to
20, as the lengths of the submain and the number of
branches received by it increase ; such drains, they say, are
self-scouring, the run of water through them is so concentrated that it keeps them clear of deposit; the branches
being very short, and running backward towards the drain
behind, instead of forward beneath the houses towards a
sewer in the street, have a quicker fall, and, in case of
leakage, leak into the open air, not into the houses, while
the cost is so much reduced by this method that blocks of
labourers' houses may be thoroughly drained and fitted
with sinks and soil-pans for an improvement rate of less
than 2*d.* per house per week.

" The Commissioners of Sewers, on the contrary, recommend large brick sewers under the street in front of the houses, beneath each of which they carry a long drain from back to front, strictly forbidding more than two houses being relieved by one pipe drain ; a system which, whether otherwise good or not, certainly entails an enormous increase of expense on the house-owners, and thereby redoubles the resistance on their part to sanitary improvement.

" The Commissioners of Sewers lay great stress on the independence secured by their system to each house-owner, whose premises are traversed by his own drain only, which he may take care of or neglect as he pleases, as its stoppage can only injure himself. They allege against the Board of Health plan, that it trespasses on private property, that any stoppage of the tubular submain causes all the houses higher up to suffer, and renders it necessary for workmen to enter private back-yards to search out and remedy the evil. They also deny the self-scouring property of the tubular submains, and refer in support of their view to several hundred cases of stoppages in tubular drains, collected in a report of Mr. Bazalgette.

" The Board of Health, on the other hand, declare the householders' pretended independence on this plan to be illusory, seeing that the big street sewer is, nine times in ten, an elongated cesspool, where the sewage of each stagnates to the detriment of all, while the stoppage of a branch drain running beneath a house on the old plan often causes stench in the houses on either side. Against the inconvenience of an occasional invasion of the back-yards by workmen, they set the greater inconvenience of periodical invasions of the street and stoppage of the traffic, besides tearing up of kitchen floors to get at foul drains under the houses. In reply to Bazalgette's report, they bring forward examples by hundreds of pipe sewers working admirably year after year, and attribute such stoppages as have occurred to errors incidental to the first introduction of a new system—errors which, once known, may be avoided

Among these they cite the use, at first starting, of drain pipes which were too thin, so that they broke under the pressure of superincumbent earth, the uneven laying and careless jointing of pipes by the jobbing builders often employed (as in some parts of Croydon) on this work, the use of soil-pans with ducts as large as the pipe drains to which they led, and, above all, the often scanty and intermittent supply of water to the system. They further point out that a population accustomed to open privies and cesspools, available as receptacles for solid refuse of every kind (refuse properly due to the ashpit), have to go through a transitional period of a few weeks, while they are learning that such practice leads to stoppage of the new circulating system, and must be abolished along with the old stagnant cesspools. They say that many of the tubular sewers, put down as failures in Bazalgette's report, are at this moment working perfectly well, for which reason they take all his allegations with very great reserve. They point, with some reason, to Lambeth Square, New Cut, where the Commissioners have allowed 32 houses to be drained by four tubular back drains, which act perfectly well, quite as well as 32 separate drains could act, though these would have cost eight times as much; and they ask, why this eightfold cost should be imposed upon London at large by the very same authority which sanctioned in this square the combined drainage, which is found to work well."

276. On the 2nd March, 1854, on the motion for the second reading of the " Great London Drainage Bill" (which was negatived without a division), Lord Palmerston stated he was about to re-organise the Commission of Sewers, and that Commission had lately matured a plan for the drain age of the metropolis, which, upon full consideration, he believed to be a good one, and likely t⁊ effect the objects which the scheme now before the House was intended to effect. He thought there would be great advantage in the drainage of the metropolis being effected and managed by one central department in some degree connected with the

local authorities. Moreover, if it should be possible to realise that which, according to this Bill, might be realised —namely, the connection of the drainage of the metropolis with some commercial advantages from the transformation of sewage into manure—that also ought to be in the hands of Commissioners, in order that any such profit might be applied for the public benefit in diminution or relief of the rates raised for the construction of the sewers.*

SECTION II.

Supply of Water.—Public Filters and Reservoirs, &c.

277. *Quantity* and *quality*—criteria of every-day application—have special reference to the supply of water for every congregation or community of human beings. The varied practical purposes of domestic life to which this invaluable agent is alone applicable, and the intimate connection of many of these purposes with the health, life, and well-being of humanity, at once attest the high importance of abundance and of excellence in our command of water. Rivers, springs, and surface collections have already been enumerated as the several sources of water for the use of towns, and the advisability of resorting to one or other, or combinations of these sources, has been shown to have some dependence upon the superficial contour of the town and suburbs. Facility of supply, promoting the economy of the means, will of course always have great influence in determining the source to be referred to. Rivers and their feeders—brooks or streams—may be classed, as the most abundant sources in most instances, but their applicability can seldom be realised without some expenditure of power— natural or artificial. Surface collections and springs, on the other hand, are frequently applicable by the force of

* The subsequent proceedings of the various public bodies down to 1865, relating to the Main Drainage of the Metropolis, the Utilisation of Sewage, and the Purification of the Thames by Embankment, will be found noticed in the Appendix, Nos. 3, 4, and 6; pp. 194, 205, and 225.

gravity unaided by power, and requiring only suitable channels in which the supply may be conducted from the higher lands around the town. The cost of power has, however, lost much of its importance as an element in the calculation since the steam-engine has enabled us to perform constant and easily-regulated duty in the raising and conveyance of water at a very small expense; and, therefore, the comparative abundance of the several sources at all seasons will determine the preference rather than the susceptibility of self-propulsion.

278. Where a choice is afforded as to sufficiency of supply, however, the *qualities* of the water should be allowed great influence in ruling the selection. Tracing all these forms of immediate supply to the common original one of rain water, we may readily infer, from a knowledge of its ordinary properties, and of the effect of its subsequent treatment, the particular stage of this treatment at which it will be the most desirable to convert the water to our purposes. Rain water, as already shown (Part I. paragraph 54), contains ammonia, but it is, as well known, the least impure in constitution of any water at our command. All the earthy, animal, and vegetable matters with which water becomes charged, are extracted from the soil through which, or the surfaces over which, it passes. The nature of these matters depends upon the constituents of the soil which is percolated; the amount of them will be in proportion to the time during which the water is maintained in communication with the soil, modified of course by the degree in which they may be adapted for mutual action. Hence it follows that the scale of comparative purity would stand thus:—1. Rain-water. 2. Water from surface drainage. 3. Water from soil drainage or percolation. 4. Water from rivers or brooks. 5. Water from springs and subterranean sources. Regarding No. 4, however, it is to be remarked that the collection of the water in any kind of channel allows of a partial deposition of the heavier particles which the water has imbibed, facilitated, of

E

course, by the depth of the body of water, and the slowness
of the current, or minimum of motion. And besides this,
the exposure of water to the action of the atmosphere
appears to assist the evolution of some constituents which
impair its purity. Water from streams and rivers comes
thus to be considered as next in comparative purity to rain
water immediately derived, while that taken from springs
and sources in which it has long remained in intimate
contact with soluble earths and other matters, is found to
have acquired a corresponding proportion of these impurities.

279. The different kinds of impurities contained in water
have been explained (56 and 57), and the means of testing
and correcting some of those shown (58 and 59). The
process of filtration through the soil, which the water
derived from subterranean sources undergoes, tends to
separate the animal and vegetable impurities, and thus
spring-water and well-water being comparatively clear, are
commonly reckoned as pure. The matters which these
waters nevertheless contain are, however, separable only by
chemical treatment, while the impurities of river-water
may be got rid of to a great extent by mere subsidence and
self-filtering.

280. The several purposes for which water is required in
a town, or collection of people, are—1. Ordinary domestic
uses, including drinking, washing of persons, clothes,
utensils, houses, yards, and watering gardens, &c. 2. Manufactures. 3. Supply of public buildings, baths, wash-
houses, &c. 4. Extinction of fires. 5. Cleansing and
watering of streets and thoroughfares. 6. Supply of fountains, and public gardens and pleasure grounds. 7. Miscellaneous and occasional purposes not included in the
foregoing.

281. The supply necessary for the total of these purposes may be reduced into an average quantity for each
individual of the population, and each square acre, yard, or
foot of the superficial area of the town. The latter datum

will also afford the means of estimating the proportion of the supply which will be immediately rendered in the form of rain, and the difference between the amount of which, and the total quantity required, will represent the proportion to be served by other means.

282. Adopting 24 inches as the average annual fall of rain, and half of this as remaining after evaporation, as this quantity will facilitate an approximate calculation, and be sufficiently near the truth for the purpose, (an exact average for places, years, and seasons being scarcely calculable even by the most laborious computation,) it appears that 1 cubic foot of rain-water is annually retained upon each square foot of surface, or 9 cubic feet on each square yard, equal to 43,560 cubic feet upon each square acre.

283. For the first, second, third, and fourth of the purposes enumerated (280), a daily supply of 20 gallons for each individual will be a fair average, being more than sufficient in towns having an ordinary proportion of manufacturing operations carried on within them, and nearly, if not quite so, even in towns where an excessive proportion of manufactories exist. This may be inferred from the quantities now supplied in towns. In Preston, Lancashire, the supply by the Waterworks Company is on an average 80 gallons daily to each house, including factories and public establishments, and as the service is constant and the quantity unrestricted, it is presumable that much of this quantity is wasted, and, if properly reserved, might be made to supply, partially at least, the cleansing of the streets. The tenements occupied by the labouring classes in this town are estimated to consume only 45 gallons each daily. Assuming 5 as the average number of occupants of each house, the supply to each in these cases will be 16 and 9 gallons respectively. In Ashton-under-Lyne the daily supply to each house is 55 gallons, or 10 gallons to each person; and 18 factories in this town consume 1,103,000 gallons daily. Experiments tried in the year 1847 proved that the daily consumption per head of the

tenants supplied by the Ashton Waterworks Company averaged 6·245 gallons; while the quantity supplied to the mills in the neighbourhood averaged about 7 gallons per head in addition, making a total of about 14 gallons per head per diem. In Nottingham, the "Trent Water Company" supply 17 or 18 gallons per individual, daily, including the trade consumption. The quantities supplied by four of the leading companies in the metropolis are as follows :—(for 1850).

East London	100 gallons per house per diem.		
New River	114 ,,	,,	,,
West Middlesex	150 ,,	,,	,,
Chelsea	154 ,,	,,	,,

These rates of supply will be found to corroborate the average we have assumed for each individual. Thus in the district supplied by the East London Water Company, including Spitalfields, Bethnal Green, Poplar, Limehouse, and other populous neighbourhoods filled with the poorer class of persons, it will be found the average number of persons is much above 5; 7 or 8 would probably be much nearer the truth. The New River Company also supplies populous districts. Many of their customers are similar to those just described, and the average of all would certainly give more than 5 persons to each house. In the districts supplied by the West Middlesex and Chelsea Companies, the population is mainly of another class, or rather classes, but all of which occupy larger houses than those in the Eastern and Northern parishes, and the average consumption in each house is high in comparison with the others, owing to two causes, the larger number of residents in each house, including domestics, &c., and the larger quantity consumed in baths and other means of private luxury and comfort which are beyond the command of the other classes of society.

284. Although it would thus appear that an allowance of 20 gallons per diem for each head of the population will

suffice for domestic and manufacturing purposes,* including the supply of public buildings and for the extinction of fires, we would prefer to provide for a constant service of 30 gallons, in order to make an ample provision for all possible casualties and increased demands. Water is pre-eminently so valuable, and, when properly sought, so cheap an agent, that extravagance should always be permitted rather than a deficiency be risked.

285. For the three remaining purposes, viz. :—the cleansing and watering of streets and thoroughfares, the supply of fountains, and public gardens and pleasure grounds, and such miscellaneous and occasional purposes as are not included in the six preceding classes, the average quantity of water required may be reduced, for an approximate estimate, into a given depth per diem, or annually, according to the surface occupied by the town and suburbs to be supplied. Towards this quantity, the rain may, as we have seen (282), be estimated to contribute an annual average depth of 12 in. available water. Now, allowing $\frac{1}{10}$th of an inch of depth of water to be daily required over the entire surface of the town for the several purposes stated, (and we believe this to be a liberal allowance,) we shall have an annual total depth of $365 \div 10 = 36\cdot5$ in., which may be regarded as 36 in., from which deducting the 12 in. supplied by the fall of rain, we have the remainder equal to 24 in. depth to be supplied by other means.

286. We thus derive a rule as to the quantity of water required to be supplied in any town, which calculates the total quantity upon two given data, viz. :—First, the amount of the population; and, secondly, the superficial extent of the town and neighbourhood to be provided for. Thus, by way of example, let us suppose a town having a population

* In June, 1850, it was estimated, upon official data, that there were 238,000 houses in the metropolis, of which 270,000 were supplied with water, the quantity of which was 45,000,000 gallons daily, or 167 gallons per house. For later years, see Appendix No. 3, p. 194.

of 100,000 persons, and an area of 1000 acres. The quantity required to be provided annually for this town, would be,

		Gallons.
Population 100,000 × 30 × 365 =		1,095,000,000
Area 1000 × 43,560 × 2 × 6 =		522,720,000
	Total annual quantity	1,617,720,000

allowing each cubic foot to equal 6 imperial gallons, which is sufficiently near the truth for a general calculation.

287. Having thus endeavoured to arrive at an approximate estimate of the *quantity* of water required for any town, formed upon the data of the amount of population and extent of surface to be supplied, we have now to refer to the question of *quality*, and cite such observations as we can, which have tended to exhibit the qualities of water derived from the several sources of rivers, springs, and surface collections, or superficial drainage. In these particulars it will also be useful to include such accounts of the topographical and geological features of the towns and districts referred to as we can collect from the trustworthy testimony of witnesses before public Commissioners.

288. The borough of Preston comprises an area of 1960 acres, a population (in 1841) of 50,131, and 9994 houses at the same date. The town stands principally upon a dry sand of the "recent formation," marl, clay, and gravel existing in some parts. At a depth of about 90 ft. from this surface-soil, the "new red sandstone" is found; the same rock forming the bed of the river Ribble, which through two miles of its course flows at about a quarter of a mile distance from the town, which has a general westerly slope towards the river, the highest sites being about 130 ft., and the lowest about 35 ft. above its low-water level. More than half of the town is supplied with water by the Preston Waterworks Company, which derives its supply from the "mill-stone grit" formation at Longridge, distant about seven miles eastward from Preston. The remainder of the town

is supplied from wells. The whole of the supply from both sources is described as of excellent quality, but we have no analysis to determine its ingredients. The geological influences by which water derived from such strata as are here described is affected, are certainly likely to furnish a water of good general quality and comparatively free from soluble mineral impurities, while the elevated position of the town in relation to the river would discourage a resort to it for general supply upon economical grounds.

289. Chorlton-upon-Medlock, one of the townships of the borough of Manchester, from which it is indeed separated only by the little river Medlock, comprises an area of about 700 acres. The number of houses in 1841 was 6021, and the population about 29,000. The soil is of two kinds, stiff clay over the southern part of the town, and gravel chiefly over the northern. The geological formation is the new red sandstone, which is found at depths varying from 3 to 90 ft. from the surface. A stream called " Corn Brook " which flows through the township for more than a mile, and delivers into the river Irwell at a distance of about two miles, is little better than an open drain, and keeps that part of the town near to its banks in a damp and unhealthy condition. The supply of water is derived partly by a Waterworks Company from Gorton Brook, which affords the only stream-water fit for use, and partly by pumps from wells in the gravel and sandstone. The water from these latter sources is described as being " bright and sparkling and well tasted, but hard."

290. The town of Ashton-under-Lyne is built on a gentle declivity on the north-west bank of the river Tame, above which it is elevated from 30 to 40 ft., the surrounding country being remarkable for its generally level character. The principal geological feature of the neighbourhood is the great coal deposit, the surface-soil being clay and loam, and the subsoil clay and gravel. The sub-strata are chiefly schistus and sandstone, with intermediate layers of coal. The water for the supply of the town is derived

by a Waterworks Company from springs in the higher parts
of the parish, and is of a medium quality, being such, however, that it is said to be " wonderfully " improved by filtration.

291. York, situated in the centre of an extended vale,
lies between the rivers Ouse and Foss, and immediately
above their junction. Both of these are navigable and
tidal rivers, but the tide is prevented from rising to the
city by a lock placed five miles below it. The available
water is derived from the river Ouse, from wells varying in
depth from 12 to 40 ft., and from borings from 350 to 380
feet deep from the surface. The inquiries of the Rev. W.
Vernon Harcourt, and of Messrs. Spence and White, of
York, have furnished us with much valuable and accurate
information as to the qualities of these waters, and the
geological conditions in which they are presented; and,
from the records of these inquiries, a few facts may be
advantageously quoted as illustrations of general principles,
which will be found commonly applicable to the several
sources of water for the supply of towns.

292. From these records it appears that the total of
gases contained in one gallon of river-water, from the Ouse,
amounted to 10·4 cubic in., and the average of 14 waters
from the springs, or superficial wells, amounted to 23·8
cubic in. That the total of solid contents (consisting of
carbonates of lime, magnesia, and iron, sulphates of lime
and magnesia, muriates of soda and potash, silica, and
vegetable matter,) in one gallon of river-water amounted to
9 grains,—while the average of solid contents of the fourteen
well-waters amounted to 64·96 grains per gallon, comprising
the same carbonates, sulphates, and muriates as found in
the river-water, with the addition of muriate of lime in
some specimens, and of the nitrates of lime, soda, or magnesia in all. An analysis of the water from the deep
springs, made by the Rev. W. V. Harcourt, showed the
presence of 96 grains of solid contents in one gallon, and
of this quantity about half consisted of medicinal salts;

viz., 33·9 grains of the crystals of sulphate of magnesia, and 14·4 grains of the crystals of sulphate of soda, besides a small proportion of bicarbonate of iron.

293. The causes of these differences of ingredients (which, together with considerable difference of level at which the waters are maintained in the several wells, evince their independence of each other, and of the river) are referable to the geological conditions under which they are collected. The section of an Artesian well sunk to a depth of 378 ft. in the city showed the following arrangement of strata:—clay and gravel, 18 ft.; fine river-sand, 60 ft.; sandstone rock and loose sand, 60 ft.; a thin seam of blue clay and water, and sandstone rock, 62 ft.; another thin seam of clay and water, and sandstone rock, 178 ft. The Rev. W. V. Harcourt describes this sandstone formation, and the structure of the bed of the river Ouse, as follows:— "This sandstone rock belongs to the beds of the new *red sandstone* formation, which crop out in a low line of undulating hills along the western margin of the basin of the vale of York, passing in a south-easterly direction from Rainton to Borough Bridge, and Ouseburn to Green Hammerton, and emerging again from beneath the diluvial covering of that basin at Bilbrough, within a few miles of York. The immediate substratum of the soil in this line over a considerable tract of country consists of these porous beds, and the water which falls or flows down upon it passes through them, between the seams of clay which alternate with the sandstone, along the dip of the strata, eastward to York; it is thus carried between the diluvium below the bed of the Ouse, and is dammed up under the superincumbent mass, in the reservoirs of the sandy beds, to the above-mentioned height of 15 or 20 ft. above the summer level of the river, to which height it is found to rise where the superior seams of clay are perforated by boring. The water of the Ouse consists chiefly of the contributions of the rivers which flow from the high hills on the north-west of York, (especially the Swale, the Ure, and

E 3

the Nid,) and are fed by the rains falling on their summits. The streams from this source, after percolating the *mill-stone grit*, with which those hills are capped, find their channels on the surface of the impervious beds of the sub-jacent *limestone* and *shale* along the valleys, and are con-veyed on linings of *diluvial clay* across the edge of the superior strata, and over the drift-covered plan of the *red sandstone* to York. To this account of the geological con-ditions under which York is supplied with water, is to be added:—1st. That the gritstone hills which furnish the river-water include few materials of saline impregnation. 2nd. That the beds of the red sandstone in which the deep springs run are pre-eminently saliferous. 3rd. That the rubbish of centuries accumulated in some parts of the city to the depth of three or four yards over the diluvial beds, which contain the superficial wells, is full of decomposing matters, tending to mineralize and contaminate the water. The waters of these wells, accordingly, are highly charged with solid matters, amounting, on an average, to about 60 grains held in solution in an imperial gallon. In two cases Mr. Spence found in them from 6 to 7 grains of Epsom salts, and in one 11 grains; in two others he found 31 and 38 grains of neutral salts of soda and potash. In these last an infiltration may be suspected from the deep springs; but in general there are sufficient materials in and upon the drifted beds to account for the sulphate and carbonate of lime, of which the solid contents of these waters are chiefly compounded, and which render them harder than is desirable, either for drinking or for culinary use."

294. The evidence here so well adduced is amply suffi-cient to account for the differences observed in the chemical qualities and adulterations of the water derived from the several sources; while that from the river Ouse, on the other hand, furnished by the gritstone hills, being purer at its source, and subsequently improved by exposure to the air, contains only 9 grains of solid contents in the gallon,

and presents an exhaustless source of water of excellent qualities for all the purposes of the city.

295. The materials of some soils are particularly preju- dicial in their effects upon water passing through them. Thus peat impregnates the water passing through it to so great an extent, that it becomes discoloured, and thus ex- poses the origin of its impurity. Mr. Homersham, who devoted much attention professionally to the several water- sources around Manchester, has recorded his observations on this subject, and cites the confirmatory remarks of per- sons residing in the valley of Longdendale in that locality, that "upon heavy rains following a drought in the summer time, the water flowing down the streams is about the colour of London porter, and so strongly impregnated with moss and peat, 'that it can at such time be smelled a field off.'"* When the water derived from peat lands passes through mineral rocks of particular formation, a process of natural filtration is effected by which the colouring matter is ab- sorbed, and the water emerges in a tolerably pure state. This fact was observed by Mr. Thom in examining water which flowed over or through a particular species of lava or trap-rock (amalgoiloid) in the hills above Greenock, and was found to have thus become purified equal to fine spring-water. Mr. Thom made good use of this observation by substitut- ing this rock, obtained in that neighbourhood at a nominal price, for charcoal in the subsequent process of artificial filtering.

296. The town of Nottingham, which is chiefly at a con- siderable elevation above the surrounding country, on the southern, eastern, and western sides, occupies the declivity of the southern termination of a long range of hills, and has the valley of the Trent about one mile in width at its foot. Three-fourths of the town has an elevation from 50 to 200 ft. above the valley, and stands immediately on the new red sandstone rock, which, being absorbent, remains

* " Report on the Water that can be Supplied to the Inhabitants of Man- chester and Salford, p. 85." Weale, 1848.

dry on the surface. The remaining portion of the town has a sub-stratum of similar material, but stands immediately on an alluvial deposit of gravel silt, and decayed vegetable matter, lying in the valley of the Trent or its tributary streams. By two of these, the Leen on the south, and the Beck on the east, which flows into the Leen, the waters are conveyed into the Trent. The town is supplied mainly by two water companies, whereof one derives its supply from springs, situated about $1\frac{1}{2}$ mile north of the town; and the other from the river Trent, on the banks of which a reservoir and other works have been constructed. A small part of the population is supplied by minor works, which, by means of steam-engines, raise their supply from wells sunk in the new red sandstone rock. The quality of all these waters is described as being good, but those from the sandstone contain " carbonate of magnesia in notable quantity," besides the sulphate and carbonate of lime, muriate of soda, &c. It is quite certain, therefore, that this water is, for all ordinary purposes, impaired in its purity and value.

297. Liverpool is situated partly on the side of the ridge of hills forming part of Everton, Edge-hill, &c., and partly on the crest of a minor elevation, the valley between the two having been the original streamlet or channel, which discharged into the old pool. The sub-stratum of about two-fifths of the city of Liverpool is clay. Along the banks of the intermediate valley the soil is chiefly a deposit of mud, with occasional beds of gravel, and in some parts irregular masses of rock. Between this valley and the southern and eastern boundaries of the town, a mixture of yellow sand and rock is found in small thin beds, but generally resting upon solid rock at an average depth of 15 ft. Liverpool is supplied with water by two public companies, one of which derives its supply from springs at Bootle, distant 3 miles from the town, and the other from wells in various parts of the town. These waters were analysed by Dr. Trail in 1825, and found to contain " muriate of soda

and of lime, the last in very small quantity; sulphate of soda, and possibly a minute quantity of sulphate of lime, carbonate of soda."

298. The town of Bilston has a declivity towards the brook called Bilston Brook, at its base, the fall being steep in the upper part of the town, and gentle in the lower part. "The geological character of the country is that of the coal measures overlying the Wenlock limestone. The only peculiarity is the presence of porphyritic greenstone, and occasionally compact basalt. The soil of Bilston, where collieries have not been opened, has a preponderance of aluminous earth. The subsoil is generally brick earth. The sandstone is rather an important feature in the geology of Bilston, on account of its compactness and great thickness." The water for the town is chiefly supplied by a Waterworks Company, and, being collected by land-streams which flow over beds of limestone, becomes impregnated with lime, and thus acquires a considerable amount of hardness.

299. Newcastle-under-Lyne stands partly on the old red sandstone formation, and partly on a strong mine of clay which extends into the coal formation of the Pottery district. The water springing from the former formation is somewhat hard, containing a small portion of carbonate of lime. That from the clay is much more hard, from its greater quantity of this carbonate.

300. Bath, which is built partly on the slope and lower part of a hill, rising from the right bank of the river Avon, where it forms a considerable bend round from east and west to north and south, stands upon the nearly horizontal beds of clays, limestones, sands, and sandstones, which constitute a portion of the series of rocks to which the term oolitic has been given—from the oolite or oviform grains in many of the limestones. From the interstratification of these different kinds of rocks, conditions for the occurrence of springs are numerous, and they are accordingly often met with, and from these the town is supplied

with water for domestic purposes. These springs occur at various elevations above the height of the river Avon, from 120 to 160 ft. The qualities of the water raised from the several wells vary according to the beds of limestones, clays, marls, sands, &c., in which they are formed. In the alluvial ground, on the right bank of the river and lower parts of the town, trees are sometimes met with in great abundance. These lie beneath an alluvial red loam, about 8 ft. thick, resting on gravel of about the same thickness, and this upon lias clay. The water where these trees are found is abundant, but never good. Some of the wells in the lias furnish tolerable water, but there are examples of sinkings in it to a depth of 200 ft., from which no water has been obtained. The sections of many wells sunk in the neighbourhood of Bath show that the water is retained among the various beds of clays at great depths beneath the Great and Inferior Oolites, and produces springs by cropping out on the sides of the hills.

301. While the topographical and geological character of the site of the town, and of the soil and sub-strata on which it stands, are the admitted guides as to the source or sources from which the town may be supplied with an adequate quantity of water of average goodness of quality, the criterion of quality as measured by relative *hardness* must be allowed a prevailing consideration. River waters, rendered impure chiefly by organic, animal, and vegetable matters, are susceptible of improvement by methods of filtration; whereas waters derived directly from drainage or internal springs are comparatively pure in these respects. but, on the other hand, are charged in various degrees with earthy and mineral matters, which at once render them less fitted for domestic purposes, and far less readily susceptible of purification. The economical results of the qualities of the water supplied to towns have been adverted to at some length in the first Part of the Rudimentary Treatise on Drainage. (Paragraphs 57, 58, and 59.)

302. In concluding these remarks on the qualities of

waters from various sources as subjects for consideration in estimating their comparative value, we may usefully refer to the confirmatory evidence supplied by analyses made under the direction of the Superintending Inspectors to the General Board of Health, of the waters available in the several towns of Chatham, Uxbridge, Croydon, and Dartford, reported upon by Mr. Ranger. The analyses were made by Dr. Lyon Playfair.

303. The water now used in Chatham is obtained principally from surface drainage from the upper chalk, but it varies greatly in the degrees of hardness. Adopting, as is presumed, the same measure of hardness as that used by Dr. Clark, and explained in Part I. (58), the hardness of the surface water from nine places of collection varied from 17° to 56°, the average of the nine being 27°, while the water of the River Medway has only 5½° of hardness. This water, however, contains a large quantity of a yellow deposit; and, comparing the qualities of all the waters, the Inspector recommended that the supply should be taken from the Boxley Abbey Spring, of which the hardness stands at 17°. This spring is about 5 miles from the town, and the situation being backed by elevated ground and considerably higher than any part to be supplied, is peculiarly adapted for the construction of reservoirs and filtering beds if required. The Report leads us to suppose that the reasons for preferring to bring water a distance of five miles, while that from the river is accessible to all parts of the town, is to avoid the expense of artificial raising of the latter. The relative hardness is, however, an item of great moment, and should receive full consideration. The deposit remarked in the river-water occurs, there is no doubt, from earthy matters held partly in solution, which would be readily removable by filtering.

304. Uxbridge is now supplied with water from four public pumps, from wells, and by dipping from the branch of the river Colne. The hardness of the water from three town pumps and two others varied from 26° to 52°, the

average being nearly 36°. The hardness of the water from one of the Artesian wells was found to be 34°; of that from two others $14\frac{1}{2}$° and 16° respectively. From the small degree of hardness in these two latter waters, we might conjecture some communication between these wells and the river Colne, the water of which has $15\frac{3}{4}$°, but the Report does not remark on this circumstance. The Inspector advised that the adequate supply for the town should be derived from the river Colne, at a part which would be favourable for the construction of reservoirs, filtering beds, and other necessary works.

305. The waters now supplied to the town of Croydon from springs and wells are found to have an average hardness of $25\frac{1}{4}$°. That from the river Wandle has 16°·1, and Dr. Clark reports than an expenditure of 1 lb. of burnt lime will, by his "lime-water softening process," suffice to purify 800 gallons of this water, reduce its hardness to 3°·9, and effect a saving of curd soap required to form a lather with 100 gallons of the water, of $24\frac{1}{4}$ oz. The Inspector recommended the river Wandle as the most eligible source, from its contiguity to the town, the favourable quality of its water, and its sufficiency to afford the means for a supply upon the constant system.

306. The town of Dartford is now supplied by wells and pumps, and dipping from the river Darent. The water from seven of these sources, excluding the river, has an average hardness of nearly 18°, while that from the river has only $13\frac{1}{2}$°, and was recommended by the Inspector as being the most desirable for the supply of the town.

307. The third consideration affecting the supply of water for towns is the relative expense at which this supply can be obtained. Springs and other sources of the less pure waters, are, doubtless, usually of more ready and economical adaptation than rivers. Upland streams and watercourses are generally applicable to some extent for supplying the adjacent parts of the town and suburbs, but the higher elevations frequently involve extra cost in forcing

water from these lower sources. With a great scarcity of records of the cost of works and conducting of the existing arrangements for supplying water to towns, we are driven to form estimates which can only be assumed as approximate, but will nevertheless suffice probably to indicate the relative economy of the several methods of supply which may be adopted.

308. The main items of cost of the supply of water to towns are:—1, collecting; 2, storing; 3, filtering; and, 4, conveying. If the supply be derived from surface-drainage or springs at superior level, so that no raising is required, the first of these items will comprise the construction of open channels, aqueducts, or artificial rivers with tributary or catch-water drains where necessary. If the supply be derived from a river or other source at lower level, this item for collection must be understood to include the expense of raising the water and delivering it to the storing or filtering beds, with such constructions of channels or piping as may be necessary for that purpose. The storing places or impounding reservoirs for drainage waters are sometimes so constructed as to answer also the purpose of filtration, and thus combine in one cost the items Nos. 2 and 3.

309. Mr. Robert Thom, who has successfully supplied several towns with water collected from surface-drainage and natural collections or basins, considers it desirable that the reservoirs should be large enough to hold at least four months' supply of water, this being necessary to provide against the irregularities of supply of water obtained from these and similar sources. For the storing of water taken directly from rivers and other ample sources from which an abundant quantity can at all times be commanded, reservoirs of less capacity are sufficient, and the first cost of construction is therefore reduced. The catch-water drains, in which the water is first received, are made to communicate either directly with the main reservoirs, or by the medium of aqueducts From the main reservoir the water

is conveyed by another channel or aqueduct into other reservoirs or regulating basins near to the town, and each of them so situated in elevation that the water from them shall rise above the highest desired service, and of such capacity that each will contain enough for two entire days' supply of water for the town.

310. If the water cannot be delivered into the regulating basins at sufficient elevation, artificial power will of course be required to raise it from the natural to the desired height. From the regulating basins it is delivered into two or more self-cleansing filters (as before described, Part I. paragraph 72), and from these into two distributing basins, whence the water is carried through the streets by a system of piping. Thus the town appliances are provided in duplicate, and the object of this is to enable one set of apparatus to be constantly commanded, and each to be alternately cleansed or repaired when necessary.

311. In our fifth Section of this Division we shall have to enter into the details of apparatus for conveying and distributing water. Our present purpose is to enumerate the general varieties of arrangements required according to the source from which the supply is derived.

312. The increased expense incurred in the formation of large reservoirs to hold a supply for a long period, such as four months, is certainly great, but not so when compared with the first cost of machinery and current expenses of raising water from rivers and sources of low elevation. The upper sources of springs and drainage-waters are, moreover, applicable in some cases where the others are inaccessible, or rendered so practically by the great distance and low elevation from which river-water can alone be conveyed and raised. The cost of constructing reservoirs may be estimated at about three-pence per cube yard on an average, if no extraordinary difficulties or expensive works are required. With reference to reservoirs as proportioned in capacity to the number of houses or persons supplied, the following particulars may be usefully

cited, referring to the operations in seven of the large towns in Lancashire, and reported upon by Dr. Lyon Playfair:—

Towns.	Number of Houses in Town in 1841.	Number of Houses or Tenants Supplied.	Capacity of Reservoir in Gallons.	Height of Surface of Water in Reservoirs above	
				Highest Parts of Town.	Lowest Parts of Town.
				Feet.	Feet.
Manchester .⎫ Salford . . .⎬	57,238	30,000	⎰ 2,000,000 ⎱ 249,360,000	0 0	155 122
Preston. . .	9,984	5,026	50,000,000	36	160
Bury . . .	5,260	2,980	4,181,760	50	130
Ashton . . .	4,700	4,000	100,000,000	200	260
Rochdale . .	8,266	2,800	22,781,253	6	96
Oldham . .	8,220	5,620	85,000,000	30	300

The capacity and expense of reservoirs for drainage or surface-collected water will of course be regulated with a view not only to the wants of the population, on the one hand, but also with reference to the extent of surface to be drained, and probable quantity which will thus accumulate. From some statements given in the Report by Mr. Homersham, before quoted from, we may present the following figures:—

Names and Situation of Reservoirs.	Contents of Reservoirs.	Area of Drainage Ground.	Per Acre of Area.
	Cubic feet.	Acres.	Cubic feet.
Turton and Entwistle Reservoir, 14 miles N.W. of Manchester . . .	100,000,000	2036	49,110
Belmont Reservoir, 14 miles N.W. of Manchester	78,000,000	1796	43,430
Bolton Waterworks Reservoir, 4 miles W. of Bolton	22,471,9.0	595	37,767
Ashton Waterworks Reservoir, 1½ mile N.E. of Ashton	14,436,397	378	38,453
Sheffield Waterworks— Redmires Reservoir	30,000,000	⎫ ⎬ 912 ⎭	32,894 Total.
New do. do.	22,000,000		57,050

The aqueducts for passing a supply of 20 gallons per diem for each individual of a population of 500,000 may be estimated at from 400*l.* to 600*l.* per mile, according to the ruggedness of the ground and other items of expense. The cost of filters upon the self-cleansing principle will average from 6000*l.* to 8000*l.* to supply the same quantity. That constructed by Mr. Thom at Paisley, which produces regularly every 24 hours a quantity equal to 106,632 cubic feet of pure water, cost about 600*l.*, and he estimates that the expense of a filter "to give a supply of water of the best quality *for family purposes*, to a town of 50,000 inhabitants, may be safely taken at 800*l.*" This supply, however, allows only 13 gallons to each individual. We prefer allowing a minimum of 20 gallons, as already estimated. Adopting the facts stated by Mr. Thom, as experienced in supplying four towns in Scotland, viz., Greenock, Paisley, Ayr, and Campbelltown, which are served by his system, but allowing the greater quantity stated, to each individual, and assuming the cost to increase in the same proportion, we find that the average annual expense per person will amount to no more that eight-pence, that is, for a regular daily supply of 20 gallons of good spring-water throughout the year. This expense includes wear and tear of apparatus, charge for superintendence, &c., and 5 per cent. per annum upon the capital employed. In the towns here referred to there is such declivity that allows of high reservoirs and constant high service to the buildings without any expenditure for power. Mr. Thom states that the cost for apparatus for the smallest of these towns, Campbelltown, of 7000 inhabitants, amounted to about 2500*l.*; or say 3800*l.*, being about 10·85 shillings to provide for the daily allowance per individual of 20 gallons instead of 13 gallons.

313. At Nottingham, about 8000 houses, or 35,000 inhabitants, are supplied with water raised from the river Trent by a Waterworks Company. The actual supply is found to amount daily to between 80 and 90 gallons per house on

an average, including breweries, dye-works, steam-engines,
inns, and other places of large consumption. The levels
of different parts of the town vary, perhaps, 80 ft., and the
water pumped up from the river is raised above the town,
so that an average pressure of 80 ft. is maintained, the
greatest pressure being about 120 ft. The water is drawn
from the river into a reservoir formed on its banks, and excavated in a stratum of clean gravel and sand, through
which the water percolates to a distance of 150 ft. from the
river. Besides the filtration which thus naturally occurs,
the water is still further clarified by passing through a tunnel 4 ft. in diameter, which is laid through a similar stratum
for a considerable distance up the adjacent lands, and constructed of bricks, without mortar or cement. The expenditure for the supply of these 8000 houses amounts to about
30,000l., and the average annual charge per house is about
7s. 6d., the water being supplied at any level required, even
into the attics of four- or five-story buildings. The average
daily allowance to each individual supplied is here equal to
about 20 gallons; and reducing the total expenditure and
the annual charge per house to an original cost and current
expense per individual, as we have done in reference to the
four towns supplied by reservoirs and aqueducts from surface collections and higher springs, we shall find the two
items stand thus:—original cost per individual, 17·14 shillings; current expense per individual per annum, including
per centage on capital, &c., 1s. 8d. The comparative statement for the four Scotch towns and for Nottingham will,
therefore, be this per head of the population supplied:—

	Original cost of appa- ratus, &c.	Current annual expense.
	s.	d.
Scotch towns supplied with *drainage*- and *spring-water*	10·85	8
Nottingham supplied with *river-water*	17·14	20

The qualities of the waters, their comparative hardness, &c.,
should be fully known and duly estimated as items in the
relative economy of the two sources.

314. For the supply of some towns it will be found desirable to combine the two sources, namely, a river—and springs, or perhaps upper streams, which, flowing from lands much higher than the general level of the river, preserve a greater elevation, and may thus be applied to furnish the higher parts of the town, and effect a judicious economy of artificial power in raising the required quantity.

315. The expense of public filtering of water has already been stated, Part I., p. 65 and 67, as varying from about 2000 to 9000 gallons per penny. An average rate of 6000 gallons may be safely assumed as the quantity which can be filtered at an expense of one penny. The annual expense of filtering the supply for each individual of the population thus appears to average only 1·2 penny. This calculation is quite conclusive as to the superior economy of public over private filtering, since no separate house apparatus for this purpose can possibly be maintained in working order at this insignificant rate of expense.

316. The public filtering of water, before distributing it into the mains and service pipes by which the streets and buildings of a town are supplied, is, however, palpably insufficient to secure purity in the water as used by the inhabitants, if the quantity for each house be received and stored in a separate tank or cistern, which is seldom or never emptied or cleansed. In these receptacles the minute impurities brought in with each day's supply accumulate into a mass of growing foulness, stirred up by the daily delivery, and undergoing constant decomposition, and thus contaminating the entire contents of the cistern and every pint of water which is drawn from it. This consideration, which may be confirmed by volumes of evidence, but is too palpable to require proof, leads to the desirability of dispensing with these separate household accumulations of water, by providing a constant supply in the mains and service pipes, so that any required quantity may be at all times instantly commanded. The supply rendered by the

Trent Waterworks Company to the town of Nottingham, and before referred to, is maintained upon this principle, the several advantages of which have been pointed out by the engineer to the works, Mr. Hawkesley, and since adopted as a general rule in the recommendations of the Superintending Inspectors to the General Board of Health.

317. The superiority of the constant service principle of the supply of water to towns over the occasional or intermittent principle is not greater in the comparative purity of the water thus obtained for the current use of the persons supplied than it is in the economy of the supply. The first cost of cisterns or tanks, with all the expensive and inefficient paraphernalia of ball-cocks, waste-pipes, &c., &c., is entirely obviated by keeping the mains, service and communication pipes always charged. It is well known that the due care and cleansing of the house-receptacles for water, whether tanks, cisterns, or butts, are greatly neglected, especially among those classes who are actively and incessantly engaged in their business or daily labours, and who are equally unable to command the services of others for such purposes. These receptacles are often imperfectly constructed and covered, open to the entrance of soot, dust, and dirt of all kinds, frequently exposed to the action of the sun, and neglected when repairs become indispensable. If these separate and inefficient means are superseded by keeping the water-pipes constantly charged, one large reservoir suffices for a whole town, or extended section of one, and this one reservoir may be so devised, constructed, and managed, that the combined supply shall be always maintained and delivered in the best possible condition. The economy of the system here advocated arises in many ways. The spaces occupied by the house-tanks are saved, and the damp which always arises from the evaporation of bodies of water is avoided, besides preventing accidents, leakage, and the occasional inconvenience of finding the cistern empty, or its contents reduced to a few inches in depth of foul mud. Another source of economy is the re-

duction in the sizes of main and service-pipes required, as
the delivery is distributed over a longer period than by the
intermittent supply, which limits the actual delivery for
present and prospective purposes to a few hours, or some
still shorter extent of time. Added to this diminution in
the sizes of pipes permitted by the constant supply is the
fact of their non-liability to be burst by the sudden gush of
water which compresses the air within the pipes with a
force which the strength even of iron cannot resist. The
alternate absence and presence of water within them, moreover, hastens their corrosion, as it has been found that
much oxide of iron accumulates in them under these circumstances. And beyond these advantages, the constant
supply system possesses the further one of immense economy in management. It is found at Nottingham that one
experienced man and one lad are sufficient to manage the
distribution of the supply to about 8000 tenements, and
keep all the distributory works, including cocks, main and
service-pipes, &c., in perfect repair. Under the intermittent supply system, a numerous staff of assistants would
be required to discharge similar duties.

318. The "Commissioners of Enquiry into the state of
large Towns" have quoted a statement to the following
effect :—That the expense of machinery or capital invested
in the arrangements for supplying the metropolis with
water, exclusive of the communication pipes to the houses,
the tenants' water-butts, tanks, &c., amounts to 3,310,342*l.*,
or about 3*l.* per individual supplied; that the annual income is 276,243*l.*, and the expenditure 133,724*l.*, leaving a
balance which is equal to an average dividend of 4 per
cent. The income from each individual supplied would
thus appear to be somewhere about 5*s.* annually. Now,
the metropolis is supplied mainly from the river Thames,
the river Lea, and the New River, from a spring at Amwell.
In the year 1843,*the entire supply was furnished by nine
companies, the names of which, and the sources of their
water, were as follows :—

* For later years, see Appendix No. 7, p. 235.

Companies.	Sources of Water.
Chelsea	River Thames.
West Middlesex . . .	Do. do.
Grand Junction . . .	Do. do.
Southwark	Do. do.
Lambeth	Do. do.
Vauxhall	Do. do.
East London	River Lea.
New River	Amwell and River Lea
Hampstead	Springs on Hampstead Hill.

The cost of construction for the water-supply at Nottingham, as already stated (paragraph 318), is between 17s. and 18s., and the expense attendant on the supply of water and management of the works amounts to about 44 per cent. on the income, which will be found somewhat less than the proportion of the like expense in London. The expenses both of formation of works and of current supply are evidently controlled to a considerable extent by the natural facilities for the former, and by the distance from which the water has to be conveyed and the height necessary to raise it. The expense to individuals must, of course, be also liable to be affected by the proximity, or otherwise, of the several tenants to be supplied

319. Another principle to be observed in conjunction with that of constant service of water is, that it shall be delivered from such an elevation, or with such a pressure, that the service may take place at least 20 ft. higher than the tops of the highest houses to be supplied. The vast ultimate economy and value of this provision are cheaply bought by the additional expense involved in the works, and current cost for making it. The experience of the Trent Water Company in supplying the town of Nottingham from the river Trent has shown that the expense of raising and delivering the water 50 ft. higher than at present would amount to only 5 or 6 per cent. additional upon the present cost. On the other hand, the advantages of this high service are too great to be easily calculable, saving. as

F

it does, all the expenses of force-pumps or other separate apparatus for raising the water from the lower floors, and affording means of supplying baths 'and other accessories of cleanliness, health, and comfort, to all parts of each house without restriction, labour, or cost. The total expenses of supplying water in the town referred to, with all the benefits of constant and high service, including wear and tear of engines, interest on capital for machinery and distributing pipes, expenses of management, &c., amount to 2·88 pence per thousand gallons.

320. The value of constant and high service of water to towns is strikingly important in its application for the extinction of fires, and for the occasional washing of streets, and cooling them by jets of water in warm seasons and times of drought. The bearing of these purposes upon the preservation of life and property, and the promotion of health and comfort, is too evident to need much illustration, although the details of the arrangements will claim, on account of their extended utility, some notice in our fifth section, which will be devoted to a brief account of all the essential apparatus for carrying these principles into practice.

321. Finally, let us recount the leading objects to be kept in view in the supplying of water to towns:—*First,* that the supply shall be *ample in quantity* for all the purposes of personal and domestic cleansing; for the public purposes of supplying baths, fountains, and gardens; for the extinction of fires, the thorough cleansing of streets and thoroughfares, and the occasional cooling of them in dry seasons, and for all such manufacturing purposes as may be required or permitted within towns.

Secondly, that this abundant supply should be procured of the *best possible quality* for the several uses to be made of it. That, if several sources are available at various rates of expensiveness, the economy of any one of them as compared with another, or others, is to be duly estimated, with a governing reference to the quality of the water so derivable; and that the question of adopting an inferior water shall be affirmed only in cases where the expense of the better quality amounts to a practical impossibility. That, besides the always recognised impurities of an animal and vegetable character, and those held in mechanical suspension only, with which some waters are usually adulterated, there are others of a soluble nature, which are consequently commonly imperceptible except to chemical analysis, but the presence of which deteriorates the quality of water in a high degree, and occasions a necessity for chemical processes to purify it, mere filtering being utterly inoperative for the purpose. That, generally, these soluble matters are found in spring- and drainage-waters in far larger proportion than in river-waters, which are more susceptible of being purified by a process of self-filtration, and are, therefore, commonly preferable for most purposes to waters of the former character. That the expense of raising river-water by steam-pumping is really very small, and unworthy of consideration, although often regarded as a weighty argument in favour of seeking the required supply from districts of land from which the water descends by gravity, without artificial aid.

And, *thirdly*, that the complete utility and greatest ultimate economy of the supply of water to towns can be realised only by a service of it which is constant in duration, and sufficiently high to discharge over the highest buildings in the town.*

SECTION III.
Width and Direction of Roads and Streets.—Substructure and Surface.— Paving and Street Cleansing.

322. The drainage and cleansing of the roads, streets, and thoroughfares of a town are acknowledged to be public purposes of the highest utility. The facility of effecting these purposes is dependent upon the several circumstances

* The London and Glasgow Water Supplies are further noticed in Appendix No. 7, pp. 232—238, and Appendix No. 8, pp. 239, 240.

of the dimensions and situation, and the sub- and super-structure of the thoroughfare. The width of the streets is influential in admitting or preventing the access of air and winds, by which the wholesomeness of their condition is largely affected; and also in rendering the process of cleansing by hand or other labour easy or difficult. The direction of roads and streets—vertically in their relative levels and inclinations, and, laterally, in their coincidence with or opposition to the courses of the prevalent winds— is a condition of great importance in affecting the facility and economy of the processes of drainage and cleansing. And the relative dampness and dryness and quantity of debris produced upon any public thoroughfare are mainly attributable to its construction in the subsoil, and superficial formation.

323. Courts and narrow passages, such as abound in most towns—relics of public ignorance and private cupidity, destined to be destroyed in the progress of enlightened sanatory reformation—limited in width and bounded by elevated buildings, never receive their due share of light, air, or water, and thus present the greatest combination of difficulties to the vital processes of drainage and cleansing And these purposes can never be economically and efficiently fulfilled until a minimum of width and a maximum of height of buildings are recognised as the elements of street proportion. The recorded and repeated evidence on this point is more than enough to establish the general principle, although the precision of the details requires observations of a more exact nature than have yet been made. It is certain, however, that no street should be less in width than the height of the buildings on either side of it—that is, that the angle formed by the transverse surface of the street, with a line from its extremity on one side to the summit of the buildings on the other, should never exceed 45°. And in proportion as this angle can be reduced will be the facility afforded for the desirable operation of the air and of such rain as may fall.

324. Provided this principle be strictly observed, the comparative declivity of the surface will become of minor importance. Certainly, the greater the declivity the more rapid and effective will be the action of rains in cleansing and washing down the debris upon the surface of the street; but it should be the peculiar province of the subterranean sewers constructed beneath, to compensate for the relative flatness of the surface, by affording a channel of artificial declivity, that shall at all times free the surface from these matters as quickly and effectually as possible.

325. Connected with the subject of road drainage as applicable in the suburban parts of a town, the necessity of providing *covered* drains cannot be too rigorously enforced. Open road ditches are known to become receptacles for filth and refuse matters of various kinds, and the trouble and expense of cleansing and keeping them in repair, involving a constant making-up of the banks and clearing of the beds, are commonly evaded by a total neglect, which leads to a stoppage of the channels and a constant exposure of decomposing matters, both offensive to the senses and injurious to health. These roadside ditches are frequently, moreover, adopted as the only available channels for dispersing the sewage of the suburban buildings; and being thus converted into open sewers with little or no attempt at formation, and very little care in preserving even their original rude form and capacity, the evils of retaining them are multiplied to a degree actually dangerous to the health of the inhabitants and of passengers.

326. Added to the inefficiency of open road drains or ditches is the waste of surface which they involve. Pedestrians in the suburbs of towns know well that of a narrow road nearly one-half the width is frequently occupied by a wide and sluggish ditch, and that, in the absence of any raised foot-path, they are frequently driven to a dangerous proximity to its foulness in order to escape destruction by the heedless and perhaps drunken drivers of vehicles. If

these ditches were covered and converted into active sewers
by the use of pipe-tiles, of comparatively small and yet
ample dimensions, space would be afforded for the formation of convenient footpaths on which a walk would become
a luxury instead of being a task of danger and annoyance.
Those who have "picked their way" along the unpave-
mented strados of Rome, and contrasted them with the
easy security of some of the similarly narrow streets of our
own metropolis, will readily appreciate the value of the
change which might be thus cheaply effected in our suburban roadways.

327. The quantity of surface wasted by the open road
ditches, and the corresponding area thus exposed for the
evaporation of stagnant moisture, may be readily calculated from the dimensions of the ditch. It may be safely
assumed that for each mile of road, at least half an acre of
surface is thus, on an average, misapplied.

328. The position of the *main sewers* of a town being
beneath and in the same direction as its streets, *these
afford the proper channels for discharging the waste water and
all other matters from the surface of the streets.* This doctrine
is liable to be challenged by all those practical economists
who contend that street debris is so injurious in its admixture with the excrementitious matters flowing from a town,
that it should be scrupulously kept separate, and periodically removed by hand and horse labour above ground.
But if we take into the account, on one hand, the small proportion which the solid part of this debris bears to the
total of solid and liquid excrements, house refuse, street
drainage, waters, &c., which are universally allowed to be
the proper subjects of sewer discharge, and, on the other,
form a due estimate of the inconvenience, expense, and
disgusting annoyance of removing this street refuse by
any expedient above ground, the result of the calculation
will lead, we think, irresistibly to the conviction that the
whole of these matters should be by the readiest possible

methods delivered into the sewers, and by them conveyed at once to receptacles suitable for their collection and treatment.

329. The exact proportion between the solid street refuse and the total of house-sewage and street-drainage (which may be supposed to find its way unavoidable into the sewers) is difficult to determine with any certainty approaching to exactness, but an approximate estimate may be formed from such materials as we can command. The excrementitious matters produced by each individual are generally considered to amount to an annual quantity equal to one ton in weight, and the other matters which are comprised in the total of house-sewage and street-drainage, may be supposed equal to a similar quantity. We have thus a total equal to two tons annually per head of the population. Now, in the township of Manchester, of which the population in 1841 was 164,000, the number of yards of street-surface swept in the same year was 21,500,000, and the number of loads of these sweepings removed equalled 25,029, each of which is equal to a weight of one ton. Assuming the proportion between the population and street-surface of this township to be a fair average for most towns, we have thus a total of house-sewage and street-drainage equal to 164,000 × 2 = 328,000 tons, and a total of street-sweepings equal to 25,029 tons, being $\frac{1}{13}$th of the former, or less than 7·7 per cent. This rough calculation will be quite sufficient to show the small proportion in which the manuring value of the sewage is liable to be injured by the admixture with it of the street debris in the common receptacles or sewers, and the consequent inadvisability of engaging in the expensive operations of carting and removing this debris by any combination of human and animal labour.

330. Arrangements for the purpose of discharging the street surface-drainage into any contiguous river or other watercourse, instead of allowing it to mingle with the sewage in the receiving wells or receptacles to which they

are both conducted by the sewers, may, if thought necessary, be provided as accessory apparatus in connection with
the wells, although it is highly probable that the growth of
our experience on this subject will develop preferable methods of treating and disposing of these matters by subsidence and chemical processes.

331. The amount of street debris, or the quantity removable from any extent of surface, is found to vary most materially, according to the structure of the street or roadway.
Thus, roads formed of broken granite or other similar materials are rapidly destroyed by the action of wet, which
loosens the superficial coating of the road, and passes into
the body of the materials; the finer particles also become
washed upon the surface, and act as sand in grinding
it down, by the action of the wheels upon it. Paving
formed with stones of irregular shapes and sizes is also
productive of a large quantity of debris, although less than
the unpaved surfaces just referred to; upon this inferior
class of paving, water acts destructively by washing up the
soil and dirt between the stones, by which they become
loosened, while a great proportion of these interposed materials have to be removed as they appear upon the surface
in the form of mud.   Pitch-paving formed with squared
blocks of granite, whin, or other stone of equal hardness
and durability set in lime grouting upon a substantial
foundation of concrete 9 to 18 in. in thickness, according
to the nature of the sub-stratum, forms the most permanent
construction for the carriage-ways of streets and thoroughfares, and affords a correspondingly small proportion of
materials to be removed from the surface, in order to preserve its cleanliness.   Wood-paving yields the minimum of
debris, and its economy, as a subject for the labours of the
scavenger, at any rate, is thus very great, as compared even
with the most perfect form of stone-paving.

332. By making the sewers thus directly available for one
of their proper purposes, that of receiving the waste matters from the streets and thoroughfares, the operation of

street-cleansing is reduced to mere sweeping of these matters to the side channels, which should be constructed so as to afford a ready passage for them to the sewers beneath. The economy thus obtained by dispensing with the raising and carting to distances sometimes extended may be inferred from the fact, that the average expense of sweeping and carting away the refuse of 1000 square yards (in Manchester) in 1843 was 4s. 6d. This was performed by the ordinary hand labourers or sweepers. In London, at the same date, the mere *sweeping up* of the refuse from the surface of Regent Street, and depositing it in the street in loads for another process of removal, was charged at the rate of 1s. 2d. per 1000 square yards, as executed by Whitworth's patent machine. The mere *sweeping* may be liberally estimated to cost 9d. for the same extent of surface, and thus ⅔ths of the entire expense of street-cleansing might be avoided by adopting the sewers for the purpose suggested.

333. Although we advocate the abandonment of all apparatus for carting and removing street-refuse, it may be useful to describe briefly the " Patent Street-Cleansing Machine," invented by Mr. Joseph Whitworth, which has been applied to a considerable extent in Manchester and elsewhere, and been considered a very promising contrivance. This will be best done by quoting the inventor's own description of his machine. as rendered in evidence before the " Commissioners of Inquiry into the State of Large Towns and Populous Districts," in 1843. " The principle of the invention consists in employing the rotary motion of wheels moved by horse or other power, to raise the loose soil from the surface of the ground, and deposit it in a vehicle attached. The apparatus for this purpose consists of a series of brooms suspended from a light frame of wrought iron, hung behind a common cart, the body of which is placed near the ground for greater facility in loading. As the cart-wheels revolve, the brooms successively sweep the surface of the ground, and carry the soil up an incline

or carrier-plate, at the top of which it falls into the body of the cart. The apparatus is extremely simple in construction, and has no tendency to get out of order, nor is it liable to material injury from accident. An indicator, attached to the sweeping apparatus, shows the extent of surface swept during the day, and acts as a useful check on the driver. It also affords the opportunity of working the machine over a given quantity of surface. The average rate of effectual scavengering by hand in Manchester, taken for a whole year, is from 1000 to 1500 square yards of surface daily for each scavenger. The manner of sweeping is different in London, and therefore an apparently larger amount of work is done, but not so effectually. When the machine is in operation, the horse going only 2¼ miles per hour, it sweeps during that time 4000 square yards ; thus performing in a quarter of an hour nearly the day's work of one man. The average amount of surface which can be swept by a machine during the day depends upon the distance of the places of deposit. In Manchester we have seven places of deposit, and the average number of yards swept daily, by a machine drawn by one horse, is from 16,000 to 24,000."

SECTION IV.

Main Sewers; Proportions and Dimensions, Inclinations, Forms, and Construction.—Upper and Lower Connections.—Means of Access and Cleansing.—Adaptation for Street-cleansing, &c.

334. In drainage, as in many other subjects, controversy has frequently been found to be excited upon those very details of the art which appear to be the most simple and the most readily deducible from observation, while the proper ground for discussion, in which it is really urgently needed, in order to determine general principles and mark out leading rules, has been left nearly or quite unoccupied. Thus the forms, sizes, and thicknesses of sewers have re-

ceived the most elaborate investigation, and provoked the expression of the most widely-differing opinions; while the principles of arrangement according to which the entire system should be laid out, and the great questions of the most healthful and economical disposal of the refuse of towns have, till lately, remained unsought and unasked. Misled by an instinctive adoption of the works of our forefathers, we have been content to build our sewers in old channels, and to put patch upon patch—add length to length of sluggish sewer or practical cesspool, in order to maintain ancient outfalls, while the subsidiary details of form and capacity have become the vexed questions and grounds of issue among the most practised advisers.

335. Not that the attention given to the details, and the neglect inflicted upon the general principles are here contrasted for the purpose of denying the importance of the former, but that, had the principles been first determined, the details would be found readily deducible from them in a manner and with a certitude admitting little dispute or discussion.

336. We have already, in the first section of this Division, shown the general principles upon which the drainage of towns should be arranged with reference to the inclinations of surface, and the means of discharging and disposing of the sewage. From these principles it immediately follows that the proper functions of sewers are twofold, and *twofold only*, viz., the conveyance and collection of house-drainage and of street-drainage. In the former are to be included the drainings of roofs of buildings and of yards, or other spaces attached to them. In these two purposes is thus comprised the superficial drainage of each entire town. Any attempt to add to this the drainage of the sub-formation is a mistaken and a supererogatory aim. This position will be denied by those who advocate the *under-drainage* of London as one of the purposes of its sewerage. Let us endeavour to understand the practical value of this

purpose, and thence deduce the infinitely small amount that would be mis-spent in any attempt to realise it. If the proper functions of sewers be effectually discharged, viz., the conveyance away from a town of all the rain-water that falls upon its surface, and of all the solid and liquid refuse produced in streets and buildings, what will be the amount of submoisture which it can be necessary or desirable to abstract in the form of land-drainage? The entire surface being maintained constantly dry, the only sources from which under-water can arise will be springs or water-bearing strata beneath, and wherever these may show themselves, they can be turned to good account, and the water they yield converted to useful purposes, without making expensive provision for their drainage beneath. Whatever relation the site of a town may have to the surrounding country, that is to say, whether the town be above or below the lands around, or be on a similar level, none of the drainage-water from these lands should be permitted to enter the town or to mingle with the soil beneath it. This is to be effected by constructing around the town a system of encircling catch-water drains, by which so much of the surrounding drainage as would otherwise find its way into the subsoil of the town will be intercepted and collected, either to be returned by suitable channels to the rivers, streams, and watercourses, to be made available in irrigating adjacent districts, or diverted directly from the catchwater drains into the main sewers of the town itself, and disposed of with their contents. With this auxiliary arrangement for preventing the access of surrounding drainage to the sub-formation of the town, all necessary provision for maintaining it in a dry and healthy condition will be completed, and no necessity can possibly arise for constructing a duplicate system of sewers in order to drain the subsoil of the town. With all due deference to official experience, we venture to predict that, *if ever tried*, the " system of permeable land-drains and sewers," as a separate addition to the "system of permeable drains for house and soil drainage," will be found as utterly useless in practice as it will be expensive in construction.

337. The proportions, dimensions, inclinations, forms, and construction of main and all other sewers, are all more or less affected and determinable by the general system of drainage adopted. We will first cull from the mass of recorded experience at our command (up to the year 1843) some detailed particulars of modes of construction (and their cost), many of which have been found *inefficient* in fulfilling the *self discharge* of the sewage matters of London and other towns in England.

338. The experience in the city of London led the surveyor to the Commissioners of Sewers to consider that the form of a semicircular top and bottom, with straight (or vertical) sides, "answered all the conditions of a sewer." Nevertheless, many have been constructed of an oval form. The smallest size in a long street is 4 ft. 6 in. by 2 ft. 6 in. The other sizes are 5 ft. by 3 ft. ; but several are considerably larger, where much water is expected to accrue from the outer districts. The outlet for the main sewer at South Place (Finsbury) is 6 ft. 6 in. by 4 ft. 6 in. The Fleet sewer, which *drains from the south-west of Highgate, is 18 ft. 6 in. by 12 ft. at the mouth, and 12 ft. 3 in. by 11 ft. 7½ in. at the city boundary; and, owing to the immense quantity of water flowing into it, "this sewer has often been surcharged."* The Eldon Street (Finsbury) sewer is 5 ft. by 3 ft. 2 in.; the London Wall sewer is 6 ft. by 4 ft., and the main trunk increases from 8 ft. 3 in. by 6 ft. 9 in. to 10 ft. by 8 ft. at its mouth. For courts and alleys the sizes are 3 ft. by 2 ft. 2 in., and sometimes, according to the number of houses, 4 ft. by 2 ft. 4 in. The sewers 4 ft. 6 in. by 2 ft. 6 in. are built in brickwork 14 in. in thickness throughout. Adapting the size of the smaller drains so as to admit a man to pass through them, they should be at least 2 ft. in width, and, to allow crawling through, 2 ft. 4 or 6 in. in height; to allow his crouching through, 3 ft. 6 in.; or to stoop through,

4 ft. 6 in. The thickness of brickwork of these sewers should not be less than 9 in., nor the depth from the ground less than 12 ft. at the shallowest part, in order to provide for the drainage of a basement story about 7 ft. 6 in. in height. Assuming 2 ft. 6 in. as the minimum height for a common sewer, and allowing 20 in. of deposit to exist in a public sewer before it can rise into the common sewers, the surveyor deduced a minimum height for public sewers of 4 ft. 2 in.

339. In the Westminster Division of Sewers the level of the outfalls into the river varies from 10 to 15 ft. below the level of high-water mark, and some of them have flaps. Some of the main sewers have *a fall of only half an inch to 100 ft.* The form of the sewers is that of a semicircular arch at the top, and a segmental invert with upright sides. The two sizes used are—first class, 5 ft. 6 in. high and 3 ft. wide ; and second-class, 5 ft. high and 2 ft. 6 in. wide. The three centre courses of every invert are built in cement, and the remainder of the work in Dorking lime-mortar. The walls are $1\frac{1}{2}$ brick in thickness, and the arch and invert 2 half-bricks, or 9 in. The cost for a sewer 3 ft wide was. for the brickwork, 14s. 3d. per ft., and for a sewer 2 ft. 6 in. wide, 12s. 6d.

340. The sewers throughout the Holborn and Finsbury Divisions discharge into the main sewers of the city of London, and have no outfalls of their own into the Thames. The Fleet sewer conveys the drainage of about 4444 square acres of surface in those divisions, and is calculated to receive annually from this surface about 100,000 cube yards of matter held in mechanical suspension, and carried to the Thames by the force of such waters as flow through the sewer. These waters, by the experiments of Mr. Roe, having been found to amount to about 100 times the bulk of the matters held in suspension by them, it follows, that the Fleet sewer discharges from this surface about 10,000,000 of cube yards of sewage-water and suspended matters into the river Thames annually. The total

work of this sewor comprises also the quantity it receives from the surface of the city, after passing through the district here referred to. A sewer carried up to Holloway, in this division, a length of nearly 3 miles, passes under Canonbury (Islington) at a *depth of 68 ft. from the surface*, and the drainage of the houses in that part is provided for by a *subsidiary sewer*.

341. Sewers constructed on the Kingston estate, through a very soft clay, are built of an oval form, the largest size being 3 ft. 6 in. high, and 2 ft. 6 in. wide, the radius of the side curves about 3 ft. ; half a brick thick in cement. The extent of cutting was from 16 to 18 ft., and the cost 15s. per lineal foot. The fall at the rate of 80 ft. in a quarter of a mile.

342. The practice in some of the provincial towns was reported as follows :—

Lancaster. — Flag or slate bottom. Rubblestone sides, laid in common mortar. Rough stone covers. Mains 2 ft. 6 in. × 1 ft. 4 in., 6s. per lineal yard. Branch street drains, 1 ft. 4 in. square, 4s. 6d. ditto. Yard drains, 6 or 7 in. square, 2s. ditto. All found to be very inefficient.

Fig. 71.

Nottingham. — Brick. Cylindrical sewers. Upper half built in mortar. Lower half laid dry. Half-brick thick. Diameter from 2 ft. to 2 ft. 6 in. Average cost 7s. per lineal yard.

Fig. 72.

Birmingham and Walsall.—2 ft. circular culverts laid 5 ft. deep. 7s. per lineal yard.

Chester.—Circular brick drains from 30 to 36 in. diameter. Average cost 12s. per lineal yard.

Bristol.—Four sizes of elliptical brick sewers.

Fig. 73.

	Ft.	in.		Ft.	in.	
1st.	. 4	0	×	3	0	
2nd.	. 3	3	×	2	6	Internally.
3rd.	. 2	8	×	2	0	
4th.	. 2	0	×	1	6	

All 9 in. thick.

Cylindrical drains, 1 ft. 2 in. in diameter internally, 7 in. thick.

Rate of fall from 1 in 60 to 1 in 360.

Frome. — Stone and lime cheap and abundant. Drains or "gouts" 18 in. square, covered with stone to take any weight, exclusive of digging, 2s. per lineal yard.

Culverts 2 ft. square, dry walls, with rubbed stone arch, turned in good coal-ash mortar, exclusive of digging, 4s. 9d. per lineal yard.

Swansea.--Oval drains, 3 ft. 2 in. × 2 ft., including excavation, 10s. 6d. per lineal yard.

Cylindrical drains, 2 ft. diameter, including excavation, 8s. per lineal yard.

Brecon.—Cylindrical drains, 2 ft. diameter, cost 8s. per lineal yard.

Square drains, side walls of dry masonry, with flat covering stone, from 3 to 4 in. thick.

Cost.—12 in. 2s. 6d. per lineal yard.

            15 in. 3s. 3d.   „

            18 in. 4s.   „

343. The egg-shaped or oviform section used in the Holborn and Finsbury divisions is shown in fig. 74, and the section commonly used in the Westminster division, up to the year 1843, is shown in fig. 75. The difference in expense between sewers of these sections has been estimated at 1660l. per mile, upon the following data. Brickwork at

20s. per cube yard.   Excavation 1s per cube yard.   Filling
in 3d per cube yard. Carting 2s. per cube yard. Remaking

Fig. 74.   *Fig. 75.*

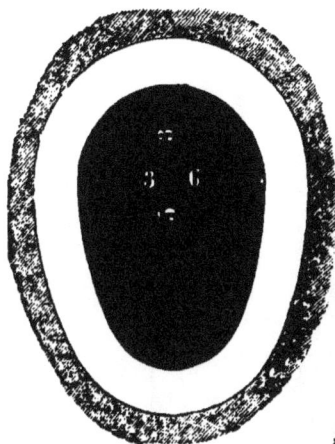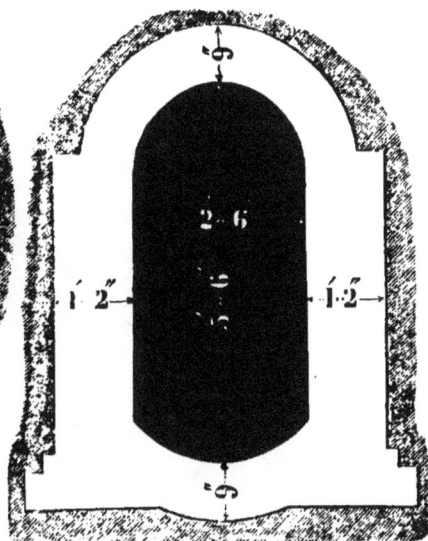

surface 1s. 6d. per superficial yard.   Average depth of excavation, 20 ft.   The quantities per mile of each sewer are
shown in the following table ; the size of the egg-shaped
sewer being 5 ft. 3 in. by 3 ft. 6 in.. and that of the upright-
sided sewer 5 ft. 6 in. by 3 ft.

	Finsbury, or Egg-shaped Sewer.	Westminster, or Upright-sided Sewer.
Bricks consumed	924,140	1,378,080
Cube yards of brickwork . .	2,272	3,388
Do. do. of excavation . .	19,555	25,420

Excess in Westminster Sewer, per mile.

1116 cube yards of brickwork at 20s. . .	£1116	0	0
5865 „ „ excavation at 1s. . .	293	5	0
5865 „ „ filling-in at 3d. . . .	73	6	3
1116 „ „ carting at 2s. . . .	111	12	0
880 super yards repairing at 1s. 6d. . .	66	0	0
Total	£1660	3	3

341. One of the Westminster sewers, built in the Har row Road, according to the section, fig. 75, failed, owing, as

Fig. 76.

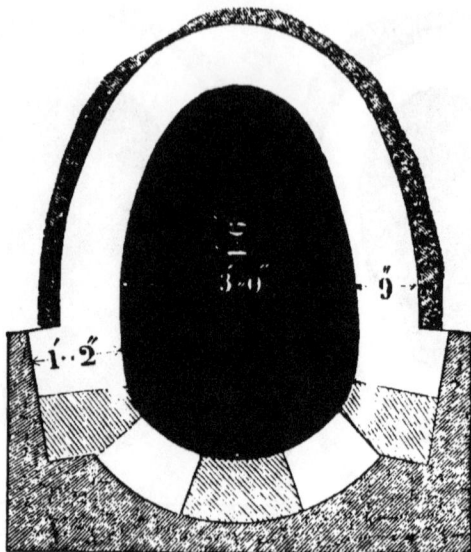

a leged, to some difficulties in the nature of the soil, and to imperfect workmanship. This was replaced by another form of sewer, which is shown in fig. 76, in which the shaded parts represent brickwork in cement, the invert and springers being bedded in concrete as high as the 14-inch work, as there shown.

345. The *capacity* of sewers is determined by a con sideration jointly of the *quantity* of sewage to be conveyed through them, and of the rate of inclination or *fall* in their vertical position. The capacity will vary directly as the quantity and inversely as the fall; since the greater the fall the more rapid will be the discharge. It has been usual to prescribe another limitation as to the minimum capacity of sewers, viz., that they shall at least, under all circum- stances, be large enough for a man to pass along them. The necessity for this allowance has arisen from the fact, that sewers are found to require cleansing by hand - that it

is utterly impossible to remove the accumulations which are liable to occur within them by any other means, and thus some 10,000*l.* has been annually expended in London alone in an employment of a most disgusting and danger-ous nature. We have no hesitation in saying, that, under a thoroughly efficient and practicable system, no such process could ever be needed. and, moreover, that if deemed desi-rable for any possible purpose, it would apply only to the principal sewers, the size of which would admit of it, as determined upon the joint data of *quantity* and *fall alone* We will, therefore, dismiss this condition from the problem, and study it upon the two data named.

346. Since the quantity of sewage due to any given ex tent of surface will depend mainly upon the amount of population to be served, it follows, that in an equalised system aiming at an uniform size for the sewers of the several classes, the points of collection or receiving wells should be arranged at distances varying inversely as the density of the population. Now, the *maximum* density of the population of London is estimated at 243,000 to a square mile. Let us suppose the drainage of one quarter of a square mile of surface, populated to this extreme de-gree of closeness, to be conveyed in *one main sewer*, and endeavour to form a rough notion of the total quantity of sewage which this sewer should be fitted to convey and dis-charge. The entire bulk of sewage must consist chiefly of the house-sewage and rain-water from the surface—at least the other constituents are of too insignificant an amount to require notice in a merely approximate estimate. And similarly the entire house-sewage may be assumed as equal to the bulk of water delivered to the total population. We have calculated, in Section II., on the supply of water (par. 283), that 20 gallons are, or ough* to be, allowed to each individual of the population per diem. The annual quantity will, therefore, be $20 \times 365 = 73,000$ gallons, or say 1200 cubic feet. The population of the square quarter

cf a mile being $\dfrac{243,000}{4}$, or about 60,000, this number

multiplied by 1200 cubic feet for each person will produce 72,000,000 as the annual quantity of sewage in cubic feet arising from this population. To this is to be added the bulk of the rain-water, which we will allow to amount to 24 inches, or 2 ft., in depth annually over the surface, and that this quantity will be discharged into the sewer without further diminution by evaporation. The total quantity to be drained annually from the surface of the quarter of a square mile will thus amount to $2640^2 \times 2 = 13,539,200$ cubic feet. Adding this, which we will call $13\frac{1}{2}$ millions, to the 72 millions of .house-sewage, we obtain a total of $85\frac{1}{2}$ millions of cubic feet of sewage to be discharged per annum from the surface of a square mile of the most densely-populated part of the metropolis. If this annual quantity were in a state of constant transition along the sewer, and with equal velocity throughout, and the effect of friction was for the moment disregarded, the proportion to be passed per minute would be of course easily calculated, being 85,500,000 divided by 525,600 (the number of minutes in a year), or 162·66 cubic feet. Now a recorded fact will be a more useful datum for our calculation here than any elaborate investigation of velocities, friction, &c.; and we will, therefore, refer to the experiments of Mr. Roe, instituted for testing the value of the flushing system as applied to sewers, and which showed that the sewage passed through the river Fleet sewer with an average velocity of 83·47 ft. per minute; the run of water being spread over a surface 10 ft. in width, and the stream being only 10 in. in depth, the passage every minute, therefore, was equal to 692·8 cube feet of sewage, and the friction in this case being greater than if the same sectional area of water had been accumulated in a cylindrical drain of smaller diameter. The solid matters held in .suspension by this water amounted to the proportion of 1 in 96 of the bulk of water, and consisted, as all sewage usually does, of decomposed animal and vegetable matter, and detritus from streets and roads. At this rate of transit, it appears that a sectional area equal to two square feet would suffice to pass the entire sewage of a thickly-populated area of a square quarter of a mile, supposing the passage to be constant and uniform, and the fall of the sewer and friction of the sewage equal to that of the river Fleet sewer, on which the experiments were made.

347. In modifying this result to provide for the difference between the assumed and the real nature of the transit, we will first admit that the bulk of the sewage, consisting of that flowing from the houses, is delivered into the drains during perhaps half the real time; that is, during 12 instead of 24 hours. The sewers will, therefore, be required to discharge double the quantity estimated during each alternate period of twelve hours, and during the intervening periods of like extent to remain empty. We will, therefore, double the capacity, and allow four square feet of transverse sectional area of main sewer for the drainage of the given surface.

348. But we have another allowance to make; we have the " *storm* waters " to provide for, about which we have heard so much, because occasionally, during the rainy month of July, a smart shower is observed to cover a flat street, or form ponds on the low side of an ill-formed roadway. Let us estimate the allowance required for this phenomenon, and infer the advisability of providing for it in the sewers. We have seen that 24 inches in depth of rain falling upon our selected spot will equal a total bulk of $13\frac{1}{2}$ millions of cubic feet. We will suppose an extraordinary case, viz. that some July day the whole quantity due to a month (2 inches) falls in 20 minutes. Then, in order to prevent any flooding of the thoroughfares, this quantity, equal to $\dfrac{13,500,000}{12} = 1,125,000$ cubic feet, will have to be disposed of in 20 minutes. Assume that the velocity produced by the pressure on the water will equal 1000 ft.

per minute. What would be the capacity of the sewer equal to discharge this rain-water as rapidly as it falls from the clouds? The quantity accruing per minute being $\dfrac{1,125,000}{20} = 56,250$ cubic feet, and the velocity equal to 1000 cubic feet, the capacity of the sewer must be equal to 56¼ square feet of transverse sectional area. Now, we have found that an area of 4 ft. will suffice ordinarily for the house-sewage. Is it desirable to increase the capacity of our sewers *fourteen fold* in order to provide for an occasional shower? There can be no necessity to answer the query. Economy of the most liberal disposition would not sanction any such arrangement. If the exact area of 4 ft. be doubled, in order to make ample provision for all ordinary contingencies, it will satisfy every reasonable requirement; and then, by suitable inlets to the sewer, the deluge of rain-waters will be prevented from overcharging it, and the effects of the shower will disappear in some hour and a half, and before any very serious mischief can be produced by the water soaking into the subsoil through well-paved streets and yards.

349. In proportion as the population is more extended, the ratio of house-sewage to surface-sewage will of course diminish, and *vice versâ;* but we believe that economy and facility of drainage will be best promoted by limiting the sum of population and area to each receiving well, so that a transverse sectional area of 8 or 9 ft. shall suffice for the main sewers.

350. In the suburban districts of a town, where comparatively large surfaces exist in gardens, and where, therefore, the effect of allowing the "storm waters" to gather might be productive of mischief by saturating the soil, the diminished amount of house-sewage will tend to make the operation of the mains more effective in relieving the surface, besides which, natural declivities will usually aid the fall of the sewers; and provision might frequently be made at little cost for receiving the surface-water in auxiliary

wells or receptacles in which it could be made available for
subsequent service in irrigation, without allowing it to bur-
den the main sewers of the district.

351. Having based our calculation as to the capacity of
main sewers upon an area of the maximum density of population, we will, with the same view of providing for the
utmost necessities, consider the question of declivity or fall
as to be applied to that description of natural surface which
presents the greatest difficulties to the operation of any
system of sewage—a perfect or dead level. The wells or
receptacles for the sewage being placed half a mile distant
from one another, so that the area drained into each of
them equals a square quarter of a mile (or each side 2640 ft.
long), half of this, or 1320 ft., may be taken as the length
of each of the main drains. The longest of the main
sewers thus measuring 1320 ft., the fall is to be computed
with reference to this length. We have seen that metropolitan sewer-practice has recognised a fall of half an inch in
10 ft., or 1 in 240, as sufficient for all the purposes of good
drainage. At this rate the fall due to 1320 ft. will be 5½ ft.
But preferring to allow double this fall as proportionally
improving the system, by aiding the discharge, we should
require a fall of 11 ft. in our main sewers of the maximum
length. And preserving 5 ft. above the head of the main,
it would lie at a depth of 16 ft. at the well. This 5 ft. will
usually be found sufficient to allow all necessary fall in
house-drains and in branch sewers, to serve the *superficial*
draining of the intervening district.

352. The utmost economy of the system would be attained by multiplying the main sewers as much as possible,
as by this means the length of the branches may be reduced to a considerable extent, and the necessary depth of
the mains also reduced correspondingly. On the other
hand, by sparing the main sewers, they are required to be
laid deeper, and the branches also; or, if depth be saved,
it is at the expense of efficiency, and the whole system is
instantly filled with insuperable difficulties in vain attempts

to reconcile the relative levels of an infinite number of collateral sewers, and to adjust the details of the arrangement to the minor variations of surface above.

353. In the depths we have assumed, as deduced from the desirable rate of inclination for the main sewers, allowance is not yet made for draining the basement stories of the buildings. It must be confessed that this purpose involves the greatest difficulty in the details of the system. On the one hand, it is evident that the construction of sewers as large as rivers, and at depths varying from 20 to 70 ft. below the surface, demands a most extravagant expenditure at the outset, and, after all, puts the works in positions which are practically inaccessible. Yet, on the other hand, we shall be reminded that the deep basements and kitchens must be provided for, and that our branch sewers must be sunk low enough to serve even the lowest of these. In order to provide for these, the main sewers will need to be laid some 8 or 10 ft. lower than the depths we have given, viz., 13 or 15 ft. at the head, and 24 or 26 ft deep at the well. Rather than permit the evils caused by sinking sewers at these depths, it will probably be preferable to reduce the distance between the wells, or even admit (although highly objectionable) some diminution in the rate of fall. We are satisfied, however, that the fullest investigation into this subject will establish the principle *that no sewage matters of any kind whatever should be allowed to be discharged into a drain from any rooms or apartments below the surface of the ground.* The difficulties which would attend any attempt to carry this principle into effect in London and similarly ill-constructed cities, may be too formidable to be now encountered, but they must be overcome before the sewerage of such towns can be reformed upon the most efficient plan which our present knowledge and experience suggest.

354. The dimensions of the branch sewers are to be determined upon the same two elements of population and surface to be served that we have referred to in estimating

the required capacity for the mains; and, according to the varying extent and proportion of these elements, a scale of sizes may be determined for the several lengths, distances apart, &c., of the branch sewers.

355. By the system of district collections here recommended, one great difficulty felt in planning sewers for concentrated discharge is at once obviated. In forming sewers which are intended at the time to serve a certain district, but which may hereafter be treated as trunks, and called upon to discharge constant accessions of sewage from an extending neighbourhood, no calculation can possibly be made as to the sufficiency or otherwise of the section it is proposed to adopt. Thus, as truly remarked by the Surveyor to the City Sewers Commission, " the sewer from Moorfields to Holloway appears to measure upon the map about *three* lineal miles. In process of time, and as buildings increase, it may throw out branches in all directions, and the *three* miles may become *thirty*. Not only all the atmospheric waters which may, upon an average, fall within the valley south-eastward of Highgate (or at least a large portion of them), but all the artificial supplies which the wants of its yet future inhabitants, as well as of those intermediate between Islington and Moorfields, may require, will have to be carried off by the City sewers." The necessary consequence of which doubtful condition is, either that the sewers are at first constructed in a most extravagant manner as to dimensions and depth, or that they are afterwards found to be utterly inadequate to their increased duty, and have to be reconstructed at greater depth and of enlarged capacity. Whereas, if district collections are adopted, each main sewer is at once properly devised as to size, form, and construction, and continues to perform its services efficiently; and, rs new districts are formed, each of them is provided with another system of sewers adapted to its defined limits, and made sufficient for all the work it will be ever expected to perform.

356. In the *form* of sewers two conditions have to be

u

fulfilled, viz. *strength*, as obtained with economy of cost, and *efficiency of action*. A hollow channel embedded in the subsoil is evidently liable to be pressed upon and against by the weight and bulk of the surrounding solid materials, and it therefore becomes necessary that the form of this channel be such as will enable it to resist effectually this pressure from without. We all know that a curved form of construction, in which the convex surface is opposed against the pressure, is stronger in this way than a plane surface, because the pressure applied to any point of the convex surface is immediately distributed to all the surrounding points on that surface, and their combined resistance is thus brought into action against the external force. And since the complete co-operation of all parts of the surface in resisting *uniform* pressure from the exterior is obtained only when all those parts have a common centre, the circle is the most perfect figure for this purpose.

357. But the pressure upon all parts of a sewer is not uniform. The top of it will be subject to the entire weight of the mass above it, minus only the friction and structural tenacity by which that mass is prevented from moving freely downward from the surrounding portions. The sides of the sewer are pressed against by the soil with forces inversely proportional to the tenacity of the material; that is to say, the less the tenacity or power of self-support, the greater will be the pressure against the sewer. The bottom of the sewer may be regarded as free from external pressure, except such as is due to the resistance with which the soil below meets the downward pressure exerted by the sewer itself, and transmitted by it from the load above.

358. The greatest pressure being thus vertically from above, a form of uniform strength would require to act with greater resistance in this direction. Hence an elliptical form, the longest diameter being placed vertically, would appear to answer the conditions better than a circle, and is, doubtless, the least imperfect form that can be

adopted. Practically it has been deemed desirable to combine, as far as possible, a considerable capacity with the means of making a reduced flow active in its passage through the sewer; and these requirements appear to be fulfilled by a form of section that differs from an ellipse in having the upper curve of larger radius than the lower one, resembling the outline of an egg standing on its smaller end, and to which the name of egg-shaped, or oviform, has therefore been applied.

359. The value of the curved bottom of reduced radius depends upon the well-known law, that the passage of fluids through channels is retarded by the friction between the water and the surface of the channel with which it is in contact. And it is an evident result of this law, that the greater the surface of contact the greater the friction. Hence any given bulk of water will flow the most rapidly in that form of channel in which this surface of contact is reduced to the minimum. The form necessary to fulfil this condition is presented at once by the well-known geometrical principle, that the circle includes a greater area within its perimeter than any other figure of equal perimeter. And as the necessity for aiding the flow by diminishing the friction occurs chiefly when only a small stream exists within the sewer, it follows that the radius of curvature should be proportionally reduced within practical limits.

360. The exact proportion which the radii of the upper and lower curves of the oviform section of sewer should bear to each other, (adopting circular curves as preferable in practice to elliptical or any indefinite curves,) would depend on the precise average minimum of water to be provided for, calculated jointly with the density and tenacity of the soil, and the depth at which the sewer is laid. As it is manifestly impossible to determine all these elements with exactness in evolving any general rule for the proportions of the section, they may be disregarded, since the main form is established by the conditions stated in

G 2

paragraph 358, and a practical rule may be formed which
will be found to answer all real purposes.

361. In the forming of sewers, as in all works of a
similar class, which are often necessarily entrusted, to a
great extent, to the charge of workmen, who cannot be
expected to pay much attention to the refinements of
geometrical principles, *simplicity* is evidently an object of
the first importance. The proportions of the several curves
required in marking out the section and forming the
moulds and gauges to be used in constructing and testing
the work should be such as can be readily understood and
exactly remembered; and in proportion as these rules are
observed in designing the form, will be the probability of
that form being preserved, and exactness attained in the
construction of the sewer.

362. In seeking this object we have worked out a diagram
of proportions, shown in fig. 77, that we can venture to

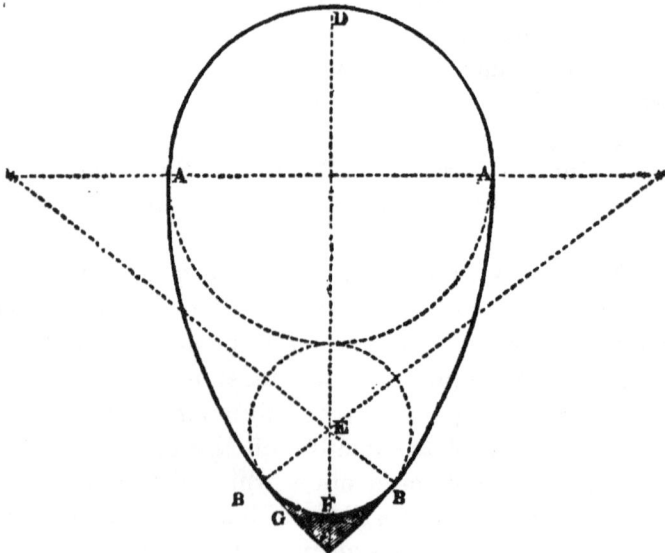

Fig. 77.

recommend on the score of simplicity and sufficiency,
which we hope will be made evident by the figure and the.

following description. In this section let the diameter A A, of the upper semicircle A D A equal 1 ; that of the lower arc B B will equal ·5. The entire height of the section D F will equal 1·5, and the radius C A A, of the side arcs A B (truly tangential to the upper and lower arcs) will also equal 1·5. The centres C C being upon the produced diameter of the upper arc, that arc will equal a semicircle, and the lower arc B D will equal 120°, the points for the meeting of the curves being at B B, found by drawing the radial lines, C B, through the centre E of the lower arc. Suppose the greatest diameter A A be determined at 3 ft., the several dimensions will be thus :—

	Ft.	In.
Diameter of upper arc	3	0
Do. of lower arc	1	6
Height of section	4	6
Radius of side arcs	4	6

And the area may be taken (as a very close approximation to the truth) as equal to that of a semicircle of 3 ft. diameter, added to the area of a circular segment whose radius is 4 ft. 6 in., and versed sine 1 ft. 6 in.; the area thus given being in excess only the small space shown in shaded lines at G.

363. The *construction* of sewers is varied according to their size, and should be also considered with reference to the economy with which different materials may be obtained according to the locality of the district, and also the nature of the soil in which the work is constructed. For the smaller sewers the " glazed stone-ware " pipes are found efficient substitutes for those built up of brickwork. They have the advantages of being much more quickly laid than the others can be built, and of presenting a very superior surface for the rapid passage of the sewage. They are also constructed in various forms of bends and junction pieces, and thus afford the means of ensuring proper form in these points. From their comparative thinness, pipes of this kind afford a much larger capacity with a given quantity of

excavation for laying them, than sewers formed of brick-work, which, even for the smallest diameter, cannot be less than half a brick, or 4½ in. in thickness. In laying them care must, of course, be taken with the joints, which are formed by a socket on one end of each length of pipe in which the plain end of the adjoining length is received. The prices at which these pipes may be procured in London are as follows:—

Straight Tubes with Socket Joints.

	Inches Bore.	Price per Lineal Foot.
		s. d.
In 3 ft. lengths	2 .	0 4
,, ,, ...	3 .	0 5
2 ft. ,, ...	4 . .	0 6
,, ,, ...	6 . .	0 8
,, ,, ...	9 . .	1 1½
,, ,, ...	12 . .	1 10
,, ,, ...	15 . .	3 0
,, ,, ...	18 . .	4 0

Diameter of Bore.	Bends Each.		Junctions Each.		Double Junctions Each.	
Inches.	s.	d.	s.	d.	s.	d.
2	1	0	1	0	1	4
3	1	3	1	3	1	8
4	1	9	1	6	2	0
6	2	3	2	0	2	8
9	3	6	3	6	4	6
12	5	6	5	6	7	0

Egg-shaped tubes are also prepared of the same material in 2 ft. lengths with socket joints, at the following prices·—

Size Inside.				Price per Lineal Foot.	
Ft.	In.	Ft.	In.	s.	d.
1	8 × 1		0	3	6
1	3 × 0		9 . .	2	3
0	9 × 0		6 .	1	1

· In Chester the following prices have been paid, including excavation of the maximum depth of 12 ft. :—

Inches.	s.	d.		
42 × 32 .	11	0	per lineal yard,	ordinary earth.
42 × 32 .	15	2	,,	rock.
36 × 28 .	9	6	,,	ordinary earth.
33 × 25 .	8	6	,,	,, ,,
30 × 22 .	7	9	,,	,, ,,
24 × 18 .	6	6	,,	,, ,,
20 × 15 .	5	6	,,	,, ,,
15 × 12	4	9	,,	,, ,,

The stone-ware tubes may be manufactured with ample strength for all purposes required in their application as minor sewers. Some experiments, made with specimen tubes of fire-clay at Glasgow, proved their power to resist a pressure equal to that of a perpendicular column of water 900 ft. in height, being three times the pressure to which it is found necessary to prove iron pipes used for the transmission of water. Drain tubes of common clay are supplied in Glasgow at the following prices:—

	s.	d.	
3 in. diameter . .	0	6	per lineal yard.
6 . ,, .	0	9	,,
9 ,, . .	1	0	,,
12 ,, . .	1	3	,,
18 ,, . .	2	0	,,

Pipes of fire-clay at Glasgow, cost:—

	s.	d.	
4 in. diameter .	1	0	per lineal yard.
6 ,, . .	1	6	,,
12 ,, .	2	3	,,

Supplied in large quantities, it is presumed that all of these prices for tubular drains would admit of considerable reduction. The following estimates for works, as ordered by the Metropolitan Commission of Sewers in the months

of April and May, 1849, contain some useful figures as to
the cost of works of this class :—

Quantities.	Localities.	Estimated Cost.
		£ s. d.
245 ft. of 12 in. pipe sewer	To be put down an open sewer in South-ampton Street, Nine Elms, Surrey .	39 0 0
400 ft. of 9 in. and 500 ft. of 12 in.	To be put down in St. Mark's Road, Kennington	131 5 0
485 ft. of 12 in.	To be put down in James Street, Kennington	78 16 3
400 ft. of 9 in. and 415 ft. of 12 in.	To be put down on the south side of Kennington Common	117 18 9
700 ft. of 4 in.	To be put down on the north side of Kennington Common	15 0 0
483 ft. of sewer, 3 ft. 6 in. by 2 ft. 3 in.	To be put down in the Wyndham Road, Camberwell	157 0 0
95 ft. of 18 in. pipe sewer	To be put down in Great Guildford Street, Borough	28 10 0
135 ft. of 15 in.		26 0 0
665 ft. of 9 in. . . In an open ditch		78 10 0
800 ft. of half-brick sewer, 3 ft. 6 in. by 2 ft. 3 in. and 158 ft. of 12 in. pipe sewer		300 0 0
240 ft. of 9 in.		15 0 0

From these estimates the average costs of supplying
and laying the pipe sewers of several sizes appear to be as
follow :—

		s.	d.	
4 in.	. . . . . .	0	5·14	per foot.
9 in.	. . . . . .	1	3	,,
12 in.	. . . . . .	3	3	,,
15 in.	. . . . . .	3	10	,,
18 in.	. . . . . .	6	0	,,

And of the egg-shaped sewer, half a brick thick, and
measuring 3 ft. 6 in. by 2 ft. 3 in., 6s. 10½d. per foot.

364. In commonly good soils brick sewers may be constructed of a single half brick, or 4½ in. in thickness, of
curved form, of considerable size. In the Finsbury division,
half brick egg-shaped sewers have been constructed, 4 ft.
6 in. by 2 ft. 9 in., and are found sufficient. Sewers of

these dimensions would be ample for the mains of properly limited and defined districts. If the soil be of a loose and uncertain character, it will be necessary to build them 9 in. in thickness, or two half-brick rings. In the small curve of the invert all brick-built sewers should be very carefully constructed, the unavoidable interstices between the bricks (if of the common square form) being filled in with pieces of slate or tile, and the whole floated in with cement to make it as one solid mass. If this be not honestly done and carefully superintended, the action of the declivity will be nullified by irregularities in the interior surface of the waterway, and a liability created to the formation of bars by the settlement of the solid portions of the sewage. Egg-shaped sewers 3 ft. 6 in. by 2 ft. 6 in., in an average excavation of 15 ft., have been executed at a cost of about 14s. per lineal foot in the neighbourhood of London. These sewers were built half a brick in thickness and in cement throughout, and the cost included excavation and refilling the soil.

365. Egg-shaped sewers formed according to the rule given (362), built half a brick in thickness and with inverts in cement, in an average excavation of 10 ft., may be estimated to cost per lineal foot as follow :—

	Ft.	In.	Ft.	In.	s.	d.	
Class 1 .	4	0	× 2	8	. 10	0	per lineal foot.
„ 2 .	3	6	× 2	4	8	6	,,
„ 3 .	3	0	× 2	0	. 7	0	,,
„ 4 .	2	6	× 1	8	5	6	,,
„ 5 .	2	0	× 1	4	. 4	6	,,

366. In forming the connections of drains with each other, viz. those of the house drains with the branch drains or sewers, and of these with the mains, through the several classes of sizes which it may be necessary to adopt, two rules should be in all cases imperatively insisted upon,— first, that all junctions shall be formed with curves, and of as large radii as possible in the direction of the current;

and, secondly, that wherever a minor drain discharges into a larger one, the bed of the former shall be kept as much as possible *above* that of the latter as the relative sizes of the two sewers will admit

367. The importance of the first of these rules has been long recognised and admits of proof, both theoretical and practical. It is found that in a sewer of 2 ft. 6 in. in width, a stream of water, flowing with a velocity equal to 250 ft. per minute, meets a resistance in suffering a change of direction, the amount of which depends upon the directness with which that change is made; the resistance occasioned being three times as great by a right angle as by a curve of 20 ft. radius, and double that produced by a curve of 5 ft. radius. The resistance thus diminishes as the radius of curvature of the junction is increased. The effect of junctions in which considerable resistance is opposed to the free passage of the sewage is, that the solid particles become deposited, and, being left by the flowing water, they accumulate until a bar is formed, which still further impedes the progress of the sewage, and eventually arrests it altogether.

368. The practical value of keeping the mouths of minor sewers above the level of the bed of the mains into which they discharge, arises from the prevention by this means of a return of the sewage up the minor drains, supposing a deficient declivity or any untoward circumstance should produce a retrograde movement within the main. The connection should also be formed in the most perfect manner, so that the mingling of the currents shall not have the effect of impeding either of them. The mouth of the minor drain should be spread into a bell-form, and the whole surface of the junction made solid and even with good cement.

369. The upper connections of the minor sewers, viz with the house drains, are small works, requiring the greatest care and circumspection. They are frequently disregarded and carelessly executed, because they appear

individually trivial matters; and, moreover, are troublesome and tedious, and correspondingly expensive. But it
is clear that the efficiency of the entire arrangement of any
system of town drainage is primarily dependent upon the
completeness with which the individual drains of houses
convey the separate contributions of sewage into the minor
or branch drains. If these tributaries fail, the trunk of
course remains idle, and all care bestowed on the larger
works is thrown away. Supposing the house drains to be
formed with clay or stone-ware pipes, and the receiving
branch sewer to be of the same material, lengths of the
latter should be introduced at intervals having sockets into
which the ends of the house drains may be fitted. If the
branch sewer be of brickwork, the junction of the house
drains should be carefully made good with a ring of cement,
and the work nicely finished on the interior surface. It
will of course be necessary to lay these house drains and
branch sewers at the same time (if the latter are of brickwork and not large enough to admit a workman), in order
to complete this work in the best manner. And as this is
not always convenient, the stone-ware pipes offer the great
advantage of jointing without any hand work inside the
branch, by simply laying the branch sewers with sufficient
socket outlet lengths at intervals, which may be communicated with by house drains at any future time, the sockets
being temporarily plugged up with wood.

370. The lower ends of the main sewers will communicate with the receiving wells, and should be well lipped
downwards to promote the ready discharge of the sewage
the moment it arrives at the mouth. These being principal
works, and few in number, are more likely to be well attended to and carefully executed than the multiplied minor
connections. The wells, adapted in capacity to the quantity
of sewage they are intended to contain, will require substantial and sound work. Being in towns necessarily sunk
to some depth in the ground, the cylinder will be the best
form in which to construct them. Behind and around the

brickwork a backing of concrete should be filled in, the excavation being made sufficiently large for this purpose, and the whole interior surface should be lined with cement or asphalte. If this be done it will not be necessary to build the work in cement, although this would, perhaps, be a wise additional precaution. Proper economy in this matter will be best arrived at by experiments, upon which an adequate sum of money would be well expended before extensive operations are commenced.

371. Means of access to the main sewers are best afforded by side entrances, such as those which have been introduced in the Holborn and Finsbury division for the purposes of inspection and flushing. Although, if the entire system were properly constructed, no necessity could occur for artificial cleansing, it will be desirable to provide means of getting at the interior of the main sewers at intervals, and the side entrances referred to are well adapted for this purpose. The side entrance consists of a vertical square or rectangular aperture, formed in brickwork, and covered by a hinged iron cover, fitted in the foot-pavement of the street. This aperture is carried down to a level of about 2 ft. above the bed of the main sewer, and terminates in a short passage or tunnel, which opens into the side of the sewer. The vertical entrance is provided with hand-irons, built into the wall, by which descent and ascent are rendered easy.

372. We have already insisted on the necessity of so arranging and constructing the sewers of a town that they shall not require any cleansing by hand, and have denied the condition of admitting workmen as an essential one in determining the size of sewers. A sewer cannot be considered as properly constructed if it retain the matters committed to it in a quiescent condition. It should act simply as a place of passage, and instantly transfer the sewage onward towards the receiving well. Failing in this purpose, and containing all the solid matters in a constantly-growing accumulation, the sewers of a town act as

combined cesspools, and the several gully-holes serve as
the outlets for the escape into the atmosphere of some of
the deadly gases constantly engendering below. The expense of cleansing by hand is, moreover, an item of considerable importance, although, of course, never incurred
until the subterranean nuisance becomes intolerable. In
the Holborn and Finsbury division, the cost of removing
the soil from the sewers provided with man holes is about
7s. per cubic yard, and from those without, 11s. per cubic
yard, including the expense of breaking the arch and
making it good again.

373. The method of cleansing the sewers in which
matter accumulates, by flushing water through them, was
practised to a great extent in the Holborn and Finsbury
division of sewers, and has been adopted by the New
Metropolitan Commission of Sewers. The principle of
this method consists simply in retaining the sewage water
for a period of time by flushing gates fitted in the·sewer,
and periodically admitting the accumulated water to pass
by opening the gates, and thus producing an artificial rush
sufficient to carry all accumulations before it. The relative
economy of the process, as practised in the Holborn and
Finsbury division, and, as compared with the hand cleansing, was stated as follows :—

	£	s.	d.
Washing away 6688 cubic yards of deposit by board-dams (a process always performed preparatory to fixing and using the flushing apparatus) . .	644	12	7
Putting inside entrances and flushing gates .	1293	0	0
	£1937	12	7

The cost of removing these 6688 cubic yards by hand
would have amounted to 2387l. The preliminary cleansing
and providing flushing apparatus were, therefore, effected
at a saving of 449l. 7s. 5d. The current expenses of the
two methods are thus stated :—

	£	s.	d.
Annual cost of cleansing 16 miles of sewers by hand	326	17	0
Annual cost of cleansing 16 miles of sewers by working flushing gates	106	0	0
Annual saving by flushing method	£220	17	0

The cost of this method, as subsequently practised under the New Metropolitan Commission of Sewers, has been reported as being about one-third that of cleansing by hand; thus 22,400 ft. of sewers, in which the deposit varied from 6 in. to 3 ft. 6 in. (in depth) were cleansed, and 3386 double loads washed away at an expense of 500l., which process, under the old system, would have cost 1500l.

374. The method of flushing is attended with one, and that a very serious, evil consequence, and the mischief of which is the greater in proportion to the force and velocity, and corresponding efficiency of the process. This is, the violent driving forward of the foul gases with which the otherwise vacant portion of a sewer holding stagnant refuse is usually filled. The flushing of the higher part of an extended line of sewer is thus frequently productive of a rising of these gases into the house-drains connected with the lower portion of the sewer, and any imperfection in the trapping of these admits the most noisome effluvia into the houses, while the streets are always poisoned with the gases thus driven up through the gratings and gully holes. Sometimes, indeed, the flushing water is forced into the house-drains, and, of course, occasions a total suspension of the flow of the sewage in the reverse direction. Accordingly, we find that the process of flushing has been discontinued during the warm season, the very time when it is most needed as an artificial means of cleansing the sewers.

375. For the efficient cleansing of the streets and thoroughfares of a town two provisions are requisite, viz. an abundant supply of water for occasional application,

when the self-supply of rain is suspended, and a complete arrangement of sewers through which to discharge all the surface-water when its purpose of cleansing has been fulfilled. For the supply of water, the system of constant supply affords the greatest facilities, giving an instant command of the required quantity.

376. It has been ascertained in London, that one ton of water is sufficient to lay the dust over a surface of 600 square yards of gravel or macadamized roads, or of 400 square yards of granite paved streets. The average number of days per annum in which it is found, from twenty years' experience, to be necessary to apply water for this purpose, is about 120. The common charge for this work is at the rate of $\frac{3}{4}d$. per square yard for the season, the water being applied only once per diem, or 50l. per mile of a main road. The common assessment per house for watering roads twice a day is 1l. for the season. The cost of doing the same work by means of jets, supplied from the main water-pipes, is estimated at 5s. per house. At Nottingham, where the constant service of water is rendered, a charge of 7s. 6d. per annum is made for a single street plug, by which some of the proprietors of shops command a ready supply, at all times, for watering the street in front of their own premises, and often of the adjoining houses also.

377. The scouring effect of jets of water thrown upon the surface of the streets is far greater than when merely dropped or thrown from the perforated pipe of a water-cart. A single jet, supplied with a force equal to throw the water vertically upward to a height of fifty feet, will, directed at an angle of 45°, command an area of about 2000 square yards, and this surface will be really cleansed by the process, whereas the mere distribution of the water, without pressure, wets without cleansing. The mud which is formed on the surface of the streets, during certain states of the weather, is well known to have an unctuous character, which resists all cleansing action less vigorous than that of jets of water under pressure.

378. The position of the main sewers beneath the streets of a town affords ready means of directly discharging the waste waters from their surface.   The adaptation of the sewers for this purpose requires inlets, at intervals, fitted with iron gratings, by which large substances are prevented from passing into the sewers.   These inlets and gratings being situated at the sides of the carriage-way, while the sewer is beneath the middle of it, they communicate by means of transverse drains or passages, which should be formed with sufficient declivity to prevent any accumulation of surface-water or road-sweepings beneath the gratings. The narrower the interstices between the bars of the gratings are, the better.   Very small spaces will suffice to admit the water with great rapidity, and also the mud which is formed upon the surface of the streets, and the narrow spaces are useful in preventing the admission of these matters during heavy showers with a force which might endanger the safety of the sewers.

SECTION V.

Conveyance of Water.—Piping, Aqueducts, Reservoirs.—Pumping Apparatus, Steam Draining and Pumping, &c.

379. For the conveyance of water from upper surfaces and sources to towns, open channels, or aqueducts, sometimes afford cheaper means than the laying of piping beneath the surface of the ground.   In supplying water from these sources to some of the towns in Scotland, Mr. Thom has had occasion to construct several miles of aqueducts, and in preference to adopting direct lines, which are commonly obtained at great cost in the necessary aqueduct bridges for crossing valleys and other expensive works for meeting the difficulties presented by the natural ruggedness of the country, Mr. Thom designs his aqueducts by winding along the slopes, however circuitous the course thus involved, and descending only with such a fall as will

allow the water to flow with a gentle current. Aqueducts
thus formed are simply artificial rivers, and the entire
expense is limited to that of constructing suitable banks
and bed for the channel. An aqueduct thus constructed
at Greenock, passes through very rugged ground, and has
cost not more than 400*l.* per mile. The New River, by
which a large section of London is supplied with water
from the springs of Chadwell and Amwell, with an additional supply out of the river Lea, near Chadwell, in Hert-
fordshire, is a fine example of an aqueduct of this kind.
This channel, the enterprise of Sir Hugh Middleton, was
commenced in 1609, and completed in 1613. The direct
length between its extremities is about 20 miles, but its
actual length is 39 miles. The average annual quantity
supplied by this aqueduct is 614,087,768 cubic feet. Deducting from this the larger consumers and street-watering,
together about 33,529,400 cubic feet, the remaining
580,558,368 cubic feet per annum are equal to about $46\frac{1}{2}$
cubic feet per tenement, supplied each alternate day. The
reservoirs, in which this supply is stored, are equal to contain the quantity consumed in seven days, or 11,774,000
cubic feet.

380. The city of New York is partially supplied with
water from the Croton river by an aqueduct 40 miles in
length. The receiving reservoir of these works contains 150,000,000 gallons, and the distributing reservoir
21,000,000 gallons. The supply is effected without either
pumps or water-wheels. An interesting work of this kind,
a suspension aqueduct, has been constructed for a canal
over the Alleghany river at Pittsburgh. This aqueduct
consists of seven spans of 160 ft. each, from centre to centre of pier. On the piers are pyramids rising 5 ft. above
the level of the side walk and towing-path, and measuring
3 ft. by 5 ft. on the top, and 4 ft. by 6 ft. 6 in. at the base.
The two wire cables which support the structure are placed
one on each side. Each is 7 in. diameter, perfectly solid
and compact and constructed in one piece from shore to

shore, 1175 ft. long, of 1900 wires of $\frac{1}{8}$ in. thickness. Each wire is varnished separately, and the whole cable has a close, compact, and continuous wrapping of annealed wire laid on by machinery. Transverse beams of timber, 27 ft. long, and 16 × 6 in., are placed in pairs at 4 ft. apart. Each pair of these beams is supported on each side of the aqueduct with a double stirrup of $1\frac{1}{8}$ in. round iron, mounted on a small saddle of cast iron, which rests on the cable. Into these beams, wooden posts 7 × 7 in. at top, and 7 × by 14 in. at bottom, are mortised. These posts are the side supports of the water-trunk, which is of wood, 1140 ft. in length, 14 ft. wide at bottom, and $16\frac{1}{2}$ ft. wide at top, and $8\frac{1}{2}$ ft. deep. The sides and bottom are composed of a double course of $2\frac{1}{2}$ in. white pine, placed so that each course crosses the other diagonally at a right angle. The extremities of the cables do not extend below the ground, but are connected with anchor-chains which, in curved lines, pass through the masonry of the abutments. The bars of these chains average $1\frac{1}{2}$ × 4 in., and from 4 to 12 ft. in length. They are formed of boiler scrap iron, and forged in single pieces without welds. The extreme links are anchored to cast-iron plates 6 ft. square. The total length of each cable and its chains is 1283 ft., and the weight of both cables 110 tons. The weight of water in each span (4 ft. deep in the trough) is 295 tons. The total solid section of anchor chains is 72 superficial inches. Deflection of chains, 14 ft. 6 in. Elevation of pyramids above piers, 16 ft. 6 in. The tension of each wire is 206 lbs., while its ultimate strength will be 1100 lbs.

381. Cast-iron pipes are now universally employed for the conveyance of water. They are formed with socket ends, so that all necessary motion is permitted according to the expansion and contraction of the metal, caused by variations of temperature. Until the commencement of the present century all the water supplied by companies to London was conveyed in pipes bored out of elm, and at

that time the New River Company had 400 miles of these wooden pipes in use. The general use of water-closets among the higher class of tenants, about the year 1809, led to the projection of new companies, who undertook to meet the growing want of water-supply at high service, by the use of steam power and iron pipes, a duty for which the old wooden pipes were inadequate. The bore of the wooden mains was from 6 to 7 in., and of the service pipes 3 in. The principal iron mains now vary from 12 to 30 in. in diameter; the sub-mains are 6 and 7 in., and the service pipes usually 4 in. The interior of the cast-iron pipes used for conveying water should be coated with a preparation of lime-water, to prevent corrosion and the consequent injurious effect upon the quality and flavour of the water.

382. Several methods have been adopted for forming the joints of iron water-pipes. Originally they were formed with flanges screwed together, but these were rapidly destroyed by the variations in the total length of piping produced by changes in temperature. Socket joints were then introduced, the joining parts being so formed that an annular space is left within the socket, and outside the entering pipe, for a ring of solder to be poured in, for the purpose of making the joint water-tight. An improvement has been effected in this kind of joint, by making the parts to fit each other, and turning them accurately to a conical form so that a water-tight joint is produced without any stuffing or packing of any kind, a little whiting and tallow only being used to assist the close adhesion of the parts. This kind of joint is so perfect that it has been adopted in forming the joints of a steam-engine suction pipe, 30 in. in diameter, with perfect success. Wooden plugs of suitable taper form have also been successfully and economically applied for forming the socket joints of water-pipes prepared with an annular space, in which they are driven.

383 The weights of cast-iron pipes, as applied for

water-supply, are as follows, according to the size or diameter of the bore.

In.							Cwt.	qrs.	lbs.	
3 diameter of bore		.	.	.	.	0	3	14	⎫
4	,,	,,	.	.	.	.	1	0	14	⎪
5	,,	,,	.	.	.	.	1	3	0	⎪
6	,,	,,	.	.	.	.	2	2	0	⎪
7	,,	,,	.	.	.	.	3	0	0	⎬
8	,,	,,	.	.	.	.	4	0	0	⎪
9	,,	,,	.	.	.	.	4	2	0	⎪
12	,,	,,	.	.	.	.	6	0	0	⎪
20	,,	,,	.	.	.	.	12	2	0	⎪
36	,,	,,	.	.	.	.	34	0	0	⎭

Weight of each pipe measuring 9 ft. in length.

384. In determining the proper size for pipes, according
to the quantity to be conveyed, the following formula has
been employed—

$$\frac{1}{15}\sqrt[5]{\frac{q^2\,l}{h}} = d,$$

in which q represents the number of gallons to be delivered
per hour, l the length of the pipe in yards, h the head in
feet, and d the diameter of the pipe in inches. In applying
this formula, Mr. Hawksley, Engineer to the Trent Water-
Works Company, calculates that for supplying a street of
600 yards in length, the total length should be divided into
three spaces of 200 yards each, and the quantity allowed
for each of these spaces should be respectively as follows :—

	Gallons per diem.
Final 200 yards	13,000
Middle 200 yards, 11,000 + 13,000 =	24,000
First 200 yards, 8,000 + 24,000 =	32,000

The calculation also assumes that the delivery of these
entire quantities will take place in four hours, and that the
whole of the water taken off from each length has to be
passed to the end of that length. The delivery of these
quantities respectively will require, according to the formula
quoted, pipes of the following sizes :—

Inches.

For the first 200 yards . , . 5·2 diameter.
  ,,   middle 200 yards . . 4·5 ,,
  ,,   final 200 yards . . . 3·6 ,,

Adding about half an inch to each of these for possible contraction by corrosion, the practical diameters become 6 in., 5 in., and 4 in. respectively. The difference in the size of pipes needed for the intermittent and the constant supply systems is exhibited in the following comparative statement :—

	Periodical Supply.	Constant Supply.
Mains .	20 in. diameter	. 12 in. diameter.
,, . .	7 ,,	. 5 ,,
,, ' .	6 ·,	. 4 ,,
Service pipes	3 ·,	. 2 ,,

385. Of the cost of raising water with pumps worked by steam-engines exaggerated conceptions are frequently formed, and it is therefore desirable to collect the best evidence on this subject. from which it appears that this cost is really an insignificant item, when the expense of the power is fairly compared with the quantity of water raised, as appears from the table of results (page 143), as stated by Mr. Wicksteed :—In all these cases the coals are taken at 12s. per ton, and all charges for working the engine, coals, labour, and stores, are included. but no charge is allowed for interest upon outlay, or repairs of machinery and buildings. To raise 160,000,000 of gallons 100 ft. high would cost according to the

     1st statement .   .   .   , £362
     2nd   ,,   . . 238
     3rd   ,, .   .   , 222
     4th   ,,   .   .   . 100

386. Of the performance of Taylor s pumping engine, in use at the United Mines, the late Mr. Farey made the following computation :—The average duty performed by this engine during the years 1841 and 1842 was equal to **the**

raising of 95¾ millions pounds weight of water, 1 foot high,
by the combustion of 1 bushel of coal. Each bushel of
coal weighs about 94 lbs., therefore each lb. of coal consumed by Taylor's engine raises 1,000,000 lbs. of water
1 ft. high. The unit of horse-power adopted by Mr. Watt,
viz. a force equal to 33,000 lbs., acting through a space of
1 ft. per minute, is found to be half as much more as the
average performance of a good draught horse working
8 hours a day and 6 days a week. A steam engine which
raises 94,000,000 per bushel (as Taylor's engine does) consumes only 1·98 lbs. of coal per hour for each horse power
which it exerts independently of overcoming its own friction, and that of the pumps That is, when it exerts a
power equal to that of 100 horses, it consumes only 198 lbs.
of coal per hour.

387. Mr. Hawksley has furnished a compendious statement of his experience in raising and conveyance of water
for the town of Nottingham, to this effect :—" The cost of
transmitting water to a distance of 5 miles, and to a height
of 200 ft., including wear and tear of pumping machinery,
fuel, labour, interest of capital invested in pipes, reservoirs,
engines, &c., amounts to about 2½d. per ton." The same
gentleman calculates the resistance from friction in conveying water in pipes according to the formula

$$p = \frac{q^3 \, l}{140 \, d^5},$$

in which p represents the horse-power necessary to overcome the friction, l the length of the pipe in inches, q the
quantity of water to be delivered in one second in gallons,
and d the diameter of the pipe in inches. For the transmission of 500 gallons of water per second, two mains,
each of 60 in. diameter, would be required, and the resistance arising from friction in these mains, 25 miles long,
would, according to this formula, require about 450-horse
power. The power required to *raise* this quantity to a reservoir at a height of 220 ft. would amount to that of 2000
horses nominally. The total power required to raise and

transmit a distance of 25 miles, through pipes, 500 gallons of water per second would thus equal that of 2450 horses. These figures are sufficient to show that the cost of raising and transmitting water by steam power is so small in proportion to the quantity of water thus placed at our command, that a pure but distant source may generally be economically applied in preference to supplying an inferior quality of water from more proximate sources.

Description of Engines.	Quantity of Water Raised per Diem.	Height to which the Water is Raised.	Cost of Raising 1000 Gallons 100 Feet High.	No. of Gallons Raised 100 Feet High for One Penny.
	Gallons.	Feet.	d.	
1. A single pumping engine, by Boulton and Watt, in 1809, working 10½ hours per diem, 6 days per week, mean power 29½ horses (Average of 2 years' working.)	612,360	100	·543	22,099
2. Two single pumping engines, by Boulton and Watt, in 1809, working 24 hours per diem, 7 days per week, mean power of each engine 30½ horses (Average of 10 years' working.)	2,922,480	90	·358	33,519
3. Two single pumping engines, by Boulton and Watt, one in 1816, and one in 1828, working 12 hours per diem, 7 days per week, mean power of each engine 76 horses (Average of 10 years' working.)	3,601,116	100	·333	36,036
4. One single pumping engine, by Harvey and Co., upon the expansive principle, in 1837, working 24 hours per diem, 7 days per week, mean power 95½ horses (Average of 4 years' working.)	4,107,816	110	·150	80,000

DIVISION III.

SECTION I.

Classification of Buildings.

388. THE principal classes of buildings, as subjects for water-supply and drainage, are—1, Dwellings; 2, Manufactories; and 3, Public Buildings. Each of these admits of several subdivisions, which should be briefly enumerated, in order to indicate the extent to which they are recipients of pure water and contributors of refuse matters to the sum total of town sewage

389. Dwellings are to be sub-classified according to the superficial area which they occupy, and the average number of residents whom they accommodate, and the arrangements to be provided for the joint purposes of supplying water and discharging sewage are required to be proportional to these two data combined. Upon the extent of area the quantity of rain water will depend, and this has to be entered in the account in two ways, first, as affording an integral portion of the supply, and secondly, as contributing to the sum of the sewage. The principal datum will be the number of persons for whom water is required in each dwelling, and each of whom will yield an average share of the refuse to be removed. The calculations of Water Companies are usually based upon the rental paid for each house as an index to the consumption of water within it, and in this way they recognise an almost infinite number of classes. It is clear, however, that the mere rental furnishes no exact criterion of the number of occupants of a house. Nor would the number of rooms in a dwelling

show this with much more accuracy. On the contrary, it is well known that houses of small rental and comparatively few apartments are frequently receptacles of a greater number of human beings than the more costly and capacious habitations of the wealthy classes. Nevertheless, *it is a fact*, that, with the present habits of the poorer sections of the population, the rental is generally in approximate proportion to the quantity of water consumed,—a fact to be accounted for only upon the recognised and deplorable principle that poverty and uncleanliness are mutual exponents and companions in the social condition of civilised beings.

390. We have estimated (283) 20 gallons as the average daily quantity for each inhabitant of a town, and have supposed this quantity to be sufficient to allow also for an ordinary proportion of manufacturing operations, for the supply of public buildings, and for the extinction of fires (284). This estimate is founded upon the experience had in several towns in which the supply is considered adequate. Reserving the details of the appropriation of this quantity for the next section, we now refer to this general estimate as the datum upon which the proper supply of water to dwelling-houses should be provided, and as being at least approximately correct, if the service be constant, and proper inducements be offered to all classes to cultivate habits of cleanliness. We would, therefore, subdivide the First Class of Buildings or Dwellings according to the average number of occupants of each, and provide the means of water-supply and drainage accordingly.

391. The Second Class of Buildings, or Manufactories, including all consumers beyond households, admits of a subdivision according to the operations carried on. Chemical works, including those for dyeing, calico-printing, &c., rank high as consumers of water. Factories for the making of paper, distilleries, breweries, bakehouses, malting-rooms, slaughter-houses, stables, &c., also consume large quantities. Steam-engines are among the wholesale

H

consumers. The charges made by the Nottingham Trent Water-Works Company are worth quoting in reference to the consumption of water, as their supply is constant, and provides for high-service, the two essential conditions of a complete water-supply. The charges for house service (according to rental, varying from 5l. to 100l.) are from 6s. to 60s. annually, being 10s. for 10l. rental; 20s. for 23l. or 24l. rental; 30s. for 39l. or 40l. rental; 42s. for 60l. rental; 50s. for a rental from 71l. to 75l.; and 60s. for 100l. rental. The incidental charges are as follow :—

	s.	d.
Stable and one horse .	4	0
Stable and more than one horse, for each horse .	2	6
Cows, each .	1	6
Warehouses—upwards from .	5	0
Offices .	5	0
Gardens .	2	6
Private baths in dwelling-houses .	10	0
Slaughter-houses .	5	0
Water-closets in private houses .	10	0
Water-closets in warehouses, &c.. .	20	0
Victuallers' brewhouses, two brewings per week .	20	0
Ditto ditto, less than twice per week.	16	0
Pipe for watering street in front of private house .	7	0
Boilers of high-pressure steam-engines, working 10 hours per day, per horse power .	9	0
Lace-dressing rooms, per yard in length, single frames .	0	9
Ditto ditto, double frames .	1	0
Bakehouses . 5s. to	8	0
Malt-rooms, per quarter of malt contained in steeping cistern .	2	6
Water consumed in erection of new buildings, per yard superficial on plan of each story .	0	1
Water consumed in erection of fence walls, per yard superficial .	0	0½
Mill-hands, for drinking and washing only, per individual employed .	0	3
Workhouses, including baths and washing rooms, per individual on the average of the whole year .	0	8

The supplies to dyers, &c., are estimated and charged for according to the size of the service pipes, by the following scale :—

Diameter of Pipe.	Estimated	
	Supply.	Charge.
Inches.	Gallons.	£ s. d.
$\frac{1}{2}$	50,000	1 10 0
$\frac{5}{8}$	100,000	2 12 0
$\frac{3}{4}$	200,000	4 12 0
$\frac{7}{8}$	300,000	6 10 0
1	400,000	8 6 0
1$\frac{1}{8}$	500,000	10 0 0
1$\frac{1}{4}$	600,000	11 12 0
1 and 1 in.	700,000	13 2 0
1$\frac{1}{8}$ and 1 „	800,000 ·	14 10 0
1$\frac{1}{8}$ and 1$\frac{1}{8}$„	900,000	15 16 0
1$\frac{1}{4}$ and 1 „	1,000,000	17 0 0
1$\frac{1}{4}$ and 1$\frac{1}{8}$„	2,000,000	32 0 0
1$\frac{1}{4}$ and 1$\frac{1}{4}$„	3,000,000	45 0 0

The waste water from condensing steam-engines of 500-horse power in the aggregate will amount to at least 1500 gallons per minute, or 3 gallons per minute per horse power.

392. Public buildings requiring constant service are to be divided according to the number of residents or persons to be supplied. Thus, union workhouses, prisons, lunatic asylums, &c., are to be provided at the minimum rate of 20 gallons per diem for each occupant. Baths and wash-houses require quantities in proportion to the maximum number of bathers and washers. Churches, theatres, and other places of public congregation are to be supplied for cleansing purposes according to the cubic contents of each building. In the baths, it may be estimated that a bulk of water measuring 6 ft. in length by 1$\frac{1}{2}$ ft. in width, and 1 ft. in depth will suffice for the ablution of each person. This quantity of water will equal 9 cubic ft., or about 54 gallons. The cost of supplying 1000 gallons by the Nottingham Trent Water-Works Company is, as we have seen (paragraph 319) 2·88d., or nearly 3d., and, as this quantity will be adequate to supply about 19 baths, the expense of water

H 2

per bath will be something less than one-sixth of a penny. The expense of fuel for heating 100 hogsheads of water— sufficient for 100 of these baths—from a medium temperature of 52° to 98°, including the replacing of heat lost by radiation, evaporation, and conduction, may be taken at about 540 lbs. of Newcastle coal, which at London prices may be averaged to cost 6*s.* The cost of heating each bath will thus amount to about ·75*d.*, and including the water, ·916*d.*, or less than 1*d.* If as much more be added for attendance, and a similar amount for interest on capital in building, and for incidentals, it appears that a hot bath may be well afforded at a charge of 3*d.*

SECTION II.

Supply of Water Levels.— Constant Service.—Quantity required.—Cisterns. —Reservoirs.—Filters.—Valves and Apparatus.—Piping, &c., &c.

393. The relative levels at which water is required to be supplied to buildings in a town will necessarily govern the height to which the main quantity must first be raised. But, practically, as the entire arrangements of the supply should be devised to command delivery at and above the most elevated of the buildings, the heights at which the delivery actually occurs will be found to affect only the current cost of raising, or the duty to be exacted from the power employed. And if this power be derived from steam-engines, its cost will appear to be insignificant in comparison to the space through which the water is raised. The expense of raising 1000 gallons to an average height of 80 ft. is found by the Trent Company to be, excluding interest on capital, less than 1½*d.*, and the cost, according to Mr. Wicksteed's Table of Results (page 143), of raising 1000 gallons to a height of 100 ft. with a single pumping engine on the expansive principle, excluding interest on capital and repairs of machinery, is less than one-sixth of 1*d.*, or 15*d.* Al

though the first cost of engines and pumping machinery of this class is very heavy, it will be a liberal allowance to balance this with the remaining ·85d., and we shall then have an average current cost of 1d. for raising 1000 gallons to an elevation of 100 feet. From this it will be readily inferred how small a difference will arise from diminishing or increasing this height to the extent of 20, 50, or 100 ft.

394. The necessity for constant service, great as it is in all buildings, is still more imperative in supplying those of which the demand is of a variable character. In certain seasons, when the occasion for repeated bathing of persons and cleansing of apartments is greatest, these duties require a much larger quantity of water than will suffice at other periods, and this demand of course increases in the same ratio with the number of persons and apartments to be supplied. Thus workhouses, prisons, and all public asylums vary considerably from time to time in the quantity of water required, and all methods of supply, short of constant service, and all provision for storage, fail in one way or another in securing the constant and unlimited command of fresh and pure water. Thus, house-tanks, cisterns, and reservoirs, however capacious and well designed, serve to receive only limited quantities; and if these be ample for all purposes, it follows that if the consumption be lessened the greater quantity of water will remain in a stagnant condition, to be added to but not replaced by the next delivery from the main. The lower body of water in the cistern will thus remain slightly changed, and stirred up only, and in this way a lower bed of impure water, surcharged and rendered heavy with deposited matters, gradually accumulates, suffering a slow diminution by the proportion of impurity which it imparts to each portion drawn off for immediate use. *Pure* or *fresh water* is, by this arrangement, put altogether out of the question.

395. In large public asylums, properly constructed, arrangements would be made for supplying a bath at least

once a week for every inmate. For this purpose an institution having 1000 residents would require weekly 54,000 gallons, or about 6000 cubic ft. of water. And if the supply be derived by a daily delivery, and the bathing be divided equally over 6 days in the week, a tank to hold the quantity for bathing only must have a capacity equal to 1000 cubic feet, or of the minimum dimensions of 20 ft. in length by 10 ft. in width, and 5 ft. in depth. The other purposes of cleansing would require (allowing 20 gallons per diem for each individual) 66,000 gallons weekly, or 11,000 gallons daily, and a tank to be daily emptied and refilled of the capacity of about 2000 ft., or measuring say 20 ft. in width and length, and 5 ft. in depth. For contingencies, provision should be made for about half this 'quantity in addition, and thus the entire capacity of the tanks should equal 4000 cubic ft., or dimensions of 80 ft. in length by 10 in width and 5 in depth. And if the consumption one day be reduced one-fourth, and the tanks be not emptied before the fresh delivery, which it is practically impossible to effect,—this quantity of stale water will remain in the lower part of the tanks, and each day's reduced consumption will tend to increase the impurity of the water, unless duplicate tanks be provided, and a large amount of water be wasted in their periodical cleansing.

396. In cases where the constant service of water cannot be obtained, and it consequently becomes necessary to provide cisterns for buildings, they should be so constructed and furnished as to combine the operation of filtering with the purpose of storing the water. For this purpose the best form of cistern will be that of which the bed inclines downwards, so that the discharge pipe may be inserted at the lowest point, and the water always drawn from that part of the cistern. The material used being commonly slate, the bottom may still be formed in a single slab for house cisterns (so as to avoid extra joints), declining in both directions. The filtering media, consisting of beds of sand and gravel of different degrees of fineness (as described in

Part I., p. 64), will be arranged in horizontal layers, excepting the lower one, which will lie in the bottom of the cistern, and be dressed to a level on its upper surface. The head of the discharge pipe should be protected with a fine wire-gauze cap, to prevent the gravel washing into the pipe. Below this pipe another cistern for the filtered water should be provided of proportionate capacity, and if the process be too tedious to admit of the filtration of all the water used, that for inferior purposes may be drawn from a pipe entering the cistern just above the filtering beds.

397. The superior quality of rain water in respect to its softness, as compared with water from all other sources, renders it exceedingly desirable, in an economical view, that all the supply derivable from this source should be carefully collected and preserved. In towns this water is commonly wasted, or at least allowed to subserve only the inferior purpose of assisting the flow of the drainage. Yet the quantity which might, by efficient arrangements, be commanded of this superior water is by no means insignificant. The roof of a house of the average dimensions of 20 ft. square, presenting a plane surface of 400 square ft., receives at least 800 cubic ft. of rain water annually, or about 4800 gallons. If well-constructed and capacious gutters are provided, this quantity may be collected with little loss from evaporation, and will form a reserve stock for such special household purposes as it is peculiarly adapted for. This quantity should be immediately received in a filtering tank, and the best available method be adopted of purifying it from the carbonaceous matters with which it becomes saturated in passing through a smoky atmosphere and flowing over roof-surfaces covered with a deposit of similar impurity. An economical and well-devised apparatus for effecting this purpose, and applicable to private and public buildings of all classes, is a desideratum yet wanting in the economical supply of water.

398. All valves and other apparatus for regulating the admission and use of water in buildings are required to be

constructed in the simplest and most efficient and durable
manner. Complicated contrivances are utterly inadmissible
to be entrusted to the ordinary carelessness and inattention
with which these things are treated in separate households.
Apparatus of costly construction will never receive the
sanction of landlords, nor will temporary tenants incur the
charge of expensive repairs, or devote regular attention to
keep ball-cocks and similar appendages in working order.
And in proportion as the rental of houses is less, these
difficulties are increased. Landlords become more par-
simonious, and tenants less interested and more neglectful
In this point of view the advantages of constant and high
service are rendered more conspicuous than in its application to tenements of a superior class in which a higher
rental enables the landlord to be liberal in the construction
and appliances of the building, and the tenant shares his
disposition to preserve their proper action in order to secure
his own comfort and convenience.

399. If the rain-water be not collected for household
cleansing purposes, it should at least be made as efficient
as possible for scouring the house-drains. An apparatus
for this purpose has been suggested by Mr. W. D. Guthrie,
a gentleman who has paid much attention to the subject of
town sewerage, and was one of the early advocates for the
use of small tubes in substitution for the larger drains,
constructed of brickwork, which were formerly prescribed
by Commissioners of Sewers as the only form of channels
which should be permitted access to their subterranean and
gigantic sewers or extended cesspools. Mr. Guthrie proposed that the rain-water from the roof be conducted into a
cistern, the lower part of which should be formed like an
inverted cone, and fitted with a conical valve at the head of
a pipe, discharging into the house-drain. This conical
valve is to be attached to a vertical chain above it, and connected with the short end of a lever to the other arm of
which a cord or chain is fixed, and by which the valve may
be occasionally raised from its seat, and the water dis-

charged from the cistern into the drain-pipe with a force proportional to the quantity in the cistern. From the upper part of the cistern a waste pipe is to descend externally and communicate with the drain pipe below the valve, so as to prevent the cistern overflowing, in case the water accumulates faster than it is discharged; the lower end of the waste-pipe being trapped, to prevent the effluvium in the drain-pipe passing into the cistern.

400. One of the most important of the occasional services for which a supply of water is required for application to buildings is, the extinction of accidental fires. For extensive buildings, such as warehouses, factories, and workrooms, tanks have been suggested, and, in some cases adopted, in which a considerable quantity may be constantly stored and ready for instant application for this purpose. This arrangement is, however, scarcely applicable for private buildings, and, where it is employed, the quantity commanded is of course limited, and can never be safely trusted to as affording an adequate supply for extinguishing the fire. In this application of water, again, the system of constant service offers great advantages. Thus, if the mains are kept always filled, an adequate supply is at all times at hand in every direction, and the grievous losses and dangers incurred by delay in obtaining water on these occasions are avoided.

401. The combination of high service with constant service in the supply of water also affords the means of instantly applying jets of water upon the fire until the fire or pumping-engines arrive. These jets are thus available as substitutes for the engines, and the experiments made to ascertain the height to which a jet of water will rise from the main and service-pipes under a fixed pressure, have shown considerable facility in applying jets for this purpose and a corresponding efficiency in their action. The practical limitation to this mode of delivery appears to arise from the extent of supply required, the economy of the use of jets depending upon the amount of pressure that can be

obtained, and the small number of jets which will suffice
for the extinction of the fire. The available power in this
case is found to decrease in proportion to the extent to
which it is employed, and the loss by friction in the leather
hose reduces the delivery, and, consequently, the height or
force of the jet, $2\frac{1}{2}$ per cent. for every 40 lineal feet of hose
through which the water passes. The importance of the
results of the experiments with jets here referred to will
justify a brief account of them in this place. They were
tried on the 31st of January, 1844, upon jets supplied from
the mains and services belonging to the Southwark Water-
Company, under a fixed pressure of 120 ft.

The first experiment was made over an extent of 800
yards of 7 in. main, which were connected with 500 yards
of 9 in., this length being joined to 200 yards of 12 in.,
continued by 550 yards of 15-inch main to the great main
leading from the Company's works at Battersea, the total
distance from the works to the experiment being 5500
yards. The heights to which the water was thrown from
$2\frac{1}{2}$-inch stand pipes, with 40 ft. of hose and a $\frac{7}{8}$-inch jet,
were as follows:—

With 1 stand pipe the water rose 50 ft.
"   2   "   "   45 "
"   3   "   "   40 "
"   4   "   "   35 "
"   5   "   "   30 "
"   6   "   "   27 "

When all the fire plugs on the main were closed, except
the first and one $2\frac{1}{2}$-inch stand pipe, and 160 ft. of hose
with a $\frac{7}{8}$-inch jet applied, the water rose to a height of 40 ft.

The quantity of water delivered from the same (7 in.)
main through one stand pipe, and different lengths of hose,
was as follows:—

With 40 ft. of hose .   .   .   . 96 gallons in 59 seconds.
"   80   "   112   "   65   "
" 160   "   116   "   70   "
"   40 ft. and $2\frac{1}{2}$-in. jet .   . . 118   "   27   "

The second experiment was made with a 9-inch main 1400 yards in length, joined to a 15-inch main of 1000 yards in length, and at a distance of 6650 yards from the works. The stand pipes used were $2\frac{1}{2}$ in., the hose 40 ft. long, and the jet $\frac{7}{8}$ inch, as before.

With 1 stand pipe the water rose . . . . . . 60 ft.
„ 2 „ „ . . imperceptible difference.
„ 4 „ „ . . . . . 45 ft.
„ 6 „ „ . . . . . 40 ft.

The quantity delivered with the same pipes, length of hose, and size of jet, being

With 1 stand pipe . . . . . 114 gallons in 64 seconds.
„ 4 „ „ . . . . 115 „ 75 „
„ 6 „ „ . . . . 112 „ 78 „

These experiments, with the two sizes of main-pipe, will indicate the rate at which the quantity is diminished by the friction of the water in smaller pipes, a result confirmed by another experiment made with the addition of 200 yards of 4-inch service and 200 yards of 5-inch pipe to the 9-inch main last referred to. The hose, 40 ft. long, and the jets $\frac{7}{8}$ inch, as before.

With $2\frac{1}{2}$-inch stand pipe fixed on the 4-inch service near the 5-inch
pipe, the water rose . . . . 40 ft.
With 2 do. do. do. . . . 31 ft.
With 1 do. fixed at end of service, or 200 yards from
5-inch pipe, the water rose . ˙. . . . . . 34 ft.
With 2 do. do. do. . . . 23 ft.

The quantity delivered in each of these last four cases being respectively as follows:—

112 gallons in 82 seconds
117 „ 103 „
112 „ 90 „
114 „ 118 „

402. In an interesting paper by Mr. James Braidwood, upon the means of applying water for the extinction of fires, read at the Institution of Civil Engineers, it is shown that elevated tanks for a reserve of water for this purpose should be adapted to contain 176 tons of water for each

fire-engine to be employed. This allows for six hours working of an engine having two cylinders of 7 in diameter with a stroke of 8 in., making 40 strokes each per minute, and fitted to throw 141 tons of water in six hours; and, allowing one fourth for waste, the supply required will be as stated, 176 tons. In the case of a large building, provision should be made for working ten engines for this period, and the quantity required will be 1760 tons, or 63,360 cubic feet of water. From this calculation, it will be evident that the dimensions of the tanks would be enormous. If steam engines can be commanded upon the premises to maintain the supply through the mains, the reserve may be reduced to a consumption for two hours, before the expiration of which time it may be expected that the engine could be got to work. This provision is such as may be supposed requisite in dockyards and for large stacks of warehouses, manufactories, &c.

403. In the town of Preston, the advantages of the constant and high service have led to the general use of jets and the comparative disuse of engines for the extinction of fires. For this purpose the hose is carried upon a reel, and should be fixed upon a light spring cart, by which the ladders may be also conveyed. The ladders are found to be invaluable appendages for the economical application of the hose without the engines, because the higher the water is carried upward in the hose, that is, the higher the nozzle of the hose is placed, the less is the resistance suffered from the atmosphere. If a jet forced by a pressure of 100 ft. attain a height of 50 ft. when delivered at the ground level, it will still attain an additional height of 20 or 25 ft., when the nozzle is carried up these 50 ft., and the discharge will then take place at a total height of 70 or 75 ft. from the level of the ground   And another advantage derived from carrying the hose as high as possible is, that of commanding a more effective discharge of the water than can be obtained when the direction of the jet is conducted on the ground

404. The piping for the conveyance of water to buildings has to be graduated in capacity according to the quantity required, in the same way that the mains and service-pipes are proportioned to the extent of district and number of buildings they are intended to serve. In supplying towns with drainage water collected in high reservoirs, and thence conveyed by aqueducts to " distributing basins," Mr. Thom adopted a general system of piping, which is so arranged that the water shall always flow within them in one direction, entering at the upper and passing to the lower end. At the lower end of each range of piping a cleansing cock is provided, by opening which occasionally any improper accumulation within the pipe may be removed. The pipes are kept constantly full, and laid at a minimum depth of 3 ft. below the surface of the pavement. In some cases, in order to provide very fine cold water to private houses, an iron cistern, to hold about 20 gallons, is sunk 8 or 10 ft below the bottom of the cellar, and supplied with water through a small lead pipe entering it at the top, while the water is drawn off for use through another small pipe, inserted a few inches above the bottom of the cistern. It would appear, however, that the cleansing of cisterns thus situated must be a somewhat troublesome duty, and the means of regular access to a cistern so deeply sunk in the ground must involve a considerable additional expense in construction.

SECTION III.

Varieties of Manufactures and best available Methods of Draining.—
Arrangement of Separate and Collective Drains.—Proportion of Area of
Drain to Cubic Contents of Dwelling-Houses.—Fall of Drains.—Mode of
Construction. Connection with Main or Collateral Sewers.—Means of
Access, &c., &c.

405. The several operations carried on within a building devoted to manufacturing purposes should afford the data

upon which to determine the extent of drainage required,
but the most ready way of estimating the amount of refuse
waters produced, will be reached by assuming this to equal
the supply of water rendered to the building. The application of the same rule to domestic buildings or dwellings
admits of a more exact calculation as to the capacity of
drains required, but these must all alike be governed by
the principle, that ample capacity for immediate discharge
is to be sought, with due regard to the fact that all passages
for the conveyance of liquid or semi-liquid matters are efficient in proportion to the narrowness of the surface over
which these matters are required to flow. This is one of
the most important results which recent inquiries have established. Sewers and drains were formerly devised with the
single object of making them *large enough*, by which it was
supposed that their full efficiency was secured. But sluggish-
ness of action is now recognised as the certain consequence
of excess of surface equally as of deficiency of declination.
A small stream of liquid matter extended over a wide surface, and reduced in depth in proportion to this width,
suffers retardation from this circumstance as well as from a
want of declivity in the current. Hence a drain which is
disproportionally large in comparison to the amount of
drainage, becomes an inoperative apparatus, by reason of
its undue dimensions; while, if the same amount of drainage is concentrated within a more limited channel, a greater
rapidity is produced, and every addition to the contents of
the drain aids, by the full force of its gravity, in propelling
the entire quantity forward to the point-of discharge.

406. There are four conditions which are to be regarded
as indispensable in the construction of all drains from all
buildings whatsoever. These conditions are—First. That
the entire length of drain is to be constructed and maintained with *sufficient declivity* towards the discharge into the
sewer to enable the average proportion and quantity of
liquid and solid matters committed to it to maintain a *constant* and *uninterrupted motion*, so that stagnation shall never

occur. Second. That the entire length of drain is to be constructed and maintained in a condition of *complete impermeability*, so that no portion of the matters put into it shall accidentally escape from it. Third. That the head of the drain shall be so efficiently trapped that no gaseous or volatile properties or products can possibly arise from its contents. And, Fourth. That the lower extremity of the drain, or the point of its communication with the sewer, shall be so properly, completely, and durably formed, that no interruption to the flow of the drainage or escape shall there take place, and that no facility shall be offered for the upward progress of the sewage in case the sewer becomes surcharged, and thus tends to produce such an effect.

407. These conditions appear so simple in their statement, that we are disposed to regard them as self-evident necessities, yet an acquaintance with the details of house-drainage as commonly regulated reveals the fact that they have been generally neglected, and that, at the best, the attention they have received has been most unwisely crippled by considerations of cheapness in first cost at the expense of permanent economy and usefulness. Thus we know that house drains are frequently laid with very imperfect fall, not sufficient indeed to propel the matters sent into them except with the aid of gushes of drainage-water; that they are often composed of defective and carelessly-built brickwork with wide joints of sandy mortar; that the head of the drain is commonly untrapped; and that the entire formation is badly designed and defectively executed. We will endeavour to show the arrangements by which the efficient action of the separate drains of houses and other buildings is most likely to be secured

408. The utmost practicable declivity being obtained in the direction of the drain, the efficiency of its action will be further much controlled by the construction adopted and the kind of surface presented to the sewage. Any roughness or irregularity in this surface will of course impede the passage of the sewage, and hence arises the neces-

sity for the greatest care in the construction, whatever the material and kind of formation. The first step in the arrangement is, to collect the whole of the drainage to one point—the head of the intended draining apparatus, and the determination of this point requires a due consideration of its relation to the other extremity of the drain at which the discharge into the sewer is to take place. In buildings of great extent this will sometimes involve a good deal of arrangement, and it will, perhaps, become desirable to divide the entire drainage into two or more points of delivery, and conduct it in so many separate drains to the receiving sewer. The length of each drain being thus reduced to a manageable extent, the necessary fall will be more readily commanded, and the efficiency of the system secured.

409. The cost of constructing these minor works, and also the main sewers with which they are connected, is so enormously aggravated by the depth to which they are frequently laid in order to accommodate the basements of buildings, that, for the sake of economy, basement-drainage should either be altogether abandoned or so modified that efficiency shall never be sacrificed in a vain attempt to reconcile the depth of the basement with the position of the sewer. In arranging the drainage of buildings, therefore, the head of the drains should be kept at the minimum depth which will suffice to sink the construction beneath the surface. We have already (353) expressed a conviction, that a thoroughly perfect and economical system of town-drainage must recognise this as a leading principle, and under this conviction we could not be satisfied to admit the difficulty now experienced to be one which should encumber our proceedings so as to involve comparative inefficiency in action and extravagant costliness in construction and repairs.

410. Although it is not within our province in this place to discuss the governmental measures which would be required to authorise and direct such an adjustment of the

details of private drainage as would be necessary to insure their conformity with the principle here advocated, we may be permitted to observe that this direction was, to a considerable extent, assumed and exercised by the old Commission of Sewers, who always declared their authority in prescribing the manner in which private drains should alone be allowed to communicate with the sewers under their jurisdiction. These prescriptions determined the rate of declivity, the relative levels, and the dimensions of the drains, and were enforced by the Commissioners' execution (by their contractors) of that portion of the work which joined with the sewers. The regulations enforced in the City of London (and which, from its independence of the new consolidated commission, are, it is supposed, still enforced) are based upon the following calculation, the stated principle being that "A house cannot be called effectually drained unless the water is taken away from the floor of its lowest story."*

	Ft.	In
Take the least height which a basement story ought to be . .	7	0
Thickness of a timber flooring on sleepers	0	9
Covering of the drain, say brick flat	0	2½
Height of drain inside	0	9
Current of drain inside the premises, say, 1 *in.* to 10 *ft.* for a house 50 ft. deep	0	5
Current outside the house, *i. e.* in the street	0	3
Height of cross-drain above the bottom of main-drain, at least .	0	6
	9	10½

This would give (9 ft. 10½ in., or) 10 ft., at the least, depth from the surface of the street to the bottom of a main drain (of 18 in. diameter), and this may be fairly assumed as the least depth at which a private house of the most ordinary description can be effectually drained; but this considers it only as for the drainage of one house.

* It should be borne in mind that this principle becomes impracticable if the lowest story of any house should, at the free will and option of its owner, have another "lower deep" excavated below it, a practice which has been indulged in in the formation of some of the leviathan warehouses in the City.

"When a series of houses, situate in a public way, inhabited by some who will use, and some who will not use, a drain fairly, is to be drained, the question has to be looked at differently.

		Ft.	In.
"For a retail shop, in which the basement story is often used as a warehouse, it cannot be unreasonable to say that the story shall not be less in height than		8	6
Flooring		0	9
Covering of drain		0	2½
Height of drain		1	3
Current inside		0	5
Current outside		0	3
Height above bottom of common sewer		1	6
		12	10½

"As it may be said that a story of less height might do as a wareroom, and, in order to keep the calculation as low as it fairly can be kept, I would assume that the bottom of a common sewer ought not in any part to be less than 12 ft. beneath the surface of the street."*

We have thus quoted these calculations at length, in order that we may be enabled to refer to the details assumed without fear of mistaking the meaning of the official provisions of the Commissioners.

411. By another of the regulations of the Commissioners of the City Sewers, affecting the details of the house drains, we find that since the year 1832 the Commissioners have required that their own tradesmen should be employed to make the whole of the drains up to the front of the buildings, these drains, 15 in. in diameter, being charged at the rate of 5s. 6d. per lineal foot. And the reason alleged for this regulation was, that the Commissioners found great difficulty in getting individuals to make the drains substantially.

412. The regulations laid down by the Commissioners of

* "Memorandum," laid before the Court of Commissioners of Sewers for the City of London, &c., by their late Surveyor.

Sewers for Westminster for the construction of private drains were as follows:—" That no drains shall be laid into a public sewer, without a special leave for that purpose from the Commissioners. That when such leave shall be obtained, the opening into the sewer shall be made, and the drain built, for a length of 3 ft. from the sewer, according to a plan and section approved by the Commissioners; the whole to be done by a workman to be employed by the Commissioners, and paid by the party requiring the drain, at the prices undermentioned:—For cutting through the springing wall of a sewer, putting in a cemented brick ring, and soundly underpinning the wall round the same, the sum of 10s. 6d. for each opening. For building a length of 3 ft. 4 in. of 9-in. barrel drain, with proper York keel stone, sound stock bricks and blue lias lime mortar, the sum of 10s. 6d. for each such length of drain. For the same length of 12-in. barrel drain, 12s. 6d. The digging to be done at the expense of the party requiring the drain; and notice to be given at the office of the Commissioners when the excavation shall have been made, in order that an officer may attend, and that a workman may be sent to do the required works. As a guide to persons about to build, it is recommended that the private drain of each house or other premises have a current not less than a quarter of an inch to each foot in length, making in the length of 60 ft. a fall of 15 in., to which, adding 13 in. for the height of the drain and brick arch over it, also 8 in. for the depth of ground and paving over the drain at the upper end, and 12 in. from the lower end of the drain to the bottom at the side of the sewer, will make, in the whole, 4 ft. from the bottom at the side of the sewer to the lowest pavement of the building, being the least height necessary to guard the premises from being flooded by water from the sewer."

413. These notices of the regulations which were enacted and enforced by two of the old Commissions of Sewers are sufficient to show that the powers which may now be required for instituting an entire system of house-drainage

under public authority, or that derived from a Commission under the Great Seal, would be no new entrenchment-upon private rights. The following order of the Westminster Commission declares its power to deny the right of draining into the public sewers if the depth of the building would require a rate of declivity less than then deemed necessary to insure the proper action of the drain:—" The Commissioners give notice, that whenever the lower floors or pavements of buildings shall have been laid so low as not to admit of their being drained with a proper current, they will not allow any sewers, or drains into sewers, to be made for the service of such buildings."

414. The regulations we have quoted are, we submit, sufficient to show also that the details thus prescribed were not calculated to contribute to a system of efficient house-drainage, being inadequate, in some of the several indispensable conditions before stated (406).

415. Thus for cylindrical drains of 9 in. in diameter, a construction composed of the ordinary rectangular bricks, with mortar joints, is essentially unsuitable and imperfect, being unavoidably permeable to a considerable extent; the irregularities which occur at every joint, moreover, impair most seriously the effectiveness of the declivity which, if only 1 in. in 10 ft., or 1 in 120, as allowed in the City of London, is, even if fully preserved, inadequate for the purpose. The Westminster allowance of a quarter of an inch in each foot, or 1 in 48, is barely sufficient to make the rapid passage of the sewage a matter of certainty. And drains are much more likely to act efficiently if laid with a fall of 1 in 20 or 30. These regulations illustrate the two alternatives to which the present system reduces the practice and the utility of house-drains. In the one division we have an utterly inefficient declivity of 1 in 120, coupled with a *minimum* depth of 12 ft. from the bottom of the common sewer; while, in the other division, the Commissioners, with an arbitrary kind of wisdom, decline to attempt the task of draining any premises with basements

"laid so low as not to admit of their being drained with a proper current." The "propriety" of the current would, however, be considerably enhanced by still increasing the fall of 1 in 48, which they adopted.

416. The common occupation of the basement stories of houses as kitchens and water-closets, has made it appear desirable to depress the drains and sewers, in order to receive the refuse matters below the level of these basements; but as this object involves one or both of the evils we have pointed out, viz. deficient declivity and consequent stagnation in the drains, and a general system of sewers sunk so deeply in the ground that incomparable expense and difficulty are created in construction, access, and repairs, the purpose of basement-draining should be abandoned, and practicable methods sought of delivering the entire drainage immediately beneath the surface of the ground. If, indeed, no practicable methods could be devised of doing this so as to render basement-draining unnecessary, it must of course be admitted as part of the purpose of house-drainage, in order to avoid the sacrifice of the healthiness of human habitations, which we all readily admit as the final object of the art of draining towns and buildings.

417. The selection of the methods to be adopted for this purpose will be dependent mainly upon the internal arrangements of the building and the occupation of its lower apartments. In the first place, *water-closets must in all cases be constructed above ground*, or, at any rate, so nearly above, that the discharge shall take place within a foot or so of the surface. However valuable the ground-floor space of any premises may be, sufficient room may and always should be reserved for this purpose, as this level is the most desirable for the situation of these accommodations. If placed higher, they cannot be so readily aided with the sewage water produced in the domestic offices, unless these occupy a similarly-elevated position; and besides this objection, is that of having an unnecessary length or extent of drains above

the ground. The most desirable arrangement, therefore, is that which collects the entire drainage at or near the ground level, and there at once and immediately delivers it into the subterranean channels. If, however, it is in any, case unavoidable that the kitchen and similar domestic offices are situated in the basement of the building, it will be still equally imperative that all the sewage water shall be delivered into the drain at or near the ground level. No sink or other apparatus for discharging refuse water should be retained in the basement, and the extra labour of carrying this water up to the surface level, or head of the drainage, must be incurred as the penalty of this misconstruction or misappropriation of the building.

418. These arrangements, although involving expense in the alteration of some of the existing buildings in towns, are not to be magnified into impracticabilities. As essential parts of a general system conducive to the health of the entire population, they should be commanded and enforced by adequate public authority, and carried into immediate effect without favour or evasion. And in the construction of new buildings, they should be regarded as imperative general orders, sanctioned by the public well-being, and, if necessary, to be obeyed under official superintendence The truth of principles and advantage of modes of action, established by experiment, should command their adoption without opposition, from the prevailing squeamish reluctance to interfere with private arrangements, which, be it remembered, are, if misdirected, really in these matters public nuisances.

419. The keeping of the basement itself of a building dry by draining is not, we submit, to be acknowledged as a proper purpose of a correct general system. Sufficient and sound construction are alone needed to maintain basement stories of any depth in a perfectly dry condition, if all sewage and rain-waters are, as they should be, collected and discharged into the sewers before they reach the basement. The draining of the surrounding subsoil to the

entire depth of the foundation of a building is a want
which cannot arise, if the entire structure up to the ground
level is *waterproof*, which we contend it should be; the
means of effecting this by the materials now at our command, being of economical and certain application

420. Discarding brickwork and all similar constructions
of small parts, as unsuitable for obtaining the impermeability and smoothness of internal surface which are
especially required in small drains, the current in which is
often reduced to a very small quantity of liquid or semi-
solid matter, we are led to seek some tubes or pipes, which
shall require only annular joining at distant points, and
thus admit of the regularity of surface which is so necessary to assist the passage of the drainage. The stone-ware
is now offered as a superior material for this purpose, admitting of much greater economy than iron, and being
entirely free from the chance of corrosion and permeability. By glazing the interior surface, moreover, tubes of
this ware are made peculiarly suitable for adoption in
forming drains; and carefully-made socket joints laid in
the direction of the current are cheaply executed, if
moulded conically and luted with a little cement of best
quality.

421. The size of the drain-pipes has to be graduated
according to the quantity to be passed through them, limited
in the minimum extreme so as to avoid stoppage from the
excessive bulk of the sewage matters, and in the maximum
extreme so as to obtain all the rapidity of progress of which
a small stream of water is capable, retarded by the friction
of the surface over which it passes. For moderate-sized
houses, say of eight rooms, and holding some five or six
persons on an average, a tube of 5 in. in diameter will
suffice for the house-drain. The area of the drain may be
proportional to the cubic contents of the house, but if so,
in diminishing ratio. That is, if a 5-in. pipe will be large
enough for an average-sized house, a pipe of double the
area of such a pipe will not be required for a house of

double the cubic contents, or holding double the average number of persons. A 6-in. pipe, laid with sufficient fall, will be ample for the most capacious private house. And from 9 to 15 in. will, under a similar condition, be sufficient to serve the average drainage of factories and other large consumers of water.

422. The trapping of the head of the drain, so as to prevent the ascent of smell and impure gas from the drain into the building, is the next indispensable requirement in the draining apparatus. So many contrivances have been applied for this purpose, that we will not attempt to make a selection; and it is beyond our limits to give any general list or detailed description of them. Simplicity of construction and permanence of action are, of course, required, with the least original outlay at which these qualities can be obtained. If only one water-closet is to be provided for, it will be desirable to gather the discharge from it and from the house-sink, &c., into one trap at the head of the drain. If two or more closets are to be served, so many separate traps will sometimes become indispensable. But for every separate inlet to the drain, which is equally an outlet for smell and gas, an efficient trapping apparatus of some kind is required.

423. The lower connection of the house-drain with the public sewer is the last point of importance to which we have to allude. A perfect construction of this portion of the work has always been recognised as an essential feature of good drainage, and the Commissioners have accordingly stipulated that its execution should be entrusted only to their own contractors, and be subject to the inspection and approval of their own officers. The level of the bed of the drain must be kept as high as possible above that of the receiving sewer. If the sewer be also constructed of the glazed stone-ware piping, lengths of it may be introduced at convenient intervals, having outlet sockets for receiving the ends of the house-drains, and those being slightly tapered or conical in form will be readily jointed with a little

of the best blue lias cement or other of equal quality. If the sewer be constructed of brickwork, a good joint will be obtained by introducing a separate socket of stone-ware to receive the house-drain pipe, and formed with a flange at the other end to surround and cover the opening in the sewer, which can then be made good with a ring of cement carefully applied.

424. Means of access to house-drains are always desirable in arranging the details of the apparatus. And this constitutes another reason against the deeply-sunk drains required to serve the basement story of houses. If the drains be constructed of glazed stone-ware pipes, carefully jointed, and laid in directions as nearly uniform as possible, the process of artificial cleansing and raking (should it ever become necessary) will be much facilitated. If any angular turns are formed in the direction of the drains, it will be worth while to consider the practicability of fitting a movable cover at the angle, by removing which, direct access should be afforded to the two branches of the drain. A long pliant rod with a stiff brush or scraper at the end could then be readily introduced into the drain, and, if necessary, these means of access by trap doors and removable covers should be afforded at intervals throughout each extended length of drain, so that thorough cleansing from the head of the drain to the outlet in the sewer could be performed as frequently as might be found requisite. Under a complete and efficient system of drainage the task of periodically examining the separate drains from the buildings would be ordered and performed with all the regularity and readiness of a necessary duty, and the drains would be maintained in a state of constant instead of intermittent cleanliness. *

* The system of *subways*, recently introduced in some parts of the metropolis, will probably supply facilities for examining the minor drains, as well as the main sewers, to a degree not hitherto practicable. (1865.)

I

SECTION IV.

Water-closets; Arrangement and Construction.—Adaptation to variou
circumstances.—Combined Arrangements for Efficient House-Drainage.—
Miscellaneous Apparatus and Contrivances.

425. The best position for a water-closet in any building
is that in which all the waste water shall be made the best
use of in scouring the contents directly through the pan of
the closet, and propelling them forward through the private
drain into the common sewer. And since the matters discharged into the closet will be, if the house drain is reserved for its proper uses, more solid and less readily conveyed than the other sewage matters, it will, moreover, be
desirable to place the closet as near as possible to the point
at which the drain discharges into the sewer. The velocity
and force of the liquid sewage are increased at the lower or
sewer end of the drain, and its effect is thus augmented in
scouring away the contributions of the closet. But if this
preferable position cannot be commanded for the closet, it
must, at any rate, be so situated with regard to the head of
the drain and the inlet for the liquid sewage, that these
shall be behind or above it. When the closet and the
house-sink are near to each other, the water from the latter
may be conducted directly into the trap or basin of the
closet, and thus secure at once a rapid discharge of its
contents and a constant supply of liquid to preserve its
action and efficiency.

426. The rudest form of domestic accommodation or
open privy over a cesspool is a contrivance which deserves
notice only on account of its several imperfections, and
which will, it may be hoped, be soon reckoned among the
obsolete mistakes of our forefathers. These cesspools are
sometimes mere pits or holes excavated in the ground, and
the contents of course rapidly permeate the surrounding
soil; by which process pits of this kind frequently are
found to drain themselves, the perviousness of the material

permitting the escape of the sewage, so that little accumulation takes place within the pit itself until the whole
neighbourhood becomes fully saturated with the drainage,
which will then ooze through and appear upon the surface,
or find its way through some defective foundation, and poison the basement of an adjoining building. Constructed
cesspools formed with brickwork of substantial quality will
prevent this saturation in proportion as their walls are carefully and imperviously built. The matters daily discharged
into these depositaries accumulate, and their decomposition
is constantly proceeding and engendering gases of the most
noisome and pestilential kind. The open privy formed
over a pit of this description affords an outlet for the escape
of these gases, which are thus regularly supplied to the
building above or adjacent to the closet. If a trap or water
basin and pan be applied to this privy, so that the pan dips
into the trap, the escape of effluvia may be prevented so
long as the trap is kept supplied with water. The supply
of water for this purpose will, however, considerably augment the bulk of the sewage, and necessitate cleansing
much more frequently than otherwise, unless some defect
in the joints of the work afford a passage for the liquid
matters in the surrounding strata, or a communication be
afforded with a drain. In this latter case of combination
of a cesspool with a drain, a waste pipe may be laid from
the former into the latter, so that the contents of the cesspool shall always be maintained at the same quantity and
depth; the trap may then be dispensed with by attaching
a vertical pipe to the lower part of the pan, so that this
pipe shall dip into the sewage, and being thus constantly
kept below its surface, no gas can pass upward through the
pipe. The cost of the pan or basin and pipe required for
this contrivance, if of stone-ware, will not exceed 13s. in
addition to that of a common privy, and its advantages in
preventing the escape of effluvia are obvious. The simplest and cheapest form of trap and basin is that in which
they are formed in one or two pieces of the stone-ware,

and may be purchased at about 7s. 6d. together. Allowing 5s. 6d. for the fixing, and also providing and fixing a short length of pipe of the same material to connect with the trap so as to dip into the sewage, this complete trapped apparatus may, at a cost of 13s., be added to a common privy over a cesspool, so as to prevent to a great extent the escape of effluvia into the house or adjacent building.

427. The great importance, however, of avoiding all sources of unwholesome and offensive effluvia, and of preserving the foundations of the buildings and the substrata of the soil of a town in a dry and clean condition, creates a severe necessity for relinquishing cesspools, and all *receptacles* for sewage, within or connected with all buildings and places whatsoever, except those to which it is conducted for the purposes of collection and treatment. *The sole purpose of all house apparatus of water-closets, sinks, and drains, and of all public constructions of branch or tributary sewers, and main sewers, should be that of affording a passage for the conveyance of the refuse waters and other matters produced in a town. This conveyance should be immediate, every particle committed to the entire ramification of passages being preserved in ceaseless motion until it arrives at the final collecting place.* ·

428. Discarding cesspools upon these grounds, we are at the same time led to the principle which should govern the whole of the details of house-draining apparatus, which should be so arranged and combined as to afford the fewest possible inlets for effluvia from the matters committed to the drains, and to make the total of the liquid refuse useful in advancing the current within the drains. The position of the water-closet being determined (425), it becomes desirable to select the most economical and efficient construction for it, and for the apparatus connected with it.

429. We have already (422) stated that the head of the drain, and every inlet to it, requires to be fitted with a trap to prevent the escape of effluvia, and this will equally

form an indispensable part of the closet apparatus. The perfect action of the trap will demand a means of supplying water on each use of the closet, and although all possible advantage should be taken of the house-sewage water in promoting the action of the drains, a separate and constantly-commanded source should be provided for this purpose. If the supply of water to the house or building be rendered upon the constant service system, a mere tap will be sufficient to afford the means of discharging a volume of water through the trap of the closet. If the water be supplied upon the intermittent system, a cistern or reservoir of some kind, provided for the house supply, must be made to communicate with the pan of the closet by a pipe with a valve and apparatus for working it. For general use it is especially desirable that economy and simplicity be combined in the whole of the apparatus of the closet. Delicacy of adjustment, requiring a complicated arrangement of parts, and a corresponding costliness of construction and repairs, and carefulness in management, is inadmissible in a design adapted for general adoption; and combinations of levers and cranks, liable to accidental derangement and injury by roughness of treatment, are therefore to be avoided as much as possible. The position of the cistern in relation to the closet will affect, in some degree, the force and efficiency of the volume of water discharged on each occasion; and, if the supply of water to the building be constant, the service-pipe should be so conducted over the closet that the tap can be conveniently placed for admitting the required quantity to the pan. If the supply is obtained from a house-cistern, this must, of course, be placed above the pan, and at such elevation that the water may acquire a sufficient impetus to flow with rapidity.

430. The glazed stone-ware basins or pans, with syphon-traps combined, before referred to (426), are the most economical and effective for general purposes. These are made in several forms: viz. with the pan and trap in one piece, and adapted to communicate either with a vertical or a hori-

zontal drain; with a separate trap, having a screwed socket on the head in which the lower part of the pan is received, being formed with a collar and screwed end; or, as a somewhat more complicated arrangement, consisting of a trap with a flanged head and a separate dip pipe having a projecting flange about its mid-length, and a spreading mouth above, into which the lower part of the pan is fitted with cement. The dip pipe, extending downwards into the trap, below the level at which its contents flow out, is secured to the head of the trap by bolts passing through the holes in the flanges. The reason for making the pan separate from the dip pipe would appear to arise from a difficulty in forming them together with the wide projecting flange so as to give sufficient steadiness to the pan above. This latter form was designed especially for prison use, under circumstances which do not allow of any fixed seat or framing above to which to secure the pan. The previous forms are found to answer all purposes in cases where this kind of support is afforded, and are preferable for their fewer number of parts. The pan in each of these forms of construction is provided with an aperture and inlet at the upper part, having a socket to receive the water-pipe. They are retailed at the price of 7s. 6d. each, and recommended by the present authorities in drainage matters.

431. The combined arrangements for efficient house-drainage comprise, besides the means of adequate water supply, as explained in section II. of this Division, the water-closets, house-sinks, and drains in which the matters committed to the closets and sinks are conveyed and delivered into the public sewers. And if the rain-water falling on the roof of the building and on the yard or space attached to the house, is not applied to any other purpose, it will have to be conducted into the drain to be discharged with the sewage. These waters being the purest of the contents should be received as near as possible to the head of the drain, and made to traverse its entire length, and thus exert all the cleansing action of which they are capable.

The house-sink or place at which the ordinary waste water of the household is discharged should communicate with the drain at a subsequent part of its course, and the closet be so placed that its contents shall traverse a minimum portion of the drain, thus reducing the liability to the escape of effluvia, and deriving the greatest scouring force from the accumulation of the rain and house waters. The drain being formed as a complete and impervious channel receiving the entire sewage and waste waters of the premises, made easy of access and examination, and provided with traps at every opening or inlet for receiving the drainage, may be graduated in size from the one extremity to the other, and, if of considerable length, it may be provided at intervals with self-acting valves or traps to prevent the possible return of any matters, waters, or gases, from the lower towards the upper end.

432. Self-acting valves or traps are constructed of the stone-ware; and the valves being hung at a slight inclination, and well fitted with a rim on the meeting surface, they remain closed against any retrograde movement of the sewage or gases, but are readily opened by a slight force of water in the outward direction of the drain. Sink traps are also formed of this material, with perforated heads or covers, and syphon bends below, which, remaining filled with the drainage water, prevent the escape of any effluvia from the drain into which they give access.

433. Beside the socket drain pipes of glazed stone-ware which we have described, another material, known under the name of " Terro-metallic," has been applied in the production of a superior quality of piping, which is manufactured in cylindrical forms both with socket ends and plain butt ends, and also in a conical form with plain ends, the cones fitting one another so that the joints are similar to conical socket joints, and may be made to fit with a great degree of exactness. The material of these pipes is of the same quality as that used in making fire-bricks, and has an extreme density with a very durable glaze upon the surface.

The prices of these pipes at the Tileries, Tunstall, Staffordshire, where they are manufactured, and in London, are as follows :—

TERRO-METALLIC DRAIN PIPES.

Diameter of Bore.	Plain Jointed Cylindrical Pipes.		Conical Pipes to Fit one another.		Cylindrical Pipes with Socket-Joints.	
	Price per Foot at Tunstall.	Price per Foot in London.	Price per Foot at Tunstall.	Price per Foot in London.	Price per Foot at Tunstall.	Price per Foot in London.
Inches.	s. d.	s. d.	s. d.	s. d.	s. d.	s. d.
2	0 2	0 3	0 2¾	0 3½	......	......
3	0 2⅝	0 4	0 3	0 5	0 3¾	0 5½
4	0 3¾	0 5	0 3¾	0 6	0 4½	0 6¾
6	0 4	0 6	0 4½	0 9	0 6	0 9
9	0 6	0 11	0 8	1 2	0 9	1 3
12	1 2	2 1½	......	......	1 4½	2 3
16	2 3	3 9	......	......	2 6	4 0

Curved and Junction pipes in the same material are charged at double the prices of the cylindrical pipes.

434. One of the most valuable improvements recently effected in the practical cleansing of buildings is a portable pumping apparatus, with hose for emptying cesspools. For conveying the sewage, this consists of a close tank mounted upon two or four wheels, according to its size, with a hose fitted to an aperture in it, and an air-pump attached, so that the chamber communicates with the interior of the tank. The hose is provided of such length that it may be laid through the passage, &c. of a house, and dipped into the cesspool, while the other end is attached to the tank at the door, into which the contents of the cesspool are rapidly transferred, without offence or nuisance, by a labourer at the pump. A small pumping apparatus with hose, but without tank, has been extensively applied for removing the contents of the cesspools into the sewers, a second hose being attached to the pump-chamber for this purpose. This apparatus, with hose complete, is furnished at the price of 15l., and the economy of its use as compared with the cost

of cleansing cesspools by the old method, effects a saving of 95 per cent. Among the instances reported by the Surveyors to the Metropolitan Commission of Sewers, the following may be quoted:—" The contents of one large cesspool, equal to 24 loads of soil, were pumped out in 3¼ hours, at a cost of 24s. Under the old system, three nights would have been occupied in emptying the cesspool, and it would have cost at least 24*l*."

435. Among the many contrivances which have been suggested for improving the house-apparatus for regulating the disposal of the water supplied, is a simple form of cistern, introduced by Mr. John Hosmer, which appears well calculated to prevent the waste of water which now frequently results from the inefficiency of the apparatus employed. The amount of this waste may be inferred from the proved fact that, in one district of the metropolis, an average quantity of *twenty-nine* gallons per house is wasted at each delivery from the works, by dribbling over the waste-pipes of the cisterns after they have become filled. Mr. Hosmer's cistern has a partition, dividing it into two spaces, one considerably larger than the other, and containing the supply for domestic use, while the smaller space is intended to contain a reserve for cleansing the drains and sewers. A two-way cock is fitted on the cistern with ball and lever, and one aperture of the cock opens into each of the spaces in the cistern. The large division of the cistern is fitted with a pipe or pipes to deliver the house supply as required, and the small division has a syphon-trapped pipe, leading into the drain and covered by a valve, the vertical rod of which is attached to the lever of the two-way delivery cock. The water from the main first fills the small division, the position of the lever being such that the valve at the lower part remains closed. The water then flows over the partition (which is kept a trifle lower than the sides of the cistern for this purpose) and fills the large division, the rising of the ball in which overcomes the pressure upon the valve in the small division, and lifts it suddenly to such

n height as to permit of a rapid discharge of water through
the syphon-trapped pipe into the drain. Similar cisterns
thus fixed and fitted, deriving their action simultaneously
from the delivery at the main, would, it is supposed, discharge streams of water at one and the same time into the
several house-drains connected with them, and thus act
with considerable efficiency in scouring these drains and
the sewer into which they discharge.

436. Although complexity of parts is to be avoided in
water-closets intended for use in the greater number of
dwellings, some of the more complete forms of apparatus
adapted for self-action, and which necessarily comprise
considerable detail of arrangement, are preferable in superior buildings in which close economy of construction is
not a first condition, and regular care and attention can be
secured for the action of the apparatus employed. In some
of these closets, the valve which opens and closes the opening into the water-pipe is attached by a rod to a lever, which,
by means of a cord or chain, is connected with the door of
the closet, so that the opening of the door opens the valve
and thus discharges a quantity of water into the pan. In
another form of apparatus, the pressure of the person on
the seat produces a similar effect. One of the most improved of these is that patented by Messrs. Bunnett and Co.,
which will be found fully described and illustrated in the
" Civil Engineer and Architect's Journal " for the month of
April, 1849. This closet is self-acting and doubly trapped,
and designed to secure a supply and force of water which
shall always be efficient and uniform without waste. It is,
moreover, so contrived, that no soil can remain in the basin
after use, and an ample supply of water being secured in
the basin so as to form a " water-lute " between that and the
syphon-trap, the rising of smell is effectually prevented.
The lower part of the pan dips into a water-pan or trap,
which is hinged and maintained in a horizontal position by
a rolling balance weight. The effect of pressure on the
seat of the closet is to depress a lever and open a valve in

the supply-box of the cistern, and thus pour a volume of water into the water-pan or trap sufficient to throw it open, and afford a passage for the soil into the lower basin, which terminates in a syphon, and is also trapped with water. When the pressure is removed from the seat, the water-pan or upper trap is immediately brought back to a horizontal position by the rolling weight, and receives sufficient water before the closing of the valve, to fill it, and thus effectually shut off all communication with the lower basin.

GENERAL SUMMARY AND CONCLUSION.

437. In the First Part of this Rudimentary Treatise, devoted to the Drainage of Districts and Lands, an attempt is made to exhibit an arranged outline of the facts which have been observed and recorded with reference to the several sources of water for agricultural purposes, and the best means of making these available. The methods of filtering and purifying water for extended purposes in districts comprising towns, are also briefly explained, and the difference pointed out between the mechanical and chemical processes required. In the sections which treat of the drainage of lands, as limited in its purpose to the discharge of superfluous water, the peculiar method to be employed is shown to depend on the united consideration of relative levels of surface and structural formation of soils. The importance of efficient draining of fens and the several works required for this purpose, are illustrated by grand instances in our own country—in the counties of Lincoln and Cambridge—and a brief description is introduced of that celebrated Dutch work by English engineers, the draining of the Lake of Haarlem. The construction of catch-water drains, and the adoption of means for aiding the supply of water to high or upland districts, are also alluded to as among the duties of the drainer. The formation of

soils is described as affording a general knowledge of their character, and aiding in the determination of the best arrangement of drains. Adopting a general classification of soils in regard to their structure, under the three leading characters of porous, retentive, and mixed, an extended notice is devoted to the several arrangements of these soils which are met with, and the modes of proceeding in each case are briefly explained. A description of the several modes of forming drains or artificial subterranean channels through lands is accompanied by practical rules as to their construction, dimensions, arrangement, and cost, and some of the best experience on this subject is quoted. A brief account of the several operations to be carried on, of contour-mapping, and of the tools employed in draining, completes the First Division of the work

438. The Second Division, of which the purport is the Drainage of Towns and Streets, aims at establishing a classification of towns as subjects for water supply and drainage according to the relative levels of the surface, and without that reference to the contiguity of rivers which has been dictated by the mistaken object of converting rivers into general sewers. An illustration of the principle here advocated is taken from the position and superficial character of our own metropolis. The value of sewage matters for agricultural purposes, and the practicability of rendering the distribution and use of these matters innocuous by chemical processes, are also stated upon the highest authority, and the evils of concentrating the sewage of large towns at few points, and misusing the channels for its conveyance, are pointed out and established upon past and extended experience. A brief notice is added of some of the general plans which have been suggested for the drainage of London, and some particulars given of the costly and inoperative works executed in the department of Metropolitan Sewerage. Upon the public supply of water to towns a mass of evidence is collected from past experience in the metropolis and the provinces, showing the effect of geologi-

cal structure upon the quality of water, and the cost of supplying, of filtering, and purifying it for the several purposes
required. The circumstances affecting the cleansing and
draining of roads and streets are also shortly noticed. The
proper functions of sewers, their arrangement, dimensions,
and construction, are deduced from the data which it is believed should be referred to, and by calculations which our
past experience enables us to form. A rule for the correct
sectional form of sewers is also given, and recommended
for its usefulness and simplicity. The cost of several descriptions of sewers is also cited from the records of experience, the stone-ware pipe sewers described, and the method
of cleansing by flushing adverted to, and its effects quoted.
The conveyance of water to towns and the several methods
adopted, with the cost of pumping by steam-power, are
described and stated. ·

439. The Third Division treats of the Drainage of Buildings as subjects of the entire system which embraces the supply of water as an accessory to the purpose of draining. It
is suggested that the classification of dwellings should be
determined by the number of persons to be served rather
than the rental paid for each house, and that larger buildings, in which human beings are congregated for manufacturing and other purposes, may be provided for according
to the cubic space inclosed by them. The arrangement,
construction, and dimensions of house-drains are described,
and the qualifications of impermeability, secure trapping at
the head and all other openings through which effluvia
might escape, and proper connection with the receiving
sewer at the lower end, are insisted upon as indispensable
to perfect and efficient construction. And in the concluding
section a general view is taken of the combined arrangements for efficient house-drainage, and the simplest construction recommended for water-closets and similar apparatus designed for general adoption.

APPENDIX No. 1.

440. THE following account of Fowler and Fry's new steam draining-plough is quoted from the "Bristol Mercury" of February 11, 1854.

"On Thursday last some experiments of a deeply interesting and important nature, in connection with the question of land-drainage, were made at Catherine Farm, on the estate of P. W. Miles, Esq.; at Kingsweston, when Fowler and Fry's new steam draining-plough was for the first time put in operation. The great utility of underground drainage upon clay soils and in marshy districts is now too generally known and appreciated to leave it a matter of debate; but the heavy expense which the adoption of the system of hand-labour entails, especially in neighbourhoods where that labour is scarce and dear, has hitherto stood in the way of its general adoption. It will be obvious that anything which has a tendency to facilitate the operation and diminish its cost, must be, in an eminent degree, beneficial to agriculture, and hence scientific and practical men have been stimulated to bend their efforts in a direction tending to those ends. The drainage of moist lands was first attempted by the simple process of digging, by spade labour, narrow trenches, and laying rude stone drains in the bottom of them. The difficulty of procuring stones and the cost of hauling them were found to stand considerably in the way of that process, and the use of draining-tiles in various forms was by degrees introduced, but without any decided improvement being effected in the mode of cutting the drain trenches. An important revolution in the process was introduced by our fellow-citizen Mr. Fowler, of the firm of Fowler and Fry, agricultural machine manufacturers, of Temple Street, in the invention of his draining-plough, by which manual labour was in a large degree superseded; but a deficiency still remained. The plough had to be worked by horse-power, four horses being employed by it in turning the windlass by which it was set in motion, and the process, although cheaper and more expeditious than spade-labour, was, nevertheless, in a degree expensive and tardy. The desirability of applying steam power to a plough upon the same principle soon became apparent, and, impressed with the importance of the object, Mr. Fowler directed his attention to it, and now, as the result of a great deal of anxiety and labour, and of no inconsiderable expense, he has perfected a steam draining-plough, which we saw in successful operation on Thursday, and which we have no doubt will speedily take rank among the most useful inventions of the day. A brief description of the machinery may prove interesting.

"The steam-engine, although mounted on wheels, and capable of being transported from point to point, is, when employed, a stationary one, and worked by a horizontal cylinder. It has connected with it two drums, which are loose on the axles. Attached to the larger drum, which draws the plough forward, is a wire rope of beautiful manufacture, the breaking strain of which is 14 tons, the working strain being 5 tons. This drum is worked by two motions off the fly-wheel shaft, which give a leverage of 22 to 1 on the plough, the drum making seven revolutions per minute. To the lesser drum, which is worked off the second shaft, is attached a rope also of wire, but of smaller calibre, which draws the plough back, when it has completed a furrow, to the side of the field from which it started and where it has to begin again. By an ingenious contrivance the drums are formed by the insides of two spur wheels, so that practically the working is effected by

ordinary spur-gearing. The drums can be instantly thrown
out of gear by clutches moving the pinions on a feather.
The larger wire rope, on being wound on to the drum for
the purpose of impelling the plough forward, works round
a sheave-wheel or pulley-block anchored to the field at such
a point as to draw the plough at right angles to the engine,
by which arrangement the necessity of shifting the engine
is obviated to so great an extent that almost any field may
be drained without once removing it from the position first
taken up by it. To the front of the plough is attached a
second sheave-wheel, round which the rope is doubled,
thereby, also, doubling the power. The coulter of the
plough is of iron, an inch in diameter at its widest point,
so that the furrow made by it upon the surface of the land
is scarcely perceptible and generally disappears after the
first storm of rain. It can be worked to a depth of four
feet, and indeed deeper, if necessary, and is so made that
it can be raised or depressed by a handwheel under the
control of the ploughman, and which works gear connected
with a rack at the back of the coulter. The boring of the
land is effected by means of a cast-iron mole or plug (the
size of which is regulated by the size of the tiles to be
laid) keyed to the bottom of the coulter, and the most
striking feature of the machine is, that as fast as it bores
the land it lays in the tile-piping, thus completing the
drain as it goes at the rate (when we saw it working) of
35 feet, and probably, under very favourable circumstances,
40 feet per minute. It should be stated, in order to the
understanding of what follows, that as the engine winds
the large rope on the large drum and draws the plough
towards it, it at the same time unwinds the small rope
which is attached to the back of the plough from the small
drum.

"The mode of operation we will now endeavour to
explain, assuming for our illustration a field of 1000 feet
square, which has to be drained by drains 10 yards apart
from east to west. The engine would be fixed at the

middle of the western edge; the plough would be placed on the eastern edge at 10 yards from the southern edge of the ground, and an anchor and sheave-wheel would be rigged exactly opposite to it on the western edge. The large wire-rope would be passed round the sheave-wheel, and thence on to the front of the plough, while the small wire-rope would be connected from the back of the plough with another anchor, &c., rigged 10 yards north of the plough—that is, at the point to which it would have to be drawn back, and from which it would have to commence again. The machinery thus arranged, the pipe-tiles are strung on ropes of fifty yards long (the length being thus limited to economise time and labour in threading), but fitted with ingeniously-contrived joints at either end, so that they can be readily and firmly joined together at any length required. These ropes are made of hemp for the sake of flexibility, while, as a matter of economy and durability, and to decrease friction as much as possible, they are coated with wire. The ropes being threaded and joined, one end is fixed immediately behind the mole, and the machine being set in motion by the steam-engine, the coulter cuts its narrow channel through the land, the mole bores and lifts the subsoil, and the pipes are drawn through the aperture, and closely and neatly put together, forming the drain. The sheave-wheels are then shifted, the plough drawn back by the small rope, and the second and succeeding drains are cut and piped in the same way. The ropes, after the tiles have been laid, are drawn out by horses, which is the only employment of horse-labour required. The plough is attended by a man, whose only duty seems to be to keep it upright where the land is out of level; but we were told by Mr. Fowler that he had perfected some self-acting level guides, which would be shortly attached, and which would enable the plough to adapt itself to any inequalities which might arise, and make it independent of any guide.

"The advantages which Mr. Fowler expects to derive from his inventions are manifold. The first and most

important is, of course, economy. The sum at present
charged by contractors, for draining on a large scale, per
acre, is from 5*l.* to 6*l.* A considerable tract of land, at
present in course of drainage in an adjacent county, is
understood to have been taken at 5*l.* 5*s.* or 5*l.* 10*s.* per acre.
Mr. Fowler considers that by his machinery land may be
drained for from 3*l.* 10*s.* to 4*l.* per acre, yielding a fair
remuneration to the contractor. One engine with ten men
and two horses will, he calculates, do as much work as
120 men, and, under favourable circumstances, as much
work as 150 men would do by the old system. A second
advantage anticipated is, the ability to drain in the summer
season, when days are long and the weather favourable, a
desideratum not now obtainable on account of labour being
at that period so fully occupied by other sources of employment. Third, drainage by the machine will be better performed. The drains will be uniform in depth and straighter,
and the tiles more closely and firmly laid, while the plough,
by lifting the land, causes the water to percolate at once,
and thus brings the drain into immediate action. Fourth,
no damage is done to the surface of the ground, which, by
the old process was often the case. With dry weather the
machinery may be erected and field drained one day, and
on the next a casual observer would be unable to perceive
that any change had taken place. Fifth, the drains when
made will be more durable.

" With regard to the capabilities of his invention, Mr.
Fowler calculates that with a single engine and plough he
shall be able to drain about 30 acres per week. At present
the machinery will only be retained in this neighbourhood
long enough to complete the drainage of about 40 acres
of land on the farm where it is now at work. This will
probably be effected by about the end of next week, after
which time it will be removed to London, in the neighbourhood of which city it will, we understand, be tried
under Government inspection, and, doubtless, will receive

the attention of many scientific and practical gentlemen interested in the advancement of agriculture

" Should the 'STEAM DRAINING-PLOUGH' meet with the approbation of those to whom it is about to be submitted, arrangements will be made for carrying it into operation upon a scale to some extent commensurate with the wants of the kingdom."

Figures 78, 79. 80, and 81, show the plough and capstan apparatus complete, as adapted to be propelled by horse-power. Figs. 78 and 79 are an elevation and plan of the capstan, and figs. 80 and 81 a corresponding elevation and plan of the plough

———

[Since the above was written, very considerable advances have been made in the practice of plough-draining by steampower, as in the application of the same mighty agency to other agricultural labours ; but the principles are nearly as above described.]

Fig. 78.   *Fig.* 79.

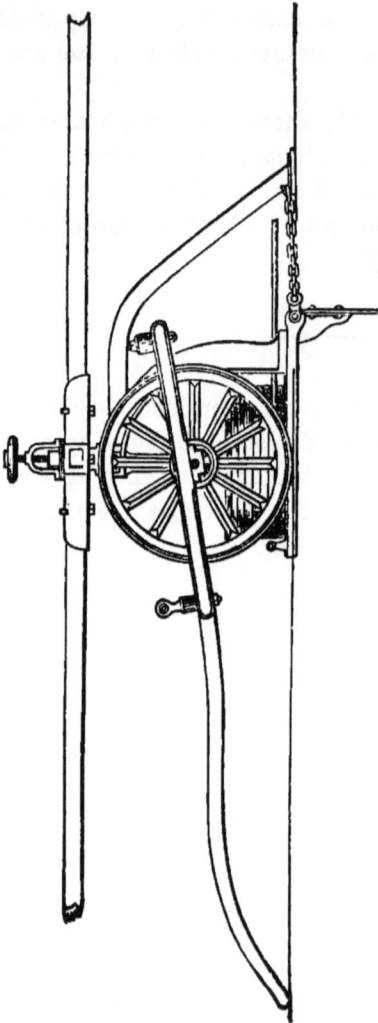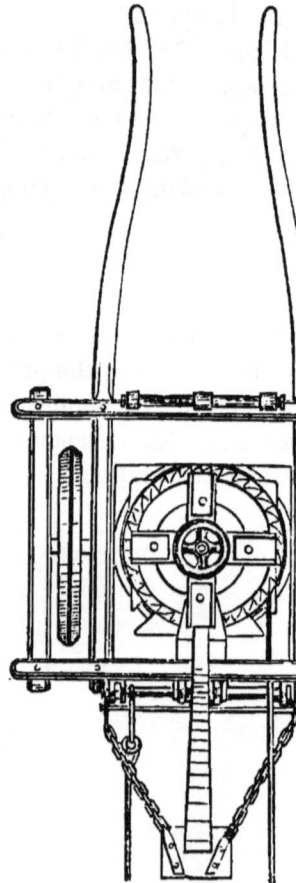

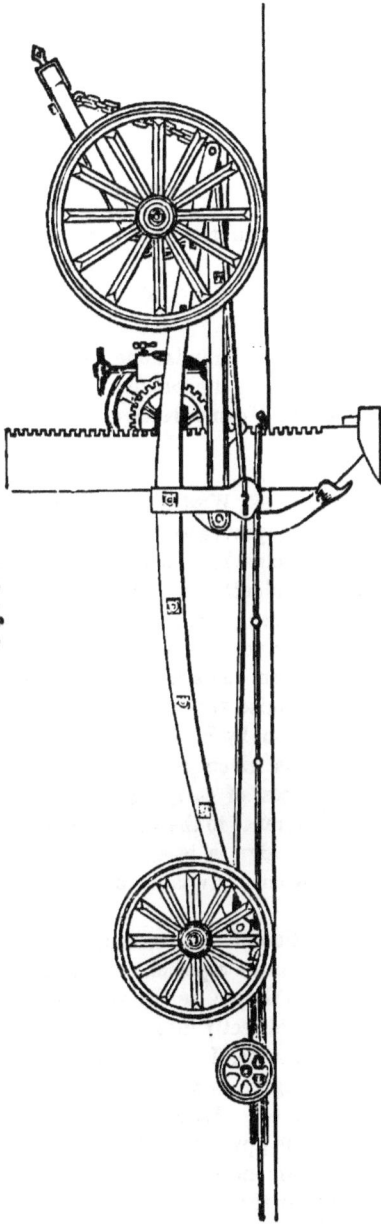

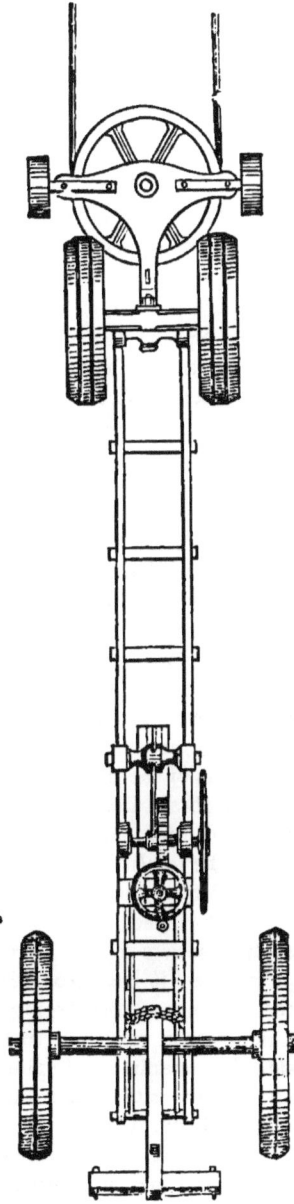

Fig. 90.

Fig. 81.

APPENDIX No. 2.

MR. STEPHENSON'S REPORT ON THE PLAN FOR THE DRAINAGE OF DIS-
TRICTS NORTH OF THE THAMES,

At a Special Court of the Metropolitan Commissioners of Sewers,
held on the 16th May, 1854, the following Report by Mr. Robert
Stephenson, M.P., on the plan proposed by Messrs. Bazalgette and Heywood, was read :—

"24, Great George-street, Westminster, May 15.
" Gentlemen,—Absence from England prevented my uniting with
Sir William Cubitt in drawing up his remarks, dated February 25,
1854, on the joint report of Messrs. Bazalgette and Heywood, on a
proposed system of intercepting drains throughout that portion of the
metropolis lying on the north side of the river Thames.

"I therefore now take leave to lay before you a few observations on
that report, and before doing so may premise that it is not my inten-
tion to revert to those points touched upon by Sir William Cubitt, as
I entirely concur in the opinion he has given respecting the general merits of the plan and the estimated cost of carrying it out.

" My observations will rather have reference to the general principle
of intercepting sewers upon which the design is based, and to one or
two of the localities where it has been deemed advisable to depart from
this system to meet special conditions.

" With respect, then, to the application of the principle of intercepting sewers to very large towns, and more especially to the metropolis, where it has become imperative to lessen as far as practicable
the nuisance of discharging the sewage directly into that part of the
Thames upon which the metropolis is situate, I believe there is considerable unanimity of opinion among engineers who have been called
upon to direct their attention to the subject. When I was a member
of your body, it fell to my lot to examine upwards of 150 suggestions
for improving the drainage of London, among which were many from
professional men of experience, and by far the majority of these, when
a comprehensive study of the subject had been made, agreed on adopting a series of intercepting sewers, as the only method affording a sim-

ple and efficient system of drainage, and at the same time freeing the Thames through London of pollution.

"The plan which was afterwards submitted by Mr. Frank Forster to the commissioners may be regarded as embodying most of the useful views which such of these plans contained as were based upon the principle of interception by main lines of sewer, carried to a point on the river so far down as to prevent the reaction of the tidal waters bringing the feculent matter back among the inhabitants of the metropolis.

"The design now produced by Mr. Bazalgette and Mr. Heywood may be regarded as the extension of Mr. Forster's views, adapted to the new features which every suburb of London is yearly presenting. They have also made some judicious modifications suggested by time and further information, one of the most important of which is the removal of the lowest intercepting drain from the shore of the river, as such a position, although chosen for the purpose of making the system of interception complete, would have been carried out with the utmost difficulty, in consequence of the loose nature of the soil, and at a cost incapable of any certain calculation beforehand.

"The construction of this low level intercepting sewer does not, however, press for immediate attention, for it is clear that, of the three intercepting sewers comprehended in Messrs. Bazalgette and Heywood's plan, the construction of the upper and middle levels should be proceeded with first. Their completion alone will greatly relieve the Thames of impurities, and by lessening the direct discharge, will render the construction of the lowest level less difficult and less costly. Indeed, it is not improbable that by the diversion of so large a portion of the sewage matter from the river as would be under the control of the middle and upper lines of interception, the formation of the lowest would become comparatively much less important, and might be postponed for some length of time, and so relieve the demand upon those who supply the funds.

"The two upper lines of interception would, in fact, leave less offensive matter to be discharged into the sewers than has taken place in any time within living recollection.

"It is the only mode of procedure which seems adapted eventually to remove, and at once to mitigate, many of the chief difficulties and annoyances attaching themselves to this complicated question.

"In this design the system of interception is applied to the whole of the metropolis on the north side of the river, excepting a portion which is designated the western district, comprising an area of about eighteen square miles, commencing a little to the east of Brompton and Chelsea, and extending westward as far as Brentford. The level of all this district is so low that it is impossible to obtain any other natural means of drainage than into the river, as at present. To obviate this objection, it is proposed to treat the western district by a different method, which appears to me well calculated to overcome the obstacles arising out of this circumstance, namely, to direct the whole of the sewage to one point on the shore of the river near the entrance into the Kensington Canal, and there take means for separating the liquid and solid portion of the sewage before the former is discharged into the river. Experiments on a large scale with this process are now being made, and with such result that its ultimate success may be fairly deemed probable.

"I agree, therefore, with the recommendations made in this report as to the method to be applied to the western district, in consequence of its extremely low level.

"With the general lines of direction which have been selected for interception I also concur; but, in anticipation of the success of the method for extracting solid manure from the sewage, and its becoming remunerative, it is suggested, 'that considerations may arise seriously affecting the scheme proposed,' and a question is raised 'whether it would not then be advisable, on the score of economy, to modify the designs presented for the middle and low-level sewers, and to abandon the chief portion of the line between the river Lea and Barking Creek.' In this I do not go to the full extent with Messrs. Bazalgette and Heywood, for, although the process for the manufacture of manure may prove successful, I do not think this probably would tend to, still less would justify, the delay of the construction of the middle level works. These seem to me to be absolutely necessary to rectify the present defective system of drainage, whatever shape the manure question may eventually take. I have already stated that the middle level is calculated, when properly dealt with, to ameliorate the evils so much complained of in so marked a manner that the execution of the lower level may perhaps safely be postponed for a time; but I can foresee no degree of improvement in the manufacture of manure in prospect that would tend to any material alteration in its construction, or position, or cost. It strikes at the root of many of the existing evils in the most direct manner, while it comprehends within its scope and influence so large and important a district, that I think no minor obstacle ought to stand in the way of its execution. The extension to Barking might, I think, be dispensed with, as suggested.

"If in extending and perfecting the drainage of London, the chief objects be the improvement of the general sanitary condition of the inhabitants and the purification of the Thames from noisome matter, I cannot help repeating that I regard the middle intercepting sewer as one of the most important features of the scheme on which I am now reporting, and, indeed, without it either object, and the last especially, could only be very partially attained, for no arrangement in the extraction of the solid matter from the sewage that I can conceive would prevent the frequent discharge of the usual amount of noxious filth into the river.

"This can only, I believe, be avoided by the intercepting system being fully carried out, and no part of the proposed plan is more effective than the middle interception, and none can be so ill spared or postponed.

"I am induced to express this opinion somewhat decidedly, because, having had my attention called to several methods that have been proposed for the extraction of manure from sewage, I have been led to the conclusion that, instead of multiplying such establishments within the precincts of towns, as some contemplate, a complete system of interception, with the concentration of the manure process on a few points, is that which is best calculated to attain success.

"With reference to the dimensions of the proposed sewers, I have not been able to go into the details of the calculations, but, having examined the tabular statements attached to the engineers' report, and having received explanations from them respecting the directions of the flow in the various sewers intended to take place, I have every confidence in the correctness of their conclusions.

"I have only further to add, that I regard the design of Messrs. Bazalgette and Heywood as comprehending, in a very practical shape, all the essentially useful suggestions which have, from time to time, been made by engineers respecting the drainage of the metropolis; and I have no doubt whatever that if the commissioners be put in a position to carry it out, it will be found effective.

"I am, gentlemen, your obedient servant,

"ROBERT STEPHENSON.

' To the Commissioners of Metropolitan Sewers."

K

194

APPENDIX No. 3.

MAIN DRAINAGE OF LONDON. THE PROCEEDINGS FROM 1854 TO 1865.

For reasons stated in the Preface, it has been considered desirable to leave Mr. Dempsey's paragraphs untouched, so far as concerns the proceedings in relation to the Main Drainage of the metropolis, and to present in this place a succinct account of what has been done in the matter between 1854 and 1865.

The first important events after 1854 in regard to this subject, were the abolition or abrogation of the *Metropolitan Commission of Sewers*, the establishment of a new body with greatly increased powers, and the settlement once for all (whether for good or bad) of a multitude of vexed questions relating to the method of draining the metropolis. On August 14, 1855, the royal assent was given to the "*Metropolis Local Management Act*," a statute of enormous length, having 251 clauses and 100 large pages of schedules. By virtue of this Act, the parishes of the metropolis elect vestries, which are bodies corporate, either singly, or two or more parishes combined, in which latter case the combined vestries form a *District Board*. The vestries and boards, together with the Common Council of the city, elect representatives from among themselves, in certain proportions; and these representatives, to the number of 45, form a *Metropolitan Board of Works*. The Secretary of State selects a chairman from three persons named by the board.

The powers of this board are chiefly the control of the main drainage of the metropolis, the naming of the streets and the numbering of the houses, the widening of narrow streets, and the facilitating of street traffic. The area over which this power extends is that which has been adopted by the Census Commissioners of 1851 and 1861, and which gives a total in the last named year of 330,237 inhabited houses, and 2,803,034 inhabitants, in the metropolis. The board, to defray the charges of any works undertaken, is empowered to levy a rate on the same principle as the county rate, and to borrow money upon the credit of that rate.

The *Parish Vestries* and the *District Boards*, besides electing members to the Metropolitan Board of Works, have extensive local powers of their own. They manage all the minor sewers and drains subordinate to the main drainage. They control the paving, lighting, watering, and cleansing of the streets, and the erection of public conveniences. They have the powers of surveyors of highways. They may appoint crossing-sweepers. They appoint *Inspectors of Nuisances*, whose duties are decided by that title. And, lastly, they appoint *Medical Officers of*

Health, to report upon the sanitary state of the several districts and parishes; to ascertain the existence of epidemics and other diseases; to suggest means for preventing the spreading of these evils; and to take cognizance of the ventilation of churches, chapels, schools, lodging-houses, and other buildings of a more or less public character. To defray the cost of these works, the vestries and district boards are empowered to levy rates on the householders, under the three forms of *Sewers rate, Lighting rate,* and *General rate.*

So far, then, as concerns the present part of our subject, the Metropolitan Board of Works, elected by the parish vestries and the district boards, is entrusted with the general drainage of the metropolis. One important clause in the Act declares that "Such board shall make such sewers and works as they may think necessary for preventing all or any part of the sewage within the metropolis from flowing or passing into the river Thames, in or near the metropolis; and shall cause such works to be completed on or before the 31st December, 1860." Of course it was right to name a limit of time in this way; but the limit has been far exceeded. Another Act, in 1856, empowered the commissioners, acting in harmony with the parish vestries and district boards, to take all the requisite measures for forming parks, pleasure-grounds, places of recreation, and healthy open spaces in the metropolis; but it did not affect the power of the commissioners in reference to the main drainage.

It was known when the Act of 1855 was passed, that the Government, and most of the influential persons concerned, were favourable to the "Intercepting" scheme, as contrasted with the local or sectional schemes. But much valuable time was spent in discussion. Not only did the remaining portion of that year, but the whole of the years 1856 and 1857, pass away without definite results; and it was not until 1858 that an Act of Parliament was obtained, empowering the commissioners to carry out a specified system of metropolitan drainage. This Act, which received the royal assent on the 2nd of August, was framed after an immense deal of trouble. Considering that 80 million gallons of liquid refuse flow from the metropolis every day; that this vast bulk of water contains 400 tons of solid refuse; that 2,000 miles of sewers convey these abominations into the Thames by about 100 mouths; that considerably more than 100 square miles of area have to be provided for in some way or other; and that upwards of 80 miles of great intercepting sewers would have to be constructed, it is no wonder that embarrassments and difficulties should arise. But there were others of a more formal kind. By the former statute (1855), the *Board of Works and Public Buildings* had a veto on the plans of the *Metropolitan Board of Works* in all that concerned main drainage; and as the rival boards had rival engineers, with rival plans, matters came to a dead lock. Moreover, by the "*Thames Conservancy Act*" of 1857, the corporation

K 2

have certain rights conceded to them over "the soil of the Thames between high and low water;" and these rights give them a voice in many drainage questions. All difficulties of this kind were, however, at length surmounted, and the Act was passed.

The basis of the system adopted was the Interception, which had been more or less before the public since 1845. The Act did not instruct the commissioners to adopt any plan defined in the Act itself, but empowered them to adopt a plan, some plan, on their own responsibility. They were to be relieved from the absurd veto of the Commissioners of Works. They were empowered to spend 3,000,000l. sterling by the 31st December, 1863; to borrow this amount under their corporate seal, and to repay the principal and interest in forty years by a levy of 3d. in the pound on the annual value of the property in the district, in the manner of a county rate.

The curious course of procedure, therefore, had been this; that, in 1855, the Metropolitan Board of Works was formed mainly with a view to the drainage of the metropolis; that, in 1856 and 1857, all parties were disputing as to the best mode of doing this; that, in 1858, an Act of Parliament was passed, empowering the commissioners to spend 3,000,000l. on some plan for preventing the sewage of the metropolis from flowing into the Thames; and that the commissioners then had to give their final vote what that plan should be.

We have now, therefore, to give some account of the greatest system of town drainage ever yet commenced. The commissioners, in 1859, finally determined upon the Intercepting plan, with Mr. Bazalgette as their chief engineer. The reader is already acquainted with the general character of the plan in the foregoing paragraphs; and all that will be necessary, therefore, is, to show how far the results are likely to be modified.

The area that will be drained under this system, both of house-sewage and of rain-fall, extends to the enormous amount of 117 square miles, nearly equal to that of a quadrangular area 12 miles long by 10 broad. All the arrangements as to the outfall, &c., are made for a prospective population of 3,500,000, about 500,000 more than the *present* population of the incluced area. The whole of this drainage will then, as heretofore, flow into the Thames, unless some plan is adopted of utilising the sewage; but nearly the whole of it will enter the river many miles below London, under circumstances which, it is hoped, will prevent a return of the pollution upwards. There will be five great lines of sewer, more or less parallel to each other and to the river; and each of these sewers will receive all the drainage from so much of the district as is on a higher level than itself. Three of the main sewers are on the north of the Thames, and two on the south. The first three will converge at a point eastward of the metropolis, and then flow into the river near

Barking Creek; the last two will converge near Deptford, and flow on together to Crossness Point, between Woolwich and Erith. We shall now describe these great lines separately, under the names by which they are generally known, and shall give a few further details concerning them collectively.

Northern High-Level.—This commences in the Gospel Oak Fields, at the foot of Hampstead, and extends to Old Ford, by way of Kentish Town, the New Cattle Market, Highbury Vale, Stoke Newington, Hackney, and Victoria Park. It absorbs on its way the old offensive open sewer called Hackney Brook, which is now emptied and closed over. It crosses beneath the Great Northern, Tottenham and Hampstead, and North London railways, and Sir George Duckett's Canal. The fall commences at about 4 ft. per mile. It lies at a depth of 20 ft. to 26 ft. below the ground, and drains 14 square miles. The sewer is 4 ft. by 2 ft. 8 in. at the western or upper end, and gradually enlarges to 12 ft. wide by 9¼ ft. high at the east end. The actual distance is about 8½ miles, but some small additional portions of sewers raise the length to 9¼ miles. The engineer's estimate for this work was about 180,000*l.*; and the whole has been admirably finished by Mr. Moxon within the estimate.

Northern Middle-Level.—This commences at Counter's Creek sewer, near Kensal Green, and extends to Old Ford. It first passes nearly south from Kensal Green and Kensington Park to Notting Hill, and then along the Bayswater Road, Oxford Street, Bloomsbury Square, Clerkenwell, Old Street, and Bethnal Green to Old Ford. The length, with branches, is about 12¼ miles, half of which is tunnelling; and for 2 miles it passes under houses and other private property. The size of sewer, as in the former case, varies from 4 ft. by 2 ft. 8 in. to 12½ ft. by 9 ft. The gradient varies from 2 ft. to 25 ft. per mile; and it lies at a depth varying from 30 ft. to 40 ft. below the surface. Mr. Bazalgette's estimate for this great work was 280,000*l.* Mr. Rowe contracted for it in February, 1860, at a price of 264,000*l.*; but he broke down after an expenditure of 12,000*l.*; and a new contract was made in December with Messrs. Brassey and Co. for 330,000*l.* This portion of the work has been beset by immense difficulties; for, owing to a paucity of plans of existing sewers, water-pipes, and gas-pipes, the excavators had to feel their way inch by inch under the roadway of Oxford Street and other places. At another place the Regent's Canal burst into this sewer, retarding the works for a considerable time.

Old Ford Storm Overflow.—The High and Middle-level sewers converge at Old Ford, where some remarkable works have been constructed. The *Penstock Chamber* consists of a large receptacle below, and machinery rooms above. The arrangements are such, that when the High and Middle-level sewers are supercharged after heavy rain, the surplus water is diverted, in the Penstock, into a short channel which flows into the river Lea, permitting only the usual average quantity to flow onwards towards the Thames. The proportion thus diverted may be varied at pleasure. This is a more important matter than would at first be supposed. Four times as much sewage passes through the London sewers at noon as at midnight; the maximum flow being about equal to a rain-fall of a quarter of an inch in depth in twenty-four hours. But besides this source of inequality, the quantity is much exceeded in rainy weather, at which time the rain-fall often greatly exceeds in bulk the sewage itself. It would have been very difficult and enormously expensive to construct the sewers large enough to carry off all the storm waters as well as the sewage; and therefore the *existing* main sewers, as well as the river Lea, are to be occasionally made use of as storm overflows, to carry the rain-fall to the Thames by channels different from those of the new sewers. The whole of the drainage collected by the Upper-level was diverted by the mechanism of the Penstock into the Lea in 1862, until the works at Barking were finished. The value of the system was further felt in 1862, when an irruption of the Fleet sewer into the Underground Railway in Victoria Street took place. The injury would have been much greater, had not the upper waters been diverted to Old Ford. It may suffice to show the formidable nature of some of these works, when we say that between Old Ford and Barking the High and Middle-level sewers have had to be carried, by iron aqueducts supported on columns, across no less than six branches of the river Lea, across the East London Waterworks Canal, and under and over four lines of railway. The works for the Old Ford Penstock Chamber were included in Mr. Moxon's contract for the High-level sewer.

Northern Low-Level.—This begins by a junction with the Ranelagh sewer at Chelsea, which it will connect with the Victoria Street sewer; then, approaching the Thames at Whitehall, it will continue near the embankment to the City; whence it will pass by way of Tower Hill, Stepney, Limehouse, and Bromley, to West Ham. At this last-named place it is to be joined by the High and Middle-level sewers. One grave difficulty has led to the delay of these works. According to the original plan, this sewer was to have been carried beneath the roadway of the Strand, Fleet Street, Ludgate Hill, &c., at a depth varying from 30 ft. to 50 ft. below the surface; but it was felt that the execution of the works would be an insupportable nuisance to the inhabitants, and that the obstructions to trade might give rise to formidable claims for compensation. It was even feared that St. Paul's Cathedral might be endangered by such deep excavations near its foundations. It was therefore determined to combine the scheme for a Low-level sewer with one for a Thames Embankment. Accordingly an Act was obtained in 1862, which enables the Metropolitan Board of Works to carry the Low-level sewer close to the proposed embankment, in the bed of the Thames itself, from Whitehall to Blackfriars Bridge. There will be difficulty enough to decide how to carry the sewer through the City; but the double scheme will at all events save the Strand and Fleet Street from the threatened disturbance. How the sewer and embankment will be combined is explained in a later page.

The Northern Outfall.—We have traced the High and Middle-level to Old Ford, and have now to describe their course over the Essex Marshes. The *Northern outfall* extends from Old Ford to Barking Creek, where the river Roding enters the Thames, a distance of about 5½ miles. The route is by way of Stratford, West Ham, Plaistow, and East Ham. It comprises two parallel channels from Old Ford to West Ham, mostly brick arches of 12 ft. by 9 ft., but partly iron tubes of great magnitude. At West Ham the Low-level sewer will join them; but as the sewage will in this be at a much lower level than in the others, it will be pumped up by powerful machinery. The three sewers will then run in three distinct and parallel channels, formed of the best brickwork, in an embankment above the general level of the ground. The three channels will be 14½ miles in all. In order to ensure permanence of structure, and to avoid a thick bed of peat which underlies the surface in the Plaistow and East Ham marshes, a concrete foundation for the embankment has been made, in some places 25 ft. below the surface. The triple sewers pass under one and over two lines of railway. For a mile and a half the embankment is built upon transverse arches, the piers of which rest upon a deeply laid bed of concrete. Mr. Bazalgette's estimate for this remarkable tripartite system of vast channels was 635,000*l.*; and Mr. Furness took the contract in 1860 for 625,000*l.* The work is considered to rank among the best specimens of brickwork ever seen—almost as highly finished as if intended to meet the eye of a connoisseur, instead of being permanently buried out of sight. It has been wisely determined that the cheapest as well as the best way in the end will be to do this work thoroughly once for all.

The Barking Reservoir and Sluices.—The sewage conducted along the three fine lines of channel just described is received at the *Outfall Station.* This consists mainly of a vast covered brick reservoir, 14 acres in area and 20 feet deep, and would, if needed, contain double the quantity of twelve hours' sewage from all the northern sewers. It is so roofed with brick arches and earth, as to prevent the escape of effluvia. From this reservoir the sewage will flow into the Thames during the first two and a half hours of ebb; that is, for two hours and a half after high water. It is calculated by the engineer (and on the correctness of this calculation depends much of the success of the whole scheme) that as the point of discharge is 12½ miles below London Bridge, the outgoing tide will carry the sewage too far away to allow it to flow up

to London again at the flood tide; especially as the bulk of water opposite the outfall is very great, and as the flow downwards is very forcible
during the first two or three hours after high water. It is supposed that
the very last gallon of each outflow will have reached 12 or 13 miles
below Barking Creek, or 25 below London Bridge, before the tide
begins to flow upwards again. The mains carry the sewage from the
reservoir far out into the bed of the river, so as to insure its mixing
with as large an amount of water as possible. Mr. Bazalgette's
estimate for the works at this outfall station was 170,000*l.*, and
Mr. Furness took the contract for 164,000*l.*, in January, 1863. It is
probable, however, that these works may be made in some way contingent
upon any future scheme for the utilisation of sewage. Mr. Bazalgette,
pending the discordant arguments on this subject, lays his plans for the
discharge of the whole of the sewage into the Thames; but the commissioners have purchased 50 acres of land adjacent, to be prepared for
any future scheme of sewage utilisation.

Western Drainage.—The western part of the Metropolitan area,
north of the Thames, is much of it at so low a level that the three great
intersecting sewers cannot well accommodate it. It is, therefore, treated
by Mr. Bazalgette as a distinct system. The districts comprised within
it are Acton, Hammersmith, Fulham, Kensington, Chelsea, Brompton,
&c. This system comprises several portions. The *Acton Branch*
extends from Acton Bottom to Notting Dale, along the Uxbridge Road.
It comprises about 1½ mile of small sewers, and descends 4 ft. per
mile, at an average depth of 15 ft. below the surface. The *Ranelagh
Storm Overflow* is a sewer commencing in the Bayswater Road, passing
through Kensington Gardens and Hyde Park, and joining the Ranelagh
sewer at Knightsbridge. It connects the Middle-level system with the
Western system. Its chief purpose is to carry off storm overflows, that
would otherwise trouble the district, and overfill the Middle-level
sewer; but another purpose is to relieve the Serpentine from foul water
that used to enter it during those overflows. This subsidiary sewer is a
little over a mile in length. The sewer is a 9 ft. barrel, at a depth of
11 to 44 ft. beneath the surface, and with a fall of 12½ ft. per mile.
These sewers vary from 3¾ ft. by 2½ to 4½ ft. by 4½. It was at first
intended to form a covered reservoir near the banks of the Thames, by
the side of the Kensington Canal, with arrangements for deodorising
the sewage before allowing it to pass into the Thames; but the inhabitants of the neighbourhood objected so strongly to the plan that the
commissioners appointed a committee to consider and report upon it.
Meanwhile these western sewers have been proceeded with, leaving the
question of the outfall in abeyance.

Southern High-Level.—The whole of the immense works hitherto
described are, or are to be, north of the Thames. We have now to

notice those on the south. Here, unlike the north, there is no *Middle-level;* the nature of the district being satisfied with a *High-level* and a *Low-level* sewer. The *High-level* extends from Clapham Common to Deptford Creek, a distance of 9¼ miles, by way of Brixton, Camberwell, and New Cross. It varies in section from 4½ ft. by 3, to 10½ ft. by 10½, and lies at a depth varying from 10 to 50 feet beneath the level of the ground. Mr. Bazalgette's estimate was 212,000*l.*; the contract was taken conjointly by Messrs. Holling & Co. and Messrs. Lee and Bowles for 217,000*l.*, and the works are completed. Connected with this High-level is a branch from New Cross to Dulwich and Norwood, superseding in its way an offensive open channel called the *Effra sewer.*

The *Southern Low-Level* extends from Putney to Deptford Creek, by way of Wandsworth, Battersea, Vauxhall, Lambeth, Southwark, and Rotherhithe, a distance of about 9¼ miles, with a Bermondsey branch from the Spa Road. In some of these works on the south side the difficulties have been most harassing. At one place the water flooded in upon the works at the rate of 8,000 gallons per minute, taxing the contractor's skill to the utmost. Although it is customary to speak only of two levels on the south of the river, it would give a better idea of the system to have regard to three—Putney to Deptford, Clapham to Deptford, and Norwood to Deptford—all rising at different levels, but all coming to the same level at Deptford.

Deptford Pumping Station.—On account of the low level of most of the ground near the south bank of the Thames, the sewage from the Low-level sewer will flow to a point far too deep to enter the river; indeed, it is stated that nearly one half of Lambeth, Bermondsey, and Rotherhithe is 6 ft. below high-water level. Hence it becomes necessary to pump the whole of it up to a higher level; and for this purpose steam-engines, boilers, furnaces, pumps, pump-wells, chimneys, coal-sheds, &c., are provided at Deptford. These works alone have cost 140,000*l.*

Southern Outfall.—This consists of about 7½ miles of sewer, 11½ ft. in diameter, at depths varying from 17 to 80 ft. below the surface, and with a fall of 2 ft. per mile. At Woolwich, it is in some places full 80 ft. beneath the surface, and of course made wholly by tunnelling. It extends from Deptford Creek to Crossness Point, in the Erith Marshes, under Greenwich and Woolwich, and across Plumstead Marshes. At Deptford it receives the contents of the High-level sewer by gravitation, and those of the Low-level by pumping.

The Crossness Reservoir and Sluices.—At the point of junction with the Thames, 14 miles below London Bridge, there is a system analogous to that at Barking Creek, but more costly to work on account of the difference of levels. · There are four double-acting condensing and rotative engines of 125 horse-power, working eight pumps, 7 ft. in diameter, with a lift varying from 10 to 34 ft.; and the works connected

K 3

with the flowing into the river of the sewage thus pumped up comprise
engine and boiler houses, reservoirs, river-walls, shafts, coal-sheds, &c.
The pumping power is equal to that of raising 25,000,000 cubic ft. per
day to the level of the outfall. These works on the south side of the
river are very costly: the Deptford and Crossness constructions, and the
connecting sewer between them, will cost no less than 800,000*l*., irrespective of the southern high and low-level sewers. The reason is, that the
whole of the sewage has to be pumped up before it can enter the Thames,
whereas at Barking it flows by gravitation to the proper level. There is
at Crossness a reservoir 5 acres in extent: connected with it are three
systems of channels, one above another—the lowest to bring down the
sewage from the sewer to the pumping-well; the highest to convey it into
the reservoir by pumping; and the middle one to discharge it from the
reservoir into the river. There is also provision for carrying off the
storm-waters by another channel. The culverts for carrying the sewage
from the great main sewer to the pumping-well are sufficiently large for
a railway train to run through, and a troop of mounted Lifeguardsmen
might do the like. The sewage passes through gratings, or strainers,
before it reaches the pump-well, formed of a kind of portcullis working
on massive hinges secured in the masonry. The strainings thus arrested
—consisting of dead cats and dogs, and all the miscellaneous bulky substances that find their way into the sewers—fall into a capacious stone
receptacle. A large wheel, carrying buckets on the periphery, rotates in
this receptacle, and dredges up the *filth* (as these solid matters are called),
and deposits it in the *filth chamber*, whence it is flushed into the river at
low tide. The bottom of the pump-room is 17 ft. below low-water mark.
This shows how formidable will be the work allotted for the pumping-
engines to perform. Indeed, Mr. Bazalgette has declared that if the
northern drainage had to be pumped up in the way which is thus necessary for the southern, he would have shrunk from incurring the vast
working costs that would be involved; and, in fact, he would not have
recommended the plan at all. When the sewage has reached the reservoir (which has a floor of Portland stone, and a roof formed of brick
arches resting upon brick piers), it will accumulate to the extent of
20,000,000 gallons (if full), and will then be let out into the Thames for
about two hours after each high-tide, day and night. The reservoir will
contain at one time much more than twelve hours' sewage from the whole
of the metropolis south of the Thames, with a sufficient extra capacity
for contingencies.

In 1861 and 1862 the Metropolitan Board of Works invited many
hundred peers, members of parliament, and other influential persons, to
visit the works, and see for themselves the nature of these very novel
operations. On the 18th July, 1863, another of these inspections took
place, by four hundred distinguished persons. Mr. Bazalgette explained

everything that could be explained in a short oral address, and then announced that all the northern drainage, except the Low-level sewer, would be finished about the end of 1863—that by the autumn of 1864 the whole of the southern drainage would be finished—and that by the end of 1866 the completion of the northern Low-level sewer and the Thames northern embankment may be expected. (The progress has not been quite so rapid as Mr. Bazalgette at that time anticipated.) After the visitors had inspected the vast works at Crossness Point they passed over by steamer to Barking Creek, where the following operation took place, as described in the daily newspapers:—"The contents of the High-level sewer, bringing down the sewage of Hampstead, Highgate, Kentish Town, Holloway, Hackney, Bethnal Green, &c., was turned on for the first time direct into the Thames. For some time past it has been discharged through the Storm Overflow into the river Lea. The complaints against the commissioners for spoiling the navigation of the pleasant little river induced the board to lay down a temporary iron duct, which conveys the sewage direct into the river Thames from the end of the great sewer, without leading it through the reservoir. To those who have faith in the value of this material as a liquid manure, it must have been grievous indeed to see this gushing, roaring, black, and stinking stream mixing in the broad waters of the Thames. Others, however, not so absorbed in considerations of the economical value of this fertilising stream, regarded it as the first great triumph of the system of main drainage."

Considering the immensity of the work, it cannot be matter for surprise that the cost will exceed the estimates. Since the commencement of the commissioners' operations, in 1858, the wages of bricklayers and the prices of bricks have risen considerably. It has been found that between 1858 and 1863 wages had risen 1s. or 1s. 6d. per day for bricklayers, and 6d. or 1s. per day for excavators and labourers. The best stock bricks, with which most of the work has been done, rose from about 26s. to 40s. per thousand; and every kind of cement had risen in price also. These augmentations of wages and prices were due mainly to the unexampled quantity of brickwork executed in and near London within the last few years. The Metropolitan or Underground Railway, the Metropolitan Extension of the Chatham and Dover Railway, the Charing Cross and Cannon Street Extensions of the South-Eastern Railway, and the Finsbury Extension of the North London Railway, besides the International Exhibition building and other public structures, and the new dwelling-houses and edifices (there were 24,000 new houses built in the metropolis between the two takings of the census in 1851 and 1861), all combined to enhance the market value of materials and labour in the building trades. The contracts taken for the main drainage in 1861, 1862, and 1863 were at higher terms than those taken in 1858,

1859, and 1860, on this account. Hence it resulted that Mr. Bazalgette found his estimates would be exceeded. The Commissioners of the Board of Works had, by the Main Drainage Act of 1858, obtained power to raise 3,000,000*l.*, to be paid by a tax or rate on the house property within the metropolitan area. The Bank of England has advanced this sum, at the low rate of 3¾ per cent. interest. In June, 1863, the commissioners announced to the Treasury that, partly owing to the rise in labour and materials, and partly to the necessity for additional works, a further addition of more than a million sterling would be necessary. After a little correspondence, the Treasury agreed to the introduction of a Bill for increased parliamentary powers; and this Bill has since become law. The commissioners may raise 1,250,000*l.*, which the Bank of England agrees to advance at 3¾ per cent. It is a curious proof of the increase in the value of property in London, that the commissioners now believe they will be able to pay off the whole 4,250,000*l.* in forty years—the same period which in 1858 they believed would be necessary for paying off 3,000,000*l.* The duty is 3*d.* in the pound on the rated rental of the property within the metropolitan area. It produced 150,389*l.* in 1858-59, and gradually rose to 157,075*l.* in 1862-63. Calculating on a similar rate of increase in future, the commissioners estimate that by the year 1898, or forty years from 1858, they will have received an aggregate total of 7,560,000*l.* This would pay off the debt of 4,250,000*l.*, together with 3,233,465*l.*, the calculated interest, and leave a small surplus. As the Treasury will guarantee the whole of the principal and interest, they will virtually claim control over the due collection and disbursement of the rate. The new Act of 1863 gave the commissioners an extension of time until the end of 1866, on the ground that it will be impossible to finish the Low-level sewer except in connection with the Thames embankment.

Should the anticipations of the engineer be realised, it will unquestionably be one of the greatest works of ancient or modern times. Even in an industrial point of view it is very remarkable; for, by the time all is finished, the works will have absorbed 300,000,000 bricks, and 800,000 cubic yards of concrete, while the excavations will have amounted to 4,000,000 cubic yards of earthwork.

Among the many difficulties which the engineer of the Main Drainage has had to meet, is that connected with new railway schemes. In 1863 there were no fewer than thirty-three distinct plans for railways in the metropolis; and when Mr. Bazalgette came to examine them, he found that they would cut his sewers all to pieces, if constructed according to the deposited plans. One would have diverted the great Fleet sewer wholly out of its present direction for a considerable distance; while others had been planned so recklessly that it would be difficult to see how the sewers could be managed at all. Happily, most of the schemes fell to the ground at one or other of the parliamentary stages; and in those for which Acts were obtained, clauses were introduced bearing reference to the safety and efficacy of the Main Drainage. Difficulties of a similar kind afterwards arose in relation to the railway projects of 1864 and 1865.

APPENDIX No. 4.

EMBANKMENT OF THE THAMES.

Although the embankment of the Thames, now (1865) legislatively sanctioned in reference to both banks of the river, is intended mainly to increase the road communication, east and west, through the metropolis, and to deepen and otherwise improve the river current, yet the intimate alliance which it has with the great intercepting main drainage scheme, on account of the line selected for the Low-level northern sewer, renders desirable some account of the matter in this place. We shall therefore rapidly trace the history of the subject, notice the extraordinary variety of projects to which it has given rise, and describe the final result to which those projects have led.

Just about a century ago plans began to be proposed for embanking one or both sides of the river at London; and they have been poured forth at intervals ever since. The object was not in the first instance to protect the river from pollution, but to improve the shores in an architectural point of view, and to facilitate traffic. So long as cesspools were used instead of sewers, the river remained comparatively clean; for the contents of such receptacles frequently found their way upon the land for manure. But when sanitary reformers urged upon the government, as a measure of public health, the desirability of encouraging the construction of sewers instead of cesspools, then did the Thames of necessity become polluted, seeing that there is no other channel through which the contents of the sewers can be conveyed out to sea. All the earlier schemes for embanking the Thames, as we have said, had other objects in view than the preservation of the water in purity; but with the present century the plans gradually became more comprehensive in their scope. A committee of the House of Commons was appointed quite early in the century to examine schemes for improving the river in various ways. Among them was one by Mr. Jessop for forming a river wall at some distance beyond the existing shore; filling in the space behind it with ballast dredged from the bed of the river; forming in this way an embankment from Blackfriars Bridge to the Tower; building wharves and warehouses on the embankment; dredging the shoals in that part of the river; and selling the reclaimed land behind the embankment to pay for the works. The plan was full of ingenuity, but nothing came of it. Twenty years afterwards, when old London Bridge was about to be, replaced by a new one, Mr. James Walker was requested by the Corporation to report on the probable effect of that change on the river and its banks. His opinion was that the effect would be rather beneficial than otherwise. Then, in 1824, came forward Sir Frederick Trench's scheme for embanking the Thames. He proposed to embank the north shore from London Bridge to Westminster Bridge, and to render the embankment available as a public thoroughfare. A bill was brought into Parliament to effect this object, by means of a public company; but without success. Next year he enlarged his plan, proposing to make an embankment from London Bridge to Scotland Yard, 80 ft. wide, with a carriage-way in the middle, and footpaths on either side; to continue this by another embankment, 110 ft. wide, from Scotland Yard to Westminster Bridge, to be surmounted by a terrace-crescent of handsome houses; to form a basin behind the embankment, 7 or 8 acres in area, for commercial purposes; and to construct roads to connect the embankment with the Strand. The scheme was one of much boldness; but, like its predecessor, it fell to the ground. In 1831 and 1832, Sir John Rennie and Mr. Mylne, at the request of the Corporation, reported on the practicability of improving the Thames by equalising in some degree the depth and the width at different points. They proposed about 12,000 ft. of quay wall on the south side, from Southwark Bridge to Westminster Bridge, and 7,000 ft. on the north side. They calculated, as other engineers had previously done, that the rental of the reclaimed land behind the river wall would pay for the whole enterprise. The Corporation, however, did nothing further in the matter. Again, nine years afterwards, engineers went over the old track, and took up again the idea of a river wall or embankment. In 1840 the Corporation, reproached for doing nothing, answered the reproach by commissioning Mr. Walker to prepare a comprehensive plan. He found on examination that, by the removal of the bulky piers of old London Bridge "the velocity of the stream had increased, the depth of water had decreased, and shoals appeared more and more above the surface; the piers of Blackfriars and Westminster bridges were becoming undermined; and the general effect had been to render the inequality of the river greater than ever, by deepening the narrow parts and shoaling the wide." Mr. Walker believed that the Thames, even at the narrowest part, is wide enough for all the traffic, if well organised; and his proposal was, to bring the river to something like an equality of width by means of embankments. The width is a minimum of 600 ft. opposite the Milbank Prison, and a maximum of 1,480 ft. opposite Buckingham Terrace. He

proposed that the maximum should nowhere exceed 870 ft. The embankment was to consist of a river wall of brick or stone, filled in behind
with soil dredged up from the bed of the river; and Mr. Walker estimated (as other engineers had estimated before him) that the rental
of the reclaimed land would pay for the works: the soil would form
good building ground behind the wall, instead of shoaling the river
itself. This well-concocted scheme fell to the ground, owing chiefly
to the opposition of wharfingers and others interested in river-side
property.

We now approach the period when Royal Commissioners began to
take up this Thames question. In 1842 a commission was appointed
"to inquire into and report upon the most effectual means of improving
the metropolis, and of providing increased facilities of communication."
As far as concerned the Thames, the commissioners examined plans by
Sir Frederick Trench, Mr. Walker, Mr. Page, Mr. Martin, and others.
Trench's plan was for an embankment, with a railway elevated above it
on columns; a promenade between the columns; a foot-pavement between
the covered walk and the river; stone landing-stairs at intervals; and a
road for vehicles on the other side of the walk. Mr. Walker's plan
comprised a continuous quay on the northern side of the river, about
4 ft. high; and the owner of every wharf was to have the portion of
quay fronting his property on certain terms to be agreed upon. Four
basins behind the quay were to be constructed for barge traffic, with
four openings through the quay itself. This plan proposed neither a
roadway nor a large area of reclaimed land behind the quay, and was in
these particulars less comprehensive than his earlier plans. Mr. Page's
plan was for a quay with numerous water-openings leading to floating
basins, or tidal docks, behind it. Every water-opening was to have a
bridge over it, so that a continuous roadway might be formed along the
quay. Mr. Martin's plan was one of the first which embraced provisions
for a great sewer, in addition to the other objects of an embankment,
and on that account it deserves notice, as a sort of precursor of the plan
now actually being adopted. He proposed the construction of a great
sewer to receive all the drainage from the adjacent parts of London, and
carry it down to Limehouse, where it would be solidified as agricultural
manure; a line of quay above the sewer for general traffic; a line of
terrace above the quay for foot passengers; and colonnaded wharves
upon the quay, at certain busy places, to land merchandise, but without
disturbing the continuity of the quay. It is worthy of remark that,
while Mr. Martin's plan resembled the one now actually adopted, in
combining a low-level sewer with an embankment, Mr. Page's resembled
it in containing a provision for paying for the works by a tax upon all
coals brought within the metropolitan area.

London was doomed to another postponement of these excellent improvements. The commissioners issued voluminous reports; but, through various causes, nothing was done—excepting the embanking of the northern shore of the river in the neighbourhood of Pimlico, chiefly through the energetic exertions of Mr. Thomas Cubitt, acting for or with the Marquis of Westminster.

In 1855 Parliament was flooded with railway schemes, many of which attacked the Thames in singular ways. Mr. Lionel Gisborne, Mr. Beaumont, Mr. Bird, Mr. Taylor, and Mr. Hawkshaw, all had schemes for combining railways in some way with embankments of the Thames, but without any arrangements for sewers. The House of Commons appointed a committee to investigate all these schemes, and this committee rejected every one of the plans that affected the Thames. And thus another year was lost. So, indeed, were lost the years 1856, 1857, 1858, and 1859, in regard to any definite plans for embanking the Thames. There were, however, proceedings in relation to the main drainage of the Metropolis, and others in relation to the temporary purification of the Thames water. The first we have already described in another part of the Appendix, and the other we will briefly notice in this place. As the water of the Thames had been getting more and more foul every year, it exhaled more and more noxious odours, especially in hot weather; until at length our judges at Westminster Hall "smell'd" the river all day, and our legislators at the Houses of Parliament all the evening, and half through the night. Mr. Goldsworthy Gurney, superintendent of the arrangements for warming and ventilating the Houses of Parliament, was requested, in 1857, to see what could be done in the matter. Mr. Gurney reported that the solid portion of sewage, being heavier than water, subsides permanently after being driven to and fro a few times by the flood and ebb tides, and forms a portion of the bed of the river; inasmuch as the black, slimy mud that we see at low water contains solid sewage as one of its constituents. No wonder, then, that such an abominable mixture should give forth offensive odours; the wonder would be if it did not. This applies to the part of the river above London Bridge, into which sewage can flow from the sewers only for a few hours before and after low water; it need not necessarily apply to the Thames about Barking and Erith, especially when the sewage flows into the river soon after high water, as in the Great Intercepting Drainage System.

Mr. Gurney conceived that if he could obtain a mastery over the banks of the river he could greatly lessen the offensive state of the water; even though the old sewers continued to pollute the river. By straightening and deepening the portion of the bed between high and low water levels, he expected to get rid of many eddies, slacks, and retrograde movements, which cause the retention of solid refuse. He proposed to deepen the shallows near the shore, and to round off projections; then

to form a breadth of 50 yards of solid, sloping banks on either side of
the river, with an inclination of one in twelve from the present shore
down to low-water level; then to dredge a deep channel, 30 yards in
width, immediately outside or beyond each of these slopes. The gravel
dredged up would furnish materials for the sloping banks. The middle
of the river he would leave untouched; the two deepened channels would
form the water ways for river steamers. The theory on which this plan
was based was, that the slopes of the banks would cause all mud and
solid refuse to flow down into these channels; that the depth of the
channels would cause a current sufficiently strong to carry down the
refuse to the sea; and that thus the Thames would gradually become
cleaner and more salubrious. Mr. Gurney further proposed, as a means
of lessening the offensive odours perceptible at and near the mouths of
the sewers, to *trap* those mouths, or to close them with valves of such
construction as to permit the passage of solid and liquid sewage, but not
gases. As the noxious gases would thus be confined within the sewers
themselves, he further recommended that special openings should be
made at spots near the mouths of the sewers for burning the gases, they
being inflammable when mixed in certain ratio with atmospheric air.
Mr. Gurney's plan, submitted to the Commissioners of Public Works in
1857, underwent their consideration during the winter; but nothing was
practically done by Mr. Gurney, except lessening the offensive odour
near the Houses of Parliament, by throwing down chloride of lime into
the sewers.

The next stage in this matter was the appointment of a committee of
the House of Commons in 1858, to investigate plans " for the purification of the River Thames, especially in the immediate vicinity of the
Houses of Parliament." Judging from the constitution of the committee, it ought to have produced good results, seeing that it comprised
the names of Lord Palmerston, Lord John Russell, Sir John Shelley,
Sir Benjamin Hall, Mr. Robert Stephenson, Mr. Joseph Locke, and Mr.
William Cubitt, besides others of less note. Mr. Gurney was the chief
witness examined, and he presented in great detail the plan just
described. Other engineers, however, handled his scheme with great
severity. Mr. James Walker believed that Mr. Gurney's sloping banks
would become covered with mud in the wide parts of the river; that the
dredged channels would fill up again; that one deepened channel in the
middle would be better than two near the sides; that the muddy deposit
could not be removed unless a system of embankment were adopted;
and that the offensive odours of the metropolis would not cease so long
as any of the sewers entered the Thames thereabouts. Mr. Bidder
expressed his opinion that the suggested channels would require constant
dredging, to remove the gravel that would otherwise silt them up; that
the navigation of the Thames would be incommoded by these artificial

irregularities of bottom; that deposits of mud would form on the sloping banks; that trapping the sewer mouths, burning the gases in shafts, and all the other parts of the plan, would involve a wasteful expenditure of money, as they would only be temporary expedients even if useful at all; and that the only good plan was a system of drainage which would carry the whole of the refuse of London into the Thames at a point far below the limits of the metropolis. Mr. Haywood, engineer to the City Commission of Sewers, contended that the ventilation of the sewers would cost annually a sum so enormous as to be insupportable; that the existing air-shafts and gulley-holes must be trapped as well as the mouths of the sewers, to give the system a fair chance; and that only a small portion of the deleterious gases could be decomposed by burning. In short, although a few civil engineers gave a favourable opinion of Mr. Gurney's plan, the balance of evidence was decidedly against it. The committee thereupon rejected it. The investigation led, however, to a strengthening of the hands of the Metropolitan Board of Works, and to an adoption of that scheme which we have already described for the main drainage of the Metropolis.

The plans for embanking the Thames remained in abeyance for some time. Mr. Gurney's deodorising of the existing sewers was abandoned; the great intercepting drainage was commenced; and the embankment schemes slept a while. It became, nevertheless, evident that the Metropolitan Board of Works were favourable to such an embankment, in connection with their Low-level sewer. There was a very general concurrence of belief that if there were a river wall, filled in behind with solid earth, the advantages would speedily become great. By narrowing the river at certain wide parts, the current would be rendered more equable; by straightening the line of shore, it would increase the scouring action of the stream; by shutting off the strip of ground between high and low water, it would prevent the formation of mud-banks; by giving adequate breadth to the embankment, a terrace roadway might be formed at the top; by enclosing and drying the space now occupied by sandbanks, new building ground might be obtained; and by having a permanent wall running along a line in advance of the present limit of the river-side houses, there would be a barrier, behind which a low-level sewer might be constructed at any desired depth, without passing under the houses and roadway of the Strand and Fleet Street. The old idea, growing in many ways since the days of Sir Christopher Wren, had been clouded from time to time by other plans; but it was pretty certain to meet with proper attention at last.

The year 1859 passed over without any decisive results in reference to the Thames embankment; but in 1860 a committee of the House of Commons collected a vast body of evidence, and published a bulky Blue Book, illustrated with many valuable maps and plans. The chief projects for embanking the Thames were eight in number, concerning each of which we will say a few words :—

1. Mr. Fowler proposed an embankment, 80 ft. wide, from Westminster to Blackfriars, to carry a road and a railway. The line would be continued (though not on an embankment) to Farringdon Street station at the one end, and to Pimlico station at the other ; and there would also be a new street from Blackfriars to Cannon Street. Three docks would be formed behind the embankment for barges, with a water area of 8 acres. The capital was to be provided, partly by the Government or by the local authorities, and partly by a company, who would work the railway.

2. Mr. Lionel Gisborne proposed to embank both sides of the river from Westminster Bridge to London Bridge, on the supposition that the southern bank would be injured if only a northern embankment were made. He looked to a large rental for the reclaimed land behind the two embankments.

3. Mr. Sewell proposed a railway on iron pillars, following nearly the low-water line, with a low-level sewer under it, and openings between the pillars for barges to reach the same water-area that is now open to them.

4. Mr. Edmeston proposed an embanked road and railway following nearly the high-water level, the whole of the barge trade to be conducted outside it.

5. Mr. Bird proposed a tunnel railway from Pimlico station to Scotland Yard ; a railway, partly elevated and partly submerged in an iron tunnel, from Scotland Yard to Queenhithe ; an embankment and roadway from Scotland Yard to Blackfriars ; and the finishing of a long-intended line of new street from Blackfriars to Cannon Street.

6. Mr. Bidder brought forward a plan in which he had been assisted by Mr. Harrison and the late Mr. Robert Stephenson. It comprised the following elements—an embankment from Westminster Bridge to Southwark Bridge ; arches to raise a roadway to a level with those bridges ; a low-level sewer beneath the embankment ; large areas of reclaimed land to be given to Somerset House and the Temple ; about 12 acres of docks behind the embankment ; an extensive surface of reclaimed land to be let or sold for warehouses and cellars ; and a double tramway for large omnibuses on the embankment. The plan also contemplated an embankment on the south side of the Thames the whole way from Battersea Park to London Bridge.

7. Mr. Page proposed an embankment from Pimlico to Queenhithe, for the most part so low as not to intercept the view of the river from the existing houses ; the road to be on the embankment·

26 acres of dock-space to be formed behind the embankment; certain portions of the reclaimed land to be laid out as pleasure-grounds; tidal gates to be constructed in the embankment, to admit barges into the docks; a low-level sewer to be made under the embankment; and the embankment to be broad enough to admit a tramway.

8. Messrs. Bazalgette and Hemans proposed an embankment from Westminster Bridge to Queenhithe; a road on the embankment 100 ft. wide, to pass *under* Hungerford, Waterloo, and Blackfriars Bridges, and inclined roads to connect the embankment with the levels of those bridges. The embankment would be formed by cylinders and sheet-piling, like new Westminster Bridge, filled in with earth; a low-level sewer would be formed beneath the embankment, and behind it would be five docks, varying from 100 to 300 ft. in width, covering 21 acres, and entered by several tidal gates.

All these plans, and many others, engaged the attention of the committee. As a result, the committee did not recommend any plan in particular, but only some plan to be executed by the Metropolitan Board of Works, under sanction of an Act of Parliament to be obtained for that purpose. They recommended a postponement for a time of any southern embankment, and a limitation of the length of the northern embankment to the space from Westminster Bridge to Blackfriars. They proposed that the cost should be defrayed out of the coal and wine duties collected in the City.

No Act for this purpose was obtained in 1861; but a royal commission was again appointed to examine and report upon all the schemes that they could get hold of. The commissioners were Sir William Cubitt, Sir Joshua Jebb, Captain Galton, Mr. Burstal, Mr. Hunt, Mr. McLean, and Mr. Thwaites. They examined no less than fifty-nine plans or schemes for the embankment of the Thames. The commissioners recommended the postponement of the southern embankment: they concocted a plan among themselves from the other fifty-nine, and they recommended that it should be carried out by a distinct commission appointed for that purpose. Mr. Thwaites, chairman of the Metropolitan Board of Works, differed from the other commissioners chiefly on this point—that he wished the embankment to be constructed by his own board, which already had the management of the main drainage.

At length, in 1862, an Act was passed for a northern embankment, after a report from *another* committee that filled 360 folio pages. The Act is the 24th and 25th Vict., c. 93. By this statute there will be an embankment from Westminster to Blackfriars. The public roadway on this embankment will be 100 ft. wide from Westminster Bridge to the eastern boundary of the Inner Temple, and 70 ft. wide from the last-named point to Blackfriars Bridge. Approach roads, 40 ft. at least in width, are to lead from Surrey Street, Norfolk Street, and Arundel Street to the embankment. There is to be another approach road, extending diagonally from Wellington Street, Strand, to the embankment near Charing Cross railway bridge, with short branches to Villiers Street and Buckingham Street; and other approach roads, with good access to the embankment, from the Adelphi, from Whitehall, and from Whitehall Yard. There are to be no docks behind the embankment—the space is to be filled in and reclaimed. The Act only marks out the general features of the plan; the Metropolitan Board of Works were to settle the details. The property is to be purchased by 1867, but the execution of the work is not limited to that period. The line of embankment is so chosen as to form a continuation of the terrace in front of the Houses of Parliament. The distance to which it will extend out into the river (at high water) is 220 ft. opposite Richmond Terrace, 400 ft. opposite Scotland Yard, 300 ft. at Charing Cross, 450 ft. opposite Buckingham Street, 300 ft. opposite Salisbury Street, and 130 ft. opposite Somerset House. The first brick pier of Charing Cross Bridge and the first pier of Waterloo Bridge will nearly denote the distances from the shore at those spots. The embankment will be solid so far only as the Temple Gardens; eastward of that point it will rest on columns, and will leave space for barge-traffic behind it. Mr. Bazalgette finds that he has to carry the foundations 30 ft. below the bed of the river. He sinks iron caissons, fills them with concrete and brickwork, and raises upon them (from below the level of low-water) a solid granite-faced embankment. The low-level sewer will be constructed behind and under the protection of the embankment wall.

Throughout the various negotiations concerning the embanking of the north side of the river, the inhabitants on the south bank strongly urged the embanking of that also; and in reference to very pressing requests, the Government appointed a commission, in 1862, to inquire into the matter. After examining about twenty plans, the commissioners, while admitting the improvement which the Thames would experience by being embanked on the south side, from Westminster Bridge to Deptford, did not feel justified in recommending, at present, such an extensive work. They proposed, however, an embankment for the distance between Westminster Bridge and Battersea Park. This embanked roadway would be about 4½ ft. above Trinity high-water mark, 70 ft. wide, and 2 miles long. It would be an ornamental viaduct opposite the Houses of Parliament, as far as Bishop's Walk; then a solid embankment as far as the London Gas Works; then on arches to Nine Elms; and then solid to Battersea Park. They further recommended the dredging of this portion of the bed of the river to a level of 5 ft. below low-water mark.

The session of 1863 witnessed the passing of an Act for this Southern Embankment. The Metropolitan Board of Works are empowered to form an embankment from Gunhouse Alley to Westminster Bridge; to enlarge the bed of the river along a portion of this distance; to connect the embankment, by approach roads, with Palace New Road, and Vauxhall Row; to reclaim the portion of foreshore behind the embankment; and to make a public footway 20 ft. wide on the embankment. It will thus be seen that this scheme is a very limited one—a first instalment of what may be a large undertaking in the course of time.

The fire-places, great furnaces, and factory stoves of London are to pay for both embankments, in the form of coal dues.

APPENDIX No. 5.

REPORT (MADE, BY ORDER, TO THE COMMISSIONERS OF SEWERS), UPON THE MOST ADVANTAGEOUS MODE OF DEALING WITH THE SEWAGE MATTER OF THE METROPOLIS, WITH A VIEW TO THE PREPARATION OF SEWAGE MANURE, BY THOMAS WICKSTEED, ESQ., CIVIL ENGINEER.

From this interesting Report (dated February 13, 1854), we present our readers with the following extracts which describe the works conducted at Leicester by Mr. Wickstead, and the results he has obtained in a successful deodorisation and disinfecting of sewer water.

"In 1845 I was called upon by the projectors of the London Sewage Company to report to them upon the practicability of carrying out a scheme for distributing the sewage water of London in agricultural districts by the application of steam power and pipes, and was further instructed, that if I found it could not be carried out thus so as to prove profitable, to suggest, if possible, some other mode for effecting the object; and it was at that period that I entered into the calculations as to the cost of such a scheme, which led me to form the opinion that it could never be made a remunerative speculation. I then considered a scheme for arresting the fertilising matter held in suspension in sewer water, and found that even had it been as valuable as the matter held in solution (which is by no means the case), the quantity to be obtained, having regard to its quality, was too small to be remunerative.

"It then occurred to me, that if by a mode much less costly than that of evaporation to dryness, a sufficient quantity of the fertilising matter held in *solution* could be separated, that the collection and extraction of fertilising matter from sewer water, might *then* be effected at a remunerative cost, or otherwise the scheme must be abandoned.

" I then consulted my friend, Mr. Arthur Aikin, the eminent chemist, at that time Professor of Chemistry at Guy's Hospital, as to whether there was any cheap substance that could be employed to separate the fertilising matter dissolved in sewer water, and which would not itself prove injurious to the manure ; and he informed me, that he had for forty years previously been in the habit of using a small quantity of lime to separate the organic matter contained in the New River water, which application had always effected its intended purpose, and that he had no doubt it might be applied with good effect to the sewer water. Mr. Aikin tried some experiments upon the London sewage water, and found, that, by the addition of about the three-thousandth part by weight of lime, the quantity of precipitate obtained was double of that which had previously resulted from the precipitation of the solid matter only, held by mechanical suspension in the sewer water. The result of these experiments showing, that not only was the weight of the manure to be obtained doubled, but that the addition being much more valuable as a fertiliser, the whole quantity was rendered superior, and it appeared to be very probable that a remunerative scheme might be formed ; accordingly I designed a plan for the projectors of the London Sewage Company, and plans were prepared and deposited for parliament, and the same was done in the following year, but at neither of the periods could the company proceed for want of the necessary capital.

" For some time after that period the state of the money market was such, that although various similar schemes were published for tunnel sewers with artificial falls, I considered my scheme would not be much advanced by bringing it in opposition to the new schemes until there was a greater chance of being able to raise capital for its execution.

" About the year 1849, I consulted my friend Mr. Robert Stephenson upon the subject, who was kind enough to enter into a thorough investigation of my scheme and an examination of the data upon which it was founded. He afterwards expressed a favourable opinion of its feasibility ; the objections to it he considered to be chiefly the large size of the reservoirs, which we agreed it would be desirable to reduce if a scheme for so doing could be devised. And as to the question of the tunnel sewer itself being constructed by a private company, instead of by a public commission, —as it is clear that the company's interest would be to make the sewer as little in excess of the size required to convey the sewer water when sufficiently impregnated with fertilising matter to render its extraction remunerative, while a

commission, not being swayed by such considerations, would naturally
consider the question of getting rid of flood waters also,—he thought,
therefore, that it would be better, if possible, to divide the scheme,
leaving the sewer to the public commission, and the process of disinfecting and utilising the sewage water to a company. Thus encouraged, I proceeded farther in the consideration of the subject, and in
the following year the idea of applying centrifugal force for the separation of the water from the deposit in the bottom of the reservoirs
suggested itself to me, and, having tried the effect practically, in
1851 I took out a patent for the manufacture of sewage manure.

" The result of experiments with the patent process was ·to show,
that, by its adoption, the size of the reservoirs might be greatly reduced; that the deposit from the bottom of the reservoir might be
abstracted without exposing it by the removal of the supernatant
water ; and that it might be rapidly reduced into a sufficiently solid
state to admit of its being packed in casks, stored in pits or heaps, or
moulded into bricks for the purpose of farther drying by natural evaporation.   In the latter part of 1851, Mr. Robert Stephenson, and
Professors Aikin and Taylor, having expressed very favourable opinions
of the practicability of my amended scheme, which opinion they allowed me to publish, parties were thereby induced to purchase my
patents for Great Britain and Ireland, and in 1852 an Act of Parliament was obtained, incorporating the " Patent Solid Sewage Manure
Company," and enabling the Company to raise capital to the extent of
100,000l.

"In February, 1852, the Directors resolved that temporary works
should be erected in Leicester for the purpose of manufacturing the
manure upon a sufficiently large and practical scale, having in view
three objects:—the first being to ascertain whether the lime process
effectually disinfected the sewer water, and if so, whether it could be
practically used upon the large scale ; the second, to ascertain whether
the removal of the precipitate from the bottom of the reservoir, and
the abstraction of the water from it by means of centrifugal force,
could be practically carried out upon the large scale and at a sufficiently small cost ; the third object being, to manufacture a sufficient
quantity of the manure to enable agriculturists to prove its commercial value, considering that their practical opinions would be a much
better test than chemical analyses only, and it was resolved that upon
the result of these trials, the question of proceeding with the Company should be decided.   Works were accordingly erected, and after
many alterations and improvements upon the original scheme, which
probably would not have suggested themselves unless the opportunity
of carrying the plan out practically had been afforded, the Directors
were so satisfied with the result, that they felt justified in entering

MR. WICKSTEED'S REPORT. 217

into a contract with the Town Council of Leicester, undertaking in return for the exclusive right to all the sewage water for a period of thirty years, to disinfect it, and discharge the water in an innoxious state into the river Soar for the same period.

"The estimated cost of the necessary works is 25,000*l*., and contracts were entered into in April last, for the erection and completion of them in last October, but owing to unfortunate and unforeseen circumstances,—first, to the delay in obtaining the land; secondly, to the unparalleled wet season which has peculiarly affected these works, the site of them being on low marsh ground; and, thirdly, to the great difficulty in obtaining labour, owing to the strikes amongst the workmen, and the delay in obtaining materials,—I am afraid these works, which, under ordinary circumstances, might easily have been completed in five months, will not be in effective operation much, if at all, before next summer.

"I will now proceed to describe the temporary works and the process, to the maturing of which I have devoted the greatest portion of my time for the last two years, and have completely satisfied myself that the scheme is not only practicable and remunerative, but may be made very profitable when carried out on a larger scale than opportunity has hitherto afforded.

"The temporary works at Leicester were erected upon ground belonging to the Town Council, upon the banks of the Leicester Navigation, near the outfall of the filthiest sewer in the town, a branch from which supplied the works with sewer water. The use of the ground was granted to the Company, at a merely nominal rate, by the Town Council, who in this, as in other instances, have afforded every facility to enable the Company to demonstrate to the public the practicability of disinfecting the water and manufacturing the manure.

"As regards the size of the temporary works, they were calculated for a population of 5,000, previous to the introduction of a supply of water into the town from the New Water Works. I do not mean to imply that we have actually deodorised the sewer water from a population of 5,000 during twelve months. for this would be inaccurate, as the constant interruption arising from the practical modifications and improvements of the machinery used in the process would of itself have prevented such a course; but the works have for different periods been kept in continuous operation day and night, that I might have the opportunity of assuring myself that the process was complete as affecting the sewer water during any of those periods, the object of these temporary works being, as I have before stated, to ascertain whether the process could be practically and remuneratively carried out by the means proposed.

"At the commencement of our operations, it was found that the

L

process of deodorising was not perfect, and it was discovered that its
partial failure was due to the sewage water being in too concentrated a
condition, the new supply of water to the town having only been introduced at Christmas last, while the operations of the Company commenced more than a year and a half ago. To prove whether this conjecture was correct, a portion of the partially deodorized water was
returned into the engine well, and when the concentrated sewage was
reduced to a quality equal in strength to that of the metropolitan
sewer water, the process was completely successful.

"Professors Aikin and Taylor ascertained the strength of the London sewer water, and determined what amount of dilution was necessary to reduce the Leicester sewer water to the requisite strength, and
by their experiments I was guided in my practical operations.

"Again, it was found that at night, when the manufactories were
not at work, and the waste water from the engines had ceased to flow
into the sewer, their contents being chiefly urine and excrementitious
matter in a state of far greater concentration than the day sewage, the
effect of the lime was only partial, but upon diluting it as in the former case, the process of deodorising was completely successful—the
effluent water from the reservoir, after the process was completed,
being perfectly free from all taste and odour, excepting occasionally
from the lime when it had been used in excess.

"A quantity of the effluent water from the reservoir was taken in
August last, by Mr. Theodore West, chemist, of Leeds, and subjected
to an analysis, and he found that there was not a trace of any other
matter than carbonate of lime, sulphate of lime, and chloride of
sodium, proving clearly that all noxious matter had been abstracted
during the process.

"The mode of operation in this process is as follows :—the water is
pumped up from the sewer, and into the pipe conveying it to the reservoir a smaller pipe is introduced, connected with the lime-pump,
which works stroke for stroke with the sewer water-pump, and the process of deodorising is so rapid that when the mixture of sewer water
and lime is discharged into the reservoir, there is no noxious odour
arising from it; the discharge takes place into the first part of the
reservoir divided into three compartments, in each of which is an
agitator worked by the engine : a thorough mixture having thus been
effected, it flows through the upper end of the reservoir, and is from
thirty to forty minutes passing through this portion, during which
time seven-eighths or more of the separated matter has been precipitated on the bottom of the reservoir; there still remains, however,
about one-eighth of solid matter, and which being lighter than the
first portion, requires a longer time for precipitation, so as to render
the water clear and bright.

"The water is, in fact, two hours in passing from the sewer to the farthest end of the reservoir, where it is discharged, and arrangements are made to enable the water to flow continuously through the reservoir with as nearly as practicable the same velocity over the whole section, the openings of the discharging gates being proportioned to the depth of water in the cross section, and thus the necessity of having two reservoirs, for the purpose of filling one while the water in the other is being cleared by deposition, is avoided, for although the stream is continuous, its velocity being only about one-fourth of an inch per second, it does not interfere with, or arrest the precipitation of the solid matter.

"The operation of removing the precipitate from the bottom of the reservoir, so as not to interfere with the continuous flow of the water in the reservoir, is performed by means of a screw, which removes the precipitated matter into an adjoining well or shaft as rapidly as it is formed, without disturbing the process of precipitation which is carried on above it.

"The bottom of the first portion of the reservoir is made to slope towards the centre, along which a culvert runs, semi-circular at bottom and open at top ; in the bottom of this the screw is laid, and the precipitate collecting upon it from the sloping sides; is, as the screw revolves, carried into the adjoining well : the practical working of this arrangement is now completely successful.

"It is the combination of these two arrangements; viz., the continuous current and the removal of the deposit without disturbing the supernatant water—that has enabled me to reduce the size of the reservoirs to so great an extent: this will be seen hereinafter when I give the sizes of the reservoirs I propose for the metropolis.

"The next operation is to raise the deposit or mud from the well or shaft, by means of a Jacob's ladder, very similar in appearance and construction to the ladder of buckets in the dredging machines used on the Thames, excepting that its position is vertical and its construction much slighter ; the mud thus raised in a semi-fluid state into a tank, flows through a pipe to the centrifugal machine, the machine is then set in motion at the rate of about 1,000 revolutions per minute, and in half an hour from the time the precipitate lay on the bottom of the reservoir, it is in a sufficiently dry state to pack in casks or to mould in the form of bricks for farther drying.

"A given bulk of the manure, when introduced into the centrifugal machine, is reduced to about one-third of its original bulk, two-thirds, as water, having been separated from it by the operation.

"The machines which are now making for the new Leicester Sewage Works, are each calculated to turn out 360 lbs. of manure in an hour, in the state of consistency previously mentioned.

"Thus it will be seen, that the whole operation of disinfection and conversion into manure is very simple, and I think it must appear evident, that after the experience of a year and a half of what may be done in works sufficient for a population of 5,000, that by simple multiplication of the means, it may be made available for any population, however great; as in this case the increased quantity of sewage water merely involves a simple increase of machinery in proportion to this increase, the increase of power for raising it being in direct proportion to the quantity.

"There is one very important conclusion that may be drawn from what has been said; viz., that an increased supply of water, by which means only the greatest and most immediate sanitary effect will be produced upon the atmosphere of dwellings in any town, does not render the plan just described abortive, on account of the enormously increased expense; but, on the contrary, within certain limits, the more the sewage is diluted, the more complete is the effect of the disinfection of the water and the precipitation of the manure, so that the sanitary objects of the Commissioners and the commercial interests of a Company carrying on the works, would not be opposed to each other as would be the case in the event of the liquid scheme being adopted.

"In the temporary works at Leicester, although the reservoir, the steam-engine and boiler, the machinery for manufacturing the manure, and the store for the manure, whether in casks or exposed in heaps for drying, are under one roof, no noxious effect has in the slightest degree been caused to the workmen employed in the process; and although the manure itself, when taken from the cask and held to the nose has a smell which an agriculturist would not object to, nevertheless, no smell whatever is perceptible at the distance of a foot or two from the manure.

"As regards, however, the proposed large works for Leicester, there will be two reservoirs, about 200 feet long, and 44 feet wide, two-thirds of the area will be covered with an iron girder and brick arch floor for the warehouses above, to economise space, and the whole will be roofed over; and as this portion of the design is intended to be carried out in all future works, however large, all chance of nuisance from exposure is avoided; but the fact is completely established, that no nuisance does arise either from the reservoirs or in the process used in manufacturing.

"Upon this point, and in corroboration of my statements, I beg leave to call your attention to the practical evidence of his Grace the Duke of Rutland, given in a letter I had the honour of receiving from him; also of Joseph Whetstone, Esq., the Chairman of the Local Board of Health; John Ellis, Esq., late M.P. for Leicester, the Town Clerk; Dr. Shaw, and other medical gentlemen of the town of Leicester. given in a certificate, which, with another from John Buck,

Esq., late Medical Officer of the Leicester Local Board of Health, who has had frequent opportunities of witnessing the operation of the patent process, are addressed to your Honourable Board, as I thought it might be more satisfactory to you to have the evidence of disinterested parties.

"Although not in a position at present to state what the 'actual commercial value of the residual manure will be, because, as before intimated, the desire of the directors of the Patent Solid Sewage Manure Company has been to leave this to be determined by the result of its practical application by the agriculturist, and at present the manure that has been so applied has been *pro tanto* inferior to what is intended to be supplied, in its having been chiefly collected from the day sewage unmixed with the richer night sewage, and also from the fact of its containing 60 or 70 per cent. of water instead of 20 per cent., which is the quantity it would have contained, if the extent of our temporary works had afforded us room for drying it in larger quantities than has hitherto been practicable; nevertheless, our experience has been quite sufficient to prove that, without the necessity of having recourse to expensive applications of artificial heat, simple exposure to atmospheric influence for a few weeks will reduce the moisture to 20 per cent., so that, bulk for bulk, the manure intended for sale will contain twice as much fertilising matter as that which has at present been forwarded to agriculturists for trial. The present results, however, show that, taking guano at 10*l.* per ton, the manure, as proposed for sale, is at least worth 2*l.* 13*s.* per ton; but as I stated to the Commissioners verbally, I have considered it safest to calculate its commercial value at 2*l.* or 2*l.* 2*s.* per ton, and this amount would yield, after deducting the cost of manufacture and repairs, a fair per centage upon the capital expended in the construction of the works; but the actual value will not be ascertained until time has afforded more extensive experience. But while it is undoubtedly of importance that the Commissioners should be satisfied that it is of sufficient value to induce capitalists to subscribe for its manufacture, its real value, if greater, must depend in some measure upon the favourable locality of the works—in relation to the agricultural districts—and hence its concentration, by reducing the cost of carriage, will increase its value, weight for weight; and again, the reduction in the cost of manufacture, which the last year's experience has already enabled me to effect, has also shown, that, with larger opportunities, further reductions may be effected, which, I need not remind the Commissioners, has generally been the case in all new manufactures.

" The cost of manufacture will be proportionably greater in small works than in larger ones: my present experience, however, enables me to state, that, upon the average, the cost of manufacture will not exceed 20*s.* per ton."

As to the application of his treatment to the sewage of the metropolis, Mr. Wicksteed had been supplied by Mr. Bazalgette, engineer to
the Metropolitan Commission, with a map and the following particulars:—

"The main sewers of the Eastern Division are capable of being
terminated at the points A, B, and D, at the river Lea, and their
contents being separately or collectively manufactured into manure in
that locality, or they can be continued so as to discharge at mean high
water at the mouth of the Roding. With this view, A and B will be
9 feet above Trinity high water at crossing over the river Lea, and
the sewage could possibly be there converted into manure without
pumping.

"The main sewers of the Western Division will have two separate
flood outlets into the river at the points E and I, the inverts being
there level with low water.

"Sewage manure works could be established at both those points,
or the sewers could be so connected as to have one establishment.

"We have taken the sewage at 5 cubic feet per head, and
the tables give the sewage thus due to the present population, and
also due to the ultimate increase of population as estimated by us.

NORTHERN SEWAGE INTERCEPTION AND DRAINAGE.

		Present Population.		Prospective Population.	
		Cubic Feet per Diem.	Cubic Feet per Minute.	Cubic Feet per Diem.	Cubic Feet per Minute.
Hackney Brook . .	A	462,240	321	2,075,040	1,441
Middle Level . . .	B	3,911,040	2,716	4,574,880	3,177
Low Level . . .	C	3,225,600	2,240	3,415,680	2,372
Total . .	D	7,598,880	5,277	10,065,600	6,990

WESTERN DIVISION.

		Present Population.		Prospective Population.	
Acton Line . .		169,920	118	662,400	460
Cheyne Walk Branch to ditto	F	224,640	156	224,640	156
Brentford Line . .	G	298,080	207	1,732,320	1,203
Fulham Branch to do.	H	37,440	26	139,680	97
Total . . .	I	730,080	507	2,759,040	1,916

"MEMORANDA.—The height of the lift, from the invert of the low level sewer at the pumping station near West Ham Abbey, to the invert of the high level sewer, is 35 feet, and to Trinity high water mark, 26 feet. The height of the lift, from the invert of the sewer at the pumping station in Fulham Meadows to Trinity high water mark, is 17½ feet."

Mr. Wicksteed's estimate for works required for treating the sewage of the metropolis, according to this arrangement of main sewers was as follows :—

PROPOSED WORKS.

"Without giving an opinion as to the actual sites that should be secured by the Commissioners, which I have no doubt you will agree with me would be premature, not to say imprudent, and in the choice of which I have no doubt you will be much influenced by your Engineers, to whom I shall be happy to give my assistance if required,—I may state, that it appears to me that four sites at least should be obtained for the proposed works : the first in the neighbourhood of the River Lea, uniting A and B, or the Hackney and the middle levels of the northern division ; the second, in the neighbourhood of the same river, to receive C, or the low level of the northern division ; the third, in the neighbourhood marked I, upon the plans supplied to me by your Engineer, uniting the Acton, Cheyne Walk, Brentford, and Fulham levels of the Western Division ; and the fourth, on the banks of the Thames, on the south side of the river, for the southern sewage.

"Supposing, in other respects, these sites to be eligible, there could be no objection to them on the part of the Patent Solid Sewage Manure Company ; but to repeat what I have hereinbefore intimated, there is nothing as regards the disinfecting and manufacturing of the manure to prevent either a concentration of, or a further subdivision of manufactories ; and if, therefore, it should appear hereafter that a different arrangement would lead to a reduction in the size and cost of the main sewers, or for causes at present not foreseen an alteration should be deemed advisable, I see no objection to its being made. However, this part of the subject does not appear to me to press in such a manner as to lead to the necessity of my delaying the completion of this Report ; should, however, the si'es herein suggested be finally determined upon by the Commissioners, the following statement will represent the principal works that would be required for each :—

I.—THE QUANTITY OF LAND REQUIRED FOR ALL PURPOSES, No. 1, AND FOR RESERVOIRS, No. 2, INCLUDED IN No. 1.

	Present Population.		Prospective Population.	
	No. 1. Land for all Purposes.	No. 2. Land for Reservoirs.	No. 1. Land for all Purposes.	No. 2. Land for Reservoirs.
	Acres.	Acres.	Acres.	Acres.
1st Site . .	8½	2¾	12½	4¼
2nd Site . .	6	2	6½	2¼
3rd Site . .	1½	½	5¼	1¾
4th Site . .	5¾	2	7	2¼
Total . .	21¾	7¼	31¼	10½

II.—FULL AND AVERAGE POWER OF PUMPING ENGINES AND COALS.

	Full Power.	Average Power.	Coals per Annum.	Full Power.	Average Power.	Coals per Annum.
	H. P.	H. P.	Tons.	H. P.	H. P.	Tons.
1st Site . .	Nil	Nil	Nil	Nil	Nil	Nil
2nd Site . .	223	148½	1545	236	157	1635
3rd Site . .	37½	25	342	141½	94⅓	981
4th Site . .	225	150	1560	270	180	1872
Total . .	485½	323½	3447	647½	431⅓	4488

"With the exception of the land and approaches, the above works would have to be provided and maintained by the Company, and in addition the Company would have to provide for manufacturing purposes—

Engine-power for present population 2,955 horses' power
Ditto for prospective ditto 4,286 " "
Coals, per annum, for present ditto 26,341 tons
Ditto, ditto, for prospective ditto 38,205 "

"To show the importance of having manufactories established on sites accessible for carriage, I may state, that the probable present and prospective annual tonnage of coals, lime, and manure, will be as follows:—

	Tons.
For present population . . .	214,000 per annum
For prospective ditto	310,000 "

"The capital required for the construction of works for the prospective population would not exceed 1,000,000*l.* sterling."

APPENDIX No. 6.

In relation to the *Main Drainage* of the metropolis, we have been able, in a former number of the Appendix, to describe a very important advance made since the publication of the second edition of this work in 1854. The *Sewage Manure* question, however, has not been so fortunate. We are still, in 1865, as we were in 1854, groping in search of a practical, if not profitable plan. Boards, committees, commissions, inspectors, and civil engineers, all have been examining and reporting; and the following brief account will show in what direction speculative ingenuity has sought for a solution of this very difficult question. For fuller details we refer to Mr. Scott Burn's Rudimentary Treatise noticed in the Preface.

In 1856 the Board of Health directed their chief superintending inspector, Mr. Henry Austin, to prepare a "Report on the Means of Deodorising and Utilising the Sewage of Towns." In 1857 that report was presented, consisting of about a hundred pages of text, and seven lithographed plans. Mr. Austin entered very fully into the chemical and agricultural relations of sewage manure, the (then) existing arrangements concerning the drainage of towns, the deodorisation of sewage, and its manufacture into solid manure. He treated the general principles of the subject, so far as they have yet been determined; and obtained evidence from various quarters as to the facts. He described various *chemical* processes for separating the solid matter of sewage, as patented or suggested by Mr. Higgs, Mr. Wicksteed, Mr. Stothert, Mr. Herapath, Mr. Dover, Dr. Angus Smith, Mr. Blackwell, Mr. Manning, and Mr. McDougall. He next similarly examined various *mechanical* processes for effecting the same end, as adopted with more or less success at Cheltenham, Uxbridge, Ely, Hitchin, and Dartmoor Prison. Mr. Austin then instituted inquiries into the utilisation of sewage in the liquid form, by *open irrigation* and by *underground pipes:* and into the relative advantages of town sewage and farm-yard manure for agricultural purposes. Mr. Austin wound up his report with a series of "Conclusions," the most practical of which were as follow :—

"That in order to avoid all further risk of injury to health, whether from discharge of the sewage into the rivers and streams, or from its application to the land, it appears desirable that the solid matter should in every case be separated from the liquid sewage at the outfall, and that a cheap portable manure should be manufactured therefrom for use in the immediate neighbourhood.

"That it should be mixed with the ashes of the town, or such other deodorising material as may be most suitable for application to the surrounding land, and prepared, if desirable, with other manuring ingredients for particular crops.

"That it appears probable that such operation will in most places pay its own expenses; but that, as some such measure is absolutely necessary for the public health, even though involving some expense, it should be the duty of local boards and other governing bodies to carry it out, just as much as arrangements devolving upon them for removal of dust or other refuse from the town. It should form, in fact, part of such service, and might be combined in the same contract.

"That the liquid portion of the sewage, thus cleared of its solid matter, but still retaining its chief value as manure, might then be applied with benefit to the neighbouring lands in any quantity; but that all land upon which this method of application of the sewage is practised should, if not naturally porous, be artificially drained, as the liquid, if allowed to become stagnant, would, as in common irrigation, be likely to engender disease in the neighbouring inhabitants, or in cattle exposed to its influence.

"That the distribution of manures in the liquid state by the hose and jet, from a system of underground pipes on the land, has been found, by the experience of several years upon farms in England and Scotland, most advantageous; and that the outlay for such works is considered by eminent agriculturists, who have had experience of their benefits, as a very profitable outlay, irrespective altogether of the question of sewage distribution.

"That upon grass lands, for which the application is best adapted, these larger quantities of the liquid sewage, deprived of its grosser particles, may be economically distributed, especially upon the lower levels, by a combination of the underground pipe system with the subsidiary open irrigation by small contour gutters.

"That the solid sewage manure, prepared and deodorised as above proposed, may be used anywhere, and any quantity of the liquid applied on absorbent or properly drained land, without any risk of injury to health, and without any of the offensiveness constantly experienced from farmyard and other solid manures applied as top-dressings.

"That in any neighbourhoods, however, where no opportunity exists for this beneficial irrigation, the liquid sewage, before being discharged into rivers or streams, should, after separation of the solid matter, be treated with lime or other deodorising and precipitating agents—a duty which should devolve upon the local board or other governing body, as a precaution in which the public health is materially concerned."

Such was the substance of Mr. Austin's recommendation. It has been a misfortune, in relation to this matter, that the same facts are described over and over again, and printed in the Blue Books at the public expense, by rival or independent boards, commissions, and committees. Instead of acting upon any of the suggestions made to the Board of Health by Mr. Austin, a new body proceeded to investigate the whole matter over again. In 1857 a *Sewage Commission* was appointed, consisting of Lord Portman, Dr. Southwood Smith, Mr. I. K. Brunel, Mr. H. K. Seymer, Mr. Rawlinson, Professor Way, Mr. Lawes, Mr. Simon, and Mr. Austin. Their duties were " to inquire into the best modes of distributing the sewage of towns, and applying it to beneficial and profitable uses." The commissioners appointed five of their number as a committee "to visit and personally examine the different localities where sewage is employed in agriculture, or treated with a view of neutralising its offensive and noxious properties;" leaving for a subsequent period in their labours, " to undertake a series of. distinct experiments to test the efficacy of existing methods, and, if possible, to improve upon them." The localities visited by the committee in which sewage was applied to the land in a liquid form were Rugby, Watford, Edinburgh, Rusholme, Mansfield, and Milan; those in which works were in operation for the purification of sewage were Croydon, Leicester, Tottenham, and Cheltenham. The committee also visited several farms, in which farm-yard liquid manure was used on a large scale; the farms selected being the Earl of Essex's, at Cashiobury; Mr. Mechi's, at Tiptree; Mr. Wheble's, at Bulmarsh Court; Mr. Kennedy's, at Myer Mill; Mr. Telfer's, at Cumming Park; the Marquis of Breadalbane's, near Luing; and Mr. Hervey's, near Glasgow. The general report of the commission, founded upon the special reports of the committee, and sent in to the Treasury in 1858, was necessarily little more than a repetition of Mr. Austin's statements, extending over nearly the same area, and arrived at nearly in the same way. It will suffice to give merely a few of the results at which the commission arrived:—

" That the methods which have been adopted with the view of dealing with sewage are of two kinds—the one being the application of the whole sewage to land; and the other that of treating it by chemical processes to separate its most offensive portions.

"That the direct application of sewage to land favourably situated, if judiciously carried out, and confined to a suitable area exclusively grass, is profitable to persons so employing it; and that, where the conditions are unfavourable, a small payment on the part of the local authorities will restore the balance.

"That this method of sewage application, conducted with moderate care, is not productive of nuisance or injury to health.

"That when circumstances prevent the disposal of sewage by direct application to land, the processes of precipitation will greatly ameliorate, and practically obviate, the evils of sewage outfalls, especially where there are large rivers for the discharge of the liquid; but that such methods of treating sewage do not retain more than a comparatively small portion of the fertilising matters; and that, although in some cases the sale of the manure may repay the cost of production, they are not likely to be successful as private speculations.

"That, considered merely as the means of mitigating a nuisance, these precipitating processes are satisfactory ; that the cost of them in any case is such as town populations may reasonably be called upon to meet; that the necessary works need not, if properly conducted, be a source of nuisance; and that, by modifications of the existing methods, even the slightest risk of nuisance may be entirely obviated.

"That the employment of the one or other method of disposing of sewage, or of both conjoined, must depend upon locality, levels, markets, and a variety of other circumstances; and that the case of each town must be considered upon its own peculiarities.

"That there is good ground for believing that the methods yet proposed for dealing with sewage are not the best that can be devised; and that further investigation will probably result in the discovery of processes more thoroughly equal to the suppression of the nuisance, and at the same time calculated to give more valuable products.

"That the magnitude of a town presents no real difficulty to the effectual treatment of its sewage, provided it be considered as a collection of smaller towns. As, however, the conditions under which the evil may be best removed will differ greatly in different localities, we think it would be desirable, before any legislation takes place on this subject, that investigation should be made into the state of the outfalls of different classes of towns, and of the condition of rivers in populous districts, with the view to advise as to the general legislative measures that might safely be adopted."

The commission also sketched a plan for dealing with the sewage of the metropolis, by combining an embankment of the Thames with a series of works for deodorising the sewage.

During the course of their labours from 1857 to 1863, the commissioners received numerous communications relating to plans for dealing with this difficult sewage question. Some proposed plans for the manufacture of " urban manure ;" some for the disinfection of drains. Mr. Beadon had a plan for the drainage, collection, and deodorisation of the sewage of the metropolis by a system of subways, in which the sewage would be conveyed to suitable districts for the purpose. Mr. Bridges Adams proposed to collect separately the solid and liquid portions of the sewage, allow them to accumulate for a given time, and finally remove them for agricultural uses. Mr. Wright suggested a plan for producing an inoffensive solid manure by using burnt clay, charcoal, and gypsum, with the sewage. Mr. Dover's suggestion was, so to employ hydrochloric acid, sulphate of iron, and common salt, as to separate the liquid from the solid portion of sewage, to render the former inodorous and inoffensive, and to manufacture the residium into a profitable, portable manure. Thirty or forty projects, more or less resembling those here noticed, were received and considered by the commissioners, but no definite course of action was adopted in reference to any of them. Two commissioners, acting as an agricultural committee, Professor Way and Mr. J. B. Lawes. have made many interesting experiments on fields near Rugby, by comparing the crops produced with and without the aid of sewage manure, Oxen were fed with grass grown on sewaged and on unsewaged land, and the fattening qualities of the two kinds were compared.

This Sewage Commission was appointed before there was a Metropolitan Board of Works, entrusted with the general drainage of the metropolis. Now, however, when there is such a board, the question is practically taken out of the hands of the commission, so far as London is concerned, seeing that it rests with the board to decide whether the sewage shall flow into the Thames, or whether the contents of the great reservoirs at Barking and Crossness shall be applied as manure. Nevertheless, the commission may still render service by encouraging plans that may be useful in other large towns.

The Report of the Metropolitan Board of Works, submitted at a meeting of the board held August 7th, 1863, emanated especially from the Main Drainage Committee, and related to tenders received from various persons for the sewage of the metropolis. Twelve parties had responded to an advertisement put forth by the committee, and the following, expressed in a condensed form, will suffice to show the nature of the tenders :—

1. Dr. Thudichum proposed by a mechanical arrangement so to separate the house-drainage as to retain that which he considered most valuable ; and he gave a sketch of his closet and drains, and tables of analysis of the constituents of fluid sewage. He estimated the cost of the works for carrying out his process at 1,500,000l. Out of the net profit he proposed that he should receive a percentage, and that one half of the balance should be paid to the board, and one half to the shareholders of a company formed for carrying out the system.

2. Mr. Curwood was not in a position to offer a tender, but suggested the separation of the solid and liquid sewage.

3. Lord Torrington, Sir Charles Fox, and Mr. Hunt, while expressing their willingness to discuss a particular plan of theirs with the Main Drainage Committee, and to enter into a provisional arrangement for presenting a Bill in Parliament during the ensuing

session, regretted their inability to comply with the conditions of
the advertisement, on the ground of the impolicy of publicly
declaring the land with which they proposed to deal.

4. The London Sewage Utilisation Company (Limited) proposed that
the board should grant them the sewage they might require at
Barking Creek for two years, at a rent of £5; that, if the experiment succeeded, the board should then grant a further term of
twenty-one years at the same rent; and that, at the end of such
further term, the rent for a lengthened period should be referred
for decision either to the Board of Trade or to the President of
the Institute of Civil Engineers.

5. Mr. Moore proposed that the board should grant the sewage for a
term of ninety years; that for fourteen years of this period the
rent should be merely pepper-corn; and that for the remainder of
the term, after deducting 10 per cent. on the capital invested, the
rent should be one half of the profits. In a subsequent tender,
made in July, 1863, Mr. Moore offered one farthing per ton for
all the sewage raised by the board to a height of 200 ft.; at
80,000,000 gallons per day, this would amount to 136,000l. per
annum. Mr. Moore stated that he had already engagements with
the occupiers of nearly 60,000 acres for the use of the sewage.
This plan of raising the sewage to so great a height seems to indicate some arrangement for allowing the liquid to flow by easy
gradients to fields at a considerable distance.

6. Mr. Shepherd, as a first communication, asked for a concession of
the whole of the sewage of London for fifty years. He proposed
to establish a company to work the plan; and that, after this
company had reached 7½ per cent. on the invested capital, he
should share the surplus profits with the board. In a second
communication, Mr. Shepherd stipulated more completely than
before for an unlimited command over the whole of the sewage.

7. Mr. Kirkman's proposal embodied the use of a patent, of which he
is proprietor, for obtaining manure from sewage by filtration and
deposit; the water after such treatment to be discharged into the
river. He proposed to erect works for this purpose at Barking,
on condition that the board convey to him the necessary land
adjoining the reservoir, deliver the sewage to him from the main
outfall sewer into his works, grant him the use of their river
frontage and wharves for the term of seven years at a pepper-corn,
seven years at a rent of one-fifth the net profits, after deducting
5 per cent. on the capital, and for any further time at a rental to
be settled by the Secretary of State. Mr. Kirkman stated that he
was prepared to guarantee a minimum rent of 10,000l. per annum
—that he could name two or more securities for the due perform-

ance of the contract, and that he would agree to the surrender of the works on two years' notice, subject to the repayment of the capital invested, and of a premium of 25 per cent. thereon, together with a royalty of 1*l.* per cent. for the use of his patent.

8. Mr. Ellis proposed to pump the sewage from the reservoirs at Barking and Crossness into certain covered tanks, and then to cause it to flow by gravitation through pipes laid along the sides of the roads adjoining the land to be irrigated. On behalf of a joint-stock company to be formed, Mr. Ellis undertook to provide and use deodorising agents. The net profits, after deducting working expenses and reserve fund, to be divided equally between the company and the board. The concession of the sewage to be in perpetuity, subject to the board's power of purchasing after fifty years, on giving three years' notice—the price to be fixed by valuers jointly chosen. Certain capitalists were prepared to back Mr. Ellis to the extent of 60,000*l.*

9. Messrs. Napier and Hope proposed to intercept the whole of the ordinary flow of the northern sewage near Abbey Mills, and to convey it by a culvert 44 miles in length to Maplin Sands on the one hand, and to Dengie Flats on the other. Both those areas are at present submerged at high water, and their redemption is part of the project, extending to 15,000 or 20,000 acres. The capital proposed to be invested was 2,000,000*l.*; the concession of the sewage to be for fifty years, subject to parliamentary authority being obtained, and to a grant of the land from the crown. The net profits to be divided equally between the company and the board, after deduction of 10 per cent. on the outlay. The board to have the power of resuming the grant of sewage, and taking the whole of the lands and works at a valuation, at the end of the term of fifty years, on giving seven years' notice; or a new concession to be granted in terms settled by the Secretary of State. The board to be represented by two directors appointed to the company. The company to be formed within two years after obtaining parliamentary powers. The company, within three months of its formation, to place in the hands of trustees a certain sum of money as security for the due fulfilment of the works.

The other three responses to the advertisement of the board did not take the form of tenders. It will suffice to show the difficulties which surround this subject to say that the committee did not feel justified in recommending *any* of the plans, and that it was resolved by the board that the committee's report, together with all the tenders, supplementary documents, and correspondence, should be printed, and copies sent to all the metropolitan vestries and district boards. From this it is pretty evident, that so far from any defined arrangement being made, the end of the year 1863 witnessed merely the commencement of another series of plans, discussions, and controversies on this much-troubled sewage question. To detail the proceedings of 1864 would be little more than going again over the same ground; the controversies and proposals were numerous, but nothing definite resulted from them. In 1865, however, the Maplin Sand reclamation scheme is brought forward in a definite way, in connection with a system of sewage utilisation.

APPENDIX No. 7.

WATER SUPPLY OF LONDON, UNDER THE ACT OF 1852.

IN addition to the information given in the text (p. 99), it may be desirable to present here a few facts concerning the supply of water to the metropolis, especially under the influence of the important statute passed in 1852. For the reason stated in the Preface, we may refer to Mr. Hughes' *Rudimentary Treatise*, for a fuller treatment of water supply generally.

In 1856 a valuable Report was presented to the General Board of Health, by the Superintending Inspectors of that Board (Messrs. Henry Austin, William Ranger, and Alfred Dickens), on the subject of the Water Supply of London. The Water companies having been required, by the terms of the Act of 1852, to make very extensive alterations and improvements in their works, it became desirable to ascertain how far the alterations had advanced. The inspector, therefore, made an exact comparison between the state of matters in 1850 and in 1856, before and after the Act of 1852 came into operation. The Act required that by August 31, 1855, no water should be taken by any of the companies (with one exception) from any part of the Thames below Teddington Lock ; that all reservoirs within 5 miles of St. Paul's should be roofed in, unless the water is filtered after leaving the reservoir ; that all the conduits or water channels within the metropolis should be covered, unless the water were filtered after leaving such channel. It may be useful to present here a few leading facts concerning the ten companies which supply the metropolis with water.

In 1850, there were 270,581 houses supplied with about 44,000,000 gallons of water daily, by nine companies; whereas in 1856 there were 328,561 houses supplied with 81,000,000 gallons per day, by ten companies : exhibiting a rise from 164 to 246 gallons per house per day. The main and branch pipes, irrespective of the private service pipes, were 2,086 miles in length, in 1856. There were 40 acres of filter beds, and 141 acres of subsiding reservoirs. The filtered water was stored in fourteen covered reservoirs, comprising an area of 15 acres, and in four uncovered reservoirs, of about 3 acres, beyond the specified distance of 5 miles from St. Paul's. The cost of the several water-works, down to the enactment of the statute in 1852, was about 5,000,000*l.*; and a further sum of 2,300,000*l.* was spent between 1852 and 1856; to which an additional large sum has been added between 1856 and 1865.

The following are a few facts relating to each of the companies, individually, in 1856.

New River.—Sources of supply, New River, River Lea, and chalk springs. Number of houses supplied, 95,083. Gross quantity supplied per day, 25,000,000 gallons. Aggregate nominal steam-power for working the pumping and other engines, 1,442 horses. Length of mains and branches, about 450 miles. Area of subsiding reservoirs, 66 acres. Area of filter beds, 9 acres. Area of covered reservoirs for filtered water, 3¾ acres. Total cost of works, about 2,000,000*l.*

East London.—Source of supply, the River Lea. Number of houses supplied, 70,000. Gross quantity supplied per day, 16,000,000 gallons. Aggregate nominal steam-power, 840 horses. Length of mains and branches, 331 miles. Area of filter beds, 12 acres. Area of covered reservoirs for filtered water, 2½ acres. Total cost of works, 1,000,000*l.*

Southwark and Vauxhall.—Source of supply, the River Thames, at Hampton. Number of houses supplied, 41,529. Gross quantity supplied per day, 10,330,000 gallons. Aggregate nominal steam-power, 1,065 horses. Length of mains and branches, 432 miles. Area of subsiding reservoirs, 8 acres. Area of filter beds, 4½ acres. Total cost of works, 650,000*l.*

Lambeth.—Source of supply, the River Thames, at Thames Ditton, Number of houses supplied, 28,541. Gross quantity supplied per day, 6,110,000 gallons. Aggregate nominal steam-power, 680 horses. Length of mains and branches, 206 miles. Area of filter beds, three-quarters of an acre. Area of open reservoirs for filtered water, 1¼ acre. Area of covered reservoirs for filtered water, 3 acres. Total cost of works, 610,000*l.*

West Middlesex.—Source of supply, the Thames, at Hampton. Number of houses supplied, 25,732. Gross quantity supplied per day, 6,900,000 gallons. Aggregate nominal steam-power, 480 horses. Length of mains and branches, 178 miles. Area of subsiding reservoirs, 16 acres. Area of filter beds, 4½ acres. Area of covered reservoirs for filtered water, 1¾ acre. Total cost of works, 800,000*l.*

Chelsea.—Source of supply, the Thames, at Seething Wells. Number of houses supplied, 25,030. Gross quantity supplied per day, 5,300,000 gallons. Aggregate nominal steam-power, 700 horses. Length of mains and branches, 198 miles. Area of subsiding reservoirs, 3¼ acres. Area of filter beds, 2 acres. Area of covered reservoirs for filtered water, 2½ acres. Total cost of works, 930,000*l.*

Grand Junction.—Source of supply, the Thames, at Hampton. Number of houses supplied, 17,221. Gross quantity supplied per day, 6,700,000. Aggregate nominal steam-power, 1,440 horses. Length of mains and branches, 117 miles. Area of subsiding reservoirs, 7¾ acres. Area of filter beds, 5¼ acres. Area of covered reservoirs for filtered water, a little over 1 acre. Total cost of works, 730,000*l.*

Kent.—Source of supply, the River Ravensbourne. Number of houses supplied, 16,077. Gross quantity supplied per day, 3,500,000 gallons. Average nominal steam-power, 500 horses. Length of mains and branches, 124 miles. Area of subsiding reservoirs, 5¼ acres. Area of filter beds, 2¾ acres. Area of open reservoirs for filtered water, 1½ acre. Total cost of works, 230,000*l.*

Hampstead.—Sources of supply, the Hampstead and Highgate Ponds, and an artesian well at Hampstead. Number of houses supplied, 6,348. Gross quantity supplied per day, 600,000 gallons. Aggregate nominal steam-power, 72 horses. Length of main and branches, 33 miles. Area of subsiding reservoirs, 35 acres. Area of filter beds, one-seventh of an acre. Total cost of works, 120,000*l.*

Plumstead and Woolwich.—Source of supply, an artesian well in the chalk. Number of houses supplied, about 3,000. Gross quantity supplied per day, 550,000 gallons. Aggregate nominal steam-power, 35 horses. Length of main and branches, 16 miles. Area of subsiding reservoirs, one-fifth of an acre. Area of covered reservoirs for filtered water, one-third of an acre. Total cost of works, 50,000*l.*

Some of the above-named companies were in 1856 without subsiding reservoirs, some without open reservoirs, and some without closed reservoirs; but very extensive additional works have been constructed between 1856 and 1865. Taking them one with another, the companies have spent about 20*l.* in works for each house supplied with water; and the interest on this amount, together with the annual working expenses, are considered in determining the annual water-rate charged upon each house.

At the Woolwich and Charlton works, established so recently as 1854, the water is softened by Dr. Clark's process, in which the chalk is rendered soluble; and the result, as stated by the inspectors to the Board of Health, is beneficial in regard to health, comfort, and economy. The inspectors made the following remarks on one still-existing source of impurity in the water supplied, notwithstanding the use of filter-beds and covered reservoirs :—" The only remaining serious cause of contamination will be the cisterns, water-butts, and other means for storing the supplies now furnished by the companies. Although considerable improvement has already taken place in the distribution, and the water formerly supplied only on alternate days is now for the most part given daily, except on Sundays, to every part of each company's district, its storage even from day to day in the private butts and cisterns of most houses, and especially in those of the poorer class, to a great extent destroys the advantages that so much pains have been taken to secure. The only complete remedy for this serious defect will be the *constant supply*—that is to say, a supply obtained at all times by direct communication of the house-service pipes with the constantly-charged mains of the companies, thus avoiding the necessity for any means of storage whatever on the private premises. The constant supply would be the means of rectifying also another serious defect to which the public is not unfrequently liable in the present system—viz. the irregularities of supply. Notwithstanding that the companies have abundant means of furnishing any quantity of water that can be legitimately used throughout their districts, loud complaints are too often heard of a want of water in certain localities. The deficiency would appear to arise not from actual lack of water, but from some irregularity from time to time in 'districting' the service, which the constant supply would obviate. We would allude also to the great advantage of constantly-charged mains in case of fire as no small consideration."

The charge for water-rate by eight out of the ten companies (omitting the Kent and the Woolwich) for the year 1856, as given in one of the parliamentary papers, was as follows:—

New River	£156,367
East London	85,286
West Middlesex	72,165
Grand Junction	52,590
Chelsea	43,071
Southwark and Vauxhall	41,914
Lambeth	34,645
Hampstead	10,580

Taking an average of the whole, this gives a gross payment, by housekeepers and manufacturers, of about 1s. for 2,800 gallons of water, of 233 gallons for 1d.

We have no concern here with the controversy respecting the Trafalgar Square Fountains as matters of taste or art; but as they are connected in a small degree with the supply of the metropolis with water, a paragraph concerning them may not be out of place. The wells which supply the fountains also supply some of the government offices. In 1843 Messrs. Easton & Amos were employed to sink wells for this purpose to the level of the springs beneath the London clay. The fountains were not intended to be merely ornamental—they were to form cooling-ponds to condense the steam of the pumping-engines, the resistance of the air to the ascending jets producing a cooling action. The Government were encouraged to undertake this work by the ascertained fact that the

interest on the cost of the new works would be less than the water-rate
paid by or for the government offices about the neighbourhood of Whitehall. The works were commenced on a piece of ground in Orange Street,
behind the National Gallery. A well was sunk to the depth of 174 ft.
A cast-iron pipe, 15 in. diameter, was then driven through 30 ft. of
plastic clay, and 10 ft. into a stratum of gravel, sand, and stones.
Within this another pipe of 7 in. diameter was driven through 35 ft. of
green-coloured sand, and 3 ft. into the chalk. Boring was then continued
to a depth of 300 ft. from the surface. The chief supply of water thence
obtained came from the chalk. A second well was then sunk in the
enclosure in front of the National Gallery, to a depth of 168 ft., and a
pipe and a boring continued nearly as in the former instance, but to a depth
of 383 ft. The springs were found to be stronger than those in the well
in Orange Street. A tunnel, 6 ft. diameter, and about 400 ft. long, was
driven to connect the two wells, at a depth of 123 ft. below Trinity high-
water mark. The works were finished in December 1844, at a cost of
8,400*l.* The water rose to within 90 ft. of the surface, and was found to
be of good quality. When the engine was pumping 110 gallons per
minute, it lowered the water only 4 ft. in the well. In 1846, a further
demand for water having arisen, a larger pump was substituted, capable
of raising 350 gallons per minute. In 1849, another well was sunk in
Orange Street, to a depth of 176 ft., and a tunnel was made to connect it
with the others. The steam-engine works one double-acting pump for
supplying the fountains, and two others for raising water from the
springs into the tanks above the building. The pumping of 600 gallons
per minute lowers the water 20 or 24 ft.; but then the level remains
permanent, however long the pumping may be continued. In the
beginning of 1859 the level of the water in the wells was found to be
nearly as it had been throughout.

In a paper by Mr. P. W. Barlow, read before the Institution of Civil
Engineers in 1855, the water-bearing strata of the London basin were
described, with a view of showing how abundant is the supply of water
available, if proper means were adopted for utilising it. This basin,
defined by a boundary running through or near Folkestone, Hythe,
Ashford, West Farleigh, Sevenoaks, Reigate, Godalming, Pewsey,
Devizes, Swindon, Wantage, Tetworth, and Cambridge, covers an area
of 8,000 square miles; and the water-bearing strata beneath this area
comprise the London clay, the chalk, the upper greensand, and the
lower greensand. The superficial area through which rain infiltrates
west of London, and from which the supply for the artesian wells of
London is obtained, is about 24 square miles. About 200 square miles
of area eastward of London, where the lower beds of clay are arenaceous
and permeable, but where the upper or impervious beds are wanting,
add very little to the supply of the ordinary London basin. Of the

3,800 square miles of chalk strata exposed to the surface, it is considered that this constitutes the great water-bearing stratum ; and that, in parts where the chalk is 60 ft. or 80 ft. below the surface, there may be several supplies of water, irrespective of each other, at different depths, and applicable to different purposes.   The district south-east of London was stated by Mr. Barlow to be peculiarly adapted for affording a water supply.   Scarcely any of the springs in that part of Kent reappear in the form of surface springs ; they mostly empty themselves into the Thames at low water, from fissures in the bed or banks.   By intercepting these springs in their course towards the Thames, a copious supply of water might be obtained from the chalk.   It was estimated that the drainage area of the water thus wasted is 190 square miles, west of the Medway, which might yield a daily supply of 60,000,000 gallons ; and that 320 square miles, east of the Medway, might yield 100,000,000 gallons per day.   Mr. Prestwich, from an investigation of the greensand, had arrived at a conclusion that those strata might be made to yield 40,000,000 gallons per day, which would probably rise to a height of 120 ft. above the surface ; and he had suggested how desirable it would be, imitating the example furnished by the artesian well at Grenelle, if a similar experiment were tried in some spot of the London basin where the lower infiltrating springs would be intersected.   Mr. Barlow quoted these speculations of Mr. Prestwich, and adduced his own experience as engineer of the South-Eastern and North Kent railways, to support the view that the metropolis ought not to be left mainly dependent on the Thames for its supply of water, seeing that that river is becoming more and more deficient, and that the strata of the London basin comprise so many beds that are water-bearing.   A lengthened discussion took place after the reading of Mr. Barlow's paper, during which many engineers adverted to the uncertainties which have marked the sinking of deep wells near London.   A well-known instance is the one on the road between Kentish Town and Highgate, sunk to obtain a supply auxiliary to that of the Hampstead Waterworks.   The well was begun at a height of 172 ft. above Trinity high-water mark, with the hope of reaching water in the chalk at about 320 ft. below the surface.   The chalk was compact, hard, free from fissures, and comparatively dry, with the flints not in layers, but distributed through the mass.   After sinking 218 ft. in the chalk, boring was commenced, and continued to a depth of 1,150 ft. below the surface.   After traversing 538 ft. of chalk, and then layers of gault and red clay, the greensand was reached ; but so discouraging was the result, owing to the scanty indications of water, that the enterprise was regarded almost as hopeless.   It was further continued, however, to a depth of 1,302 ft., and then abandoned, after two years and a half of labour, and a very heavy expense.

It does not seem very probable, now that the ten water companies havo invested such large sums in enlarging and improving the water supply of the metropolis, that any further radical changes will be made in this matter. But the future will have to settle the question. It is supposed that more than 50,000,000 gallons of water per day are at the present time (1865) taken out of the Thames near Hampton and Thames Ditton, in aid of the London supply; and it has still to be ascertained what effect that large draught will have on the general condition of the river. Hence such inquiries as those instituted by Mr. Prestwich and Mr. Barlow, and mentioned in the last paragraph, are important. We may notice, too, a project which bears relation to the chalk strata near Grays, in Essex, opposite Northfleet. In excavating the chalk for shipment, 2,000,000 gallons of water per day require to be pumped away into the river; and it has been proposed to utilise this water for the supply of towns, instead of thus allowing it to run to waste. Purfleet, Rainham, Brentwood, Dagenham, Ilford, Romford, Barking, East Ham, and other towns between Grays and London, might (it is conceived) be thus supplied. Messrs. Easton and Amos have estimated that, if 220,000*l.* were expended in the necessary works, a supply of 2,000,000 gallons every twelve hours might be obtained; 4,000,000 gallons by an expenditure of 268,000*l.*; and 6,000,000 gallons by an expenditure of 475,000*l.*

APPENDIX No. 8.

The Great Waterworks from Loch Katrine to Glasgow.

Without going into the subject of the water supply of our great towns generally, as developed between the years 1854 and 1865, we deem it desirable to notice those for the supply of Glasgow, unparalleled as they are in this country for engineering grandeur and public success.

The works were commenced in 1856, after lengthened inquiries in the preceding years. The old supply from the Clyde at Dumbarton had become quite inadequate. A plan had been proposed for bringing water from Loch Lubnaig, but had fallen to the ground. Loch Katrine was then named; and after a very favourable report from two distinguished engineers, since deceased, Mr. Brunel and Mr. Robert Stephenson, the corporation commenced that magnificent work which has become an honour to the city and to the engineer. The water comes from the mountain-lakes on the borders of Stirlingshire and Perthshire. The sources of supply are Loch Katrine, 8 or 9 miles long, with an area of 3,000 acres; Loch Vennachar, 4 miles long, with an area of 900 acres; and Loch Drumkie, with an area of about 150 acres. The three lakes would, if quite full, contain 1,600,000,000 cubic feet of water. The ba-ins which find their drainage into these lakes cover an area of 45,800 acres, and have an average rain-fall of about 80 in. per annum. The works at Loch Katrine are so managed that all the water for 4 ft. above the ordinary summer level, and 3 ft. below it, can be treated as a reservoir or store, with proper channels for drawing it off; this store is equal to 50,000,000 gallons per day for 120 days without rain. The water from Lochs Vennachar and Drumkie is chiefly appropriated to the supply of mill-owners, fishermen, and others interested in certain rivers; the supply for Glasgow being mainly obtained from Loch Katrine. At the outlets provision is made for the discharge of floods as well as for the daily regulated supply, and for securing the passage of salmon and other fish by properly constructed "salmon-ladders." Loch Katrine being 360 ft. above the river level at Glasgow, there is scope for a gentle descent the whole way, and still leaving a pressure of 70 ft. or 80 ft. above the highest summit of land within the city. The whole length of aqueduct from Loch Katrine to Glasgow is about 34 miles, 10 or 11 of which consist of ridges of very hard rock, forming spurs of Ben Lomond. Through these ridges, in a tolerably straight line, the aqueduct is carried, principally by tunnelling. The tunnels are 8 ft. in diameter, and have a fall of 10 in. to the mile. Across several deep and wide valleys the water is conveyed by cast-iron pipes, 4 ft. in diameter, with a fall of 5 ft. per mile. The average inclination of the whole is 2 ft. per mile. Near Mugdock Castle, about 8 miles from Glasgow, a reservoir has been constructed, 70 acres in extent, and capable of containing 500,000,000 gallons. From this reservoir, the top-water of which is 311 ft. above the sea, the water flows to the city through two lines of cast-iron pipes, 3 ft. in diameter. Of the 26 miles which lie between Loch Katrine and the reservoir, 13 miles are tunnelling, 3¾ miles iron piping, while the remainder, where the ground has been cut open, is an arched aqueduct, 8 ft. in diameter, with the same angle of descent as the tunnels. Where the ground has been thus excavated, it has been filled in again over the aqueduct, which is covered throughout, and the surface restored to its original condition.

It will at once be seen from this description that many of the works must be very heavy. There are 70 distinct tunnels, upon which 44 vertical shafts have been sunk. The greatest tunnel, a mile and a half long, just out of Loch Katrine, is worked through the hardest gneiss and mica slate; and five out of the twelve shafts sunk for working it are 500 feet deep. The tunnel just before entering the great reservoir is also a mile and a half long, and is worked in whinstone. In some places, where the mica slate is largely mixed with quartz veins, the rock is so obdurate that the progress did not exceed 3 linear yards in a month, although the work was carried on day and night. In tunnelling the mica slate, the progress was generally 5 yards per month. In drilling

the holes for blasting the rock, a fresh drill or chisel was required fc
every inch in depth; and 60 drills were constantly kept working at onc
There are twenty-five important iron and masonry aqueducts over rivei
and ravines, some 60 or 80 ft. in height, with arches of 30 to 90 ft. spar

It was known when the plans were first laid that the hard rock wa
likely to be free from water, and that, therefore, the working, thoug
slow, would not be retarded by harassing inbreaks of springs. N
water occurred in any of the tunnels or workings in mica slate or cla
slate. When the works emerged from the slate rocks and entered th
old red sandstone, tunnelling was avoided as much as possible; but eve
in this formation water was met with in much less quantity than had bee
anticipated.

The works were inaugurated by the Queen on the 14th of October, 185?
when her Majesty and the royal family were *en route* from Balmoral t
Windsor. The royal party drove from Holyrood to the Loch, embarke
on a small steamer, and went to the mouth of the tunnel by which th
water finds its exit, and which was, of course, gaily decked out for th
occasion. The ceremony was a very simple one. The Queen turned
small tap, which set in motion a 4-horse hydraulic engine at the mout
of the tunnel, and this raised the great iron shutters which permitte
the water to enter the tunnel on its journey of 34 miles to Glasgow. I
a few graceful words before leaving, her Majesty said:—"It is with mucl
gratification that I avail myself of this opportunity of inaugurating
work which, both in its conception and its execution, reflects so mucl
credit upon its promoters, and is calculated to improve the health an
comfort of the vast population which is rapidly increasing round th
great centre of manufacturing industry in Scotland. Such a work i
worthy of the spirit of enterprise and the philanthropy of Glasgow, an
I trust that it will be blessed with complete success."

With a wise foresight, the corporation provided for a much large
supply than is at present needed. They obtained powers to drav
50,000,000 gallons per day from the lakes, and constructed all the work
necessary thereto; but their present actual requirements very little excec
20,000,000 gallons. Mr. Bateman, the engineer, in a letter to th
Builder, in 1862, draws attention to the wonderful advantages whicl
have followed the introduction of such a copious supply of pure sof
water:—"The saving to the inhabitants in soap, and other articles o
private and trading consumption, is estimated at about 40,000*l*. pe
annum, equal to a free gift to the city of 1,000,000*l*., being a sum greatei
than the cost of the whole works. The saving of soap alone in trading
establishments, such as bleach and print works, is from one-half to five
eighths of the quantity which was previously used. These facts cannoi
be too strongly impressed on the minds of the public, as they show the
importance and economy of *soft water* supplies.

INDEX.

PRINTED BY JAMES S. VIRTUE, CITY ROAD, LONDON.

· RUDIMENTARY TREATISE

ON THE

DRAINAGE

OF

DISTRICTS AND LANDS.

BY G. DRYSDALE DEMPSEY, C.E.,

AUTHOR OF
"THE PRACTICAL RAILWAY ENGINEER,"
ETC. ETC. ETC.

THIRD EDITION,
REVISED AND GREATLY EXTENDED.

London:
JOHN WEALE, 59, HIGH HOLBORN.

1859.

LONDON:
BRADBURY AND EVANS, PRINTERS, WHITEFRIARS.

PREFACE TO THE FIRST EDITION.

A FEW years since, the subject of the following volume would have been considered scarcely a necessary theme for one of a series of works intended to bear a popular as well as a technical character. The entire subject would have been deemed sufficiently disposed of by describing the subterranean works of the navigator and the bricklayer, and the sub-aquatic and rude operations of the ditcher. Now, however, our subject occupies a prominent position in the public thought, and may be regarded as nearly a new branch of practical art, based, or to be based, upon principles of science, and essential to the health, life, and morality of our race.

Urged, almost insensibly, by the strong earnestness and foresight of a few leading minds, the British public and Legislature have been brought to discern the urgent need of reforming the substructures of their dwellings and highways, and to feel affrighted at the dangerous apathy in which they and their ancestors have hitherto innocently indulged. The rudeness of our past practice is indeed the subject of our astonishment; the facts adduced are but the pictures of our individual experience, and the simplicity of the principles now first recognised brings them home to us with all the familiarity of things known long ago. We

now wonder at the folly of digging holes beneath houses
for the accumulation of filth, till the surrounding ground
becomes overcharged, and the bulk demands periodical
removal.   We can see clearly enough that the pursuits of
the scavenger are offensive equally to common sense and
to common decency, and admit, without further proof, the
sanatory axiom, that the infusion of the refuse of a town in
the water which serves at once the libations and ablutions
of its people, is *not* adapted either to perfect the purity of
the liquid, or promote the health of the human system.
And in that great branch of the subject which is devoted to
agricultural practice, by which the farmer is endowed with
all the valuable experience of the most intelligent inquirers,
and taught the art of economising the natural resources
of his streams and watercourses, and the fructifying products of his farm-yard, Drainage has acquired a well-
recognised value in the estimation of the scientific public,
and is daily recording results of the highest practical
character.

   The general principles which are now commonly entertained upon the subject of Drainage, maintain its primary
value as a branch of sanatory science, and its claim to be
regarded among the paramount duties of every civilised
Legislature.   District Commissions are disbanded as incapable of achieving the great purposes which the health of
the people demands, and which can no longer be entrusted
in the hands of incompetent authorities.   Cleanliness and
health are now considered in the relation of cause and
effect; and the first requirements of the physician's success are admitted to consist in the constructive conditions
of the patient's dwelling.   Medical philanthropists have
explored the hidden horrors of our metropolis and towns,
and shown that the open sewer and the offal heap are the

contaminators of the rich, and the agents of death to the poor. And, akin to these public pestilences, we are now made aware that a cesspool, or an imperfect drain in a house, is to be reckoned only as a means of gathering fever and disease; and that the cleansing of the rooms above, while one of these radical abominations is sending forth its putrid gases from below, is but an illustration of the ancient error of rectifying secondary, in mistake for primary, evils.

While thus the subject of Drainage is attaining a commanding importance among the social necessities of our times, a corresponding occasion has arisen for its thorough examination as a branch of practical science. Its principles are, or must be, determined, and its rules thence deduced and embodied among the vital applications of the useful arts. The future works of the engineer, the architect, and the builder, must be regulated by considerations of the available methods of securing ample water-supply and efficient drainage; and these considerations will present themselves with that imperative character which they derive from the public will, and which cannot be countervailed by any scruples of private economy, or any opposition of corporate prejudice.

To collect carefully the records of the experience of the past, to compare results and deduce principles with critical anxiety, and upon these principles to establish practical rules, propounded and illustrated with exactness and fidelity, will doubtless become the common object of many of the students of practical art. The following elementary treatise cannot attempt so extensive and laborious a range, but it aims at accomplishing a general survey of the subject, and a brief enumeration of the details which properly belong to it.

PREFACE TO THE SECOND EDITION.

In the Introduction to the first edition of this little book-
a congratulation was ventured upon the interest in its sub,
ject taken by the public at that time. That that interest is
still a growing one (albeit scarcely yet very fruitful in good
works), is now an admitted sign of the times. Commissions and Boards—metropolitan and provincial—are hard
at work, doing their best, and stayed only by professional
discordance, or limitation of resources. The demand for
a second edition of the Rudimentary Treatise on Drainage
should be deemed gratifying to the public, as an evidence
of the progress of the subject, rather than complimentary
to the author as an acknowledgment of his ability in
treating it.

Care has been taken to incorporate a record of the most
recent results, by which the book is considerably augmented
in size, while the indicial headings to the pages may, it is
hoped, assist in referring to the facts and details.

DRAINAGE.

1. DRAINAGE is the collecting and conveying away refuse
waters, and other matters, from *lands, towns,* and *buildings.*
It ascertains the means and methods of accomplishing
these purposes in the most complete manner; and, as *water*
is the principal agent in all cleansing processes, the means
required for insuring its supply are among the necessary
provisions of efficient drainage. By simply extending the
same means, the supply of water may be made adequate to
satisfy all other purposes; and it hence becomes desirable
to include among the objects of drainage the entire supply
of water for towns and buildings, and for the irrigation of
lands. *Sewers* are among the essential means of town
drainage, and therefore have to be so considered, and their
positions, forms, sizes, and modes of construction duly ascertained. Our subject thus embraces several matters which
may be treated separately, but which are properly branches
of the art of draining, and cannot be consistently studied
and usefully applied without a full appreciation of their
several and intimate connections.

2. Beyond the limits of the subject of draining as defined (1), it is also to be extended to the ultimate disposal
of the refuse matters which it has first to remove from
streets and dwellings: and one of its most important duties
is to effect this disposal in such a manner that human health
shall not be thereby impaired; and, moreover, that the

B

matters removed shall be made available to the utmost in
promoting the fertility of the land, and effecting all chemical purposes for which they are the best fitted.

3. The synopsis of the several heads under which we
propose to arrange our facts, principles, and rules, is the
following :—

DRAINAGE.

DIVISION I.—DRAINAGE OF DISTRICTS AND LANDS.
DIVISION II.—DRAINAGE OF TOWNS AND STREETS.
DIVISION III.—DRAINAGE OF BUILDINGS AND DWELLINGS.

DIVISION I.

SECTION I.—Sources of Water.—Natural and Artificial Sup-
ply.—Rain, Ocean, Rivers, Streams, Springs, &c.—Seasons,
Evaporation, Temperature, &c.—Quantity required.—Na-
ture of Soils and Crops, and position of Districts.—Quali-
ties of Water.—Rain Water, Sea Water, River and other
Waters—Four kinds of Impurities.—Modes of Purifying.—
Subsidence. — Filtration. — Chemical Process. — Natural
Filters.

SECTION II.—Upper and Lower Districts.—River-watered
and Sea-coast Districts.—Reclamation of Land.—Modes of
Draining, Pumping, &c. — Water-wheels, as applied for
Draining and supplying Upland Districts.

SECTION III. — Means of conveying, distributing, and
discharging Water.—Drains and Watercourses ; Forms,
Sizes, and Methods of Construction.—Implements employed.—Shallow and Deep Draining.—Stone, Tile, Earthenware, and Brick Drains, &c.

DIVISION II.

SECTION I.—Classification of Towns according to Position and Extent.—Varieties of Surface Levels and Inclinations.

SECTION II.—Supply of Water.—Public Filters and Reservoirs, &c.

SECTION III. — Width and Direction of Roads and Streets; Substructure and Surface.—Paving and Street Cleansing.

SECTION IV.—Main Sewers; Proportions and Dimensions, Inclinations, Forms, and Construction.—Upper and Lower Connections.—Means of Access and Cleansing.— Adaptation for Street Cleansing, &c.

SECTION V.—Conveyance of Water.—Piping, Aqueducts, Reservoirs. — Pumping Apparatus, Steam Draining and Pumping &c.

DIVISION III.

SECTION I.—Classification of Buildings.

SECTION II.—Supply of Water Levels.—Constant Service.—Quantity required.—Cisterns.—Reservoirs.—Filters. Valves and Apparatus.—Piping, &c., &c.

SECTION III.—Varieties of Manufactures, and best available Methods of Draining.—Arrangement of Separate and Collective Drains.—Proportion of Area of Drain to Cubic Contents of Dwelling-Houses.—Fall of Drains.—Mode of Construction.—Connection with Main or Collateral Sewers. —Means of Access, &c., &c.

SECTION IV.—Water-Closets; Arrangement and Construction.—Adaptation to various circumstances.—Combined Arrangements for efficient House Drainage.—Miscellaneous Apparatus and Contrivances.

GENERAL SUMMARY AND CONCLUSION.

DIVISION I.

DRAINAGE OF DISTRICTS AND LANDS

SECTION I.

Sources of Water.—Natural and Artificial Supply.—Rain, Ocean, Rivers, Streams, Springs, &c.—Seasons, Evaporation, Temperature, &c.—Quantity required.—Nature of Soils and Crops, and position of Districts.—Qualities of Water.—Rain Water, Sea Water, River and other Waters.—Four Kinds of Impurities.—Modes of Purifying.—Subsidence.—Filtration.—Chemical Process.—Natural Filters.

1. Water is indispensable to animal existence and health. The means of obtaining, treating, and economizing this vital liquid are therefore among the most important objects of human art. The several sources, primary and secondary, of water, are the ocean, rivers, streams, lakes, subterranean collections or springs, and rain. Some or other of these sources are at our command, to some extent, in every region of the habitable globe. The applicability of the first-named four sources is limited by the geographical position of the district; the latter two of them are obtainable nearly everywhere. The ceaseless cycle of operations by which the waters on the earth and of the ocean mingle with the atmosphere by the medium of evaporation, and, descending in the forms of rain and dew, sprinkle the surface, and again unite through streams and rivers in their common reservoir, is one of the most beautiful and interesting illustrations of the compensating principle of the economy of Providence.

2. In adopting the terms *natural* and *artificial* supply as contradistinguished, it may appear that the former should apply commonly to all the sources enumerated, except the subterranean. We would, however, limit the term *natural* supply to rain and dew, since all the other sources require more or less of artificial means before they are generally available for the purposes of man. Thus the water of the ocean must undergo chemical change or distillation,

and that of rivers requires artificial channels and conduits for its distribution. These, therefore, like the subterranean, for which wells and borings are necessary, have to be classed among the artificial sources of water.

3. The quantity of evaporation from land-surface is evidently more limited than that from water-surface, the one depending upon the retentive power of the super-soil, and the facility for capillary action, while the other arises from a source comparatively inexhaustible. The rate of the process is controlled by temperature, and accelerated in proportion to the heat acting upon the surface; the temperature being affected by the elevation, and reduced in proportion as the elevation increases. The joint result of these conditions is, that proximity to the sea, the river, or the lake, promotes the natural supply of water in the form of rain. The geography of the district, therefore, affects the facility or difficulty of the natural supply.

But another consideration also affects this supply, viz., the superficial features of the district. Thus a mountainous character, augmenting the surface exposed to oblique showers, increases the quantity received on the one side, and diminishes that on the other; and the sides of a valley, in like manner, receive more or less than the quantity due to a level district.

4. The natural supply is, moreover, modified in effect by the structure of the surface on which it falls. Thus, upon a rock-surface (such as that presented by mountains), which resists percolation, the rain collects in masses, floods itself through a fissure, or wears a channel along the line of the most pervious formation, and reaches the lower plains in the formidable rush of a mountain torrent. And as, generally, the effect of the natural supply of water is in proportion to the comparative impermeability of the soil, it follows, that the value of this supply in any district is further conditional on the structural character of the adjacent districts. Thus, from a higher impermeable district it will receive, and to a lower more permeable district it will give.

Natural supply is hence, in effect, determined by the geographical situation, the superficial character, and the geological structure of a district, modified also by the structure of the surrounding district.

5. The quantity of rain that falls annually at several places, has been observed, and recorded as follows :—

In England, the mean annual depth of the eight years 1836 to 1843, both included, was 26·61 inches, having varied between the extremes of 21·1 and 32·1 inches. The average annual fall at some other places has been recorded as follows:—

> South Carolina   .   .   . 50 inches.
> Bombay (mean of 10 years)   .   . 78
> Brazil (in 1821)   .   .   . 280
> Cumana   8

Humboldt has assigned the fall of rain to vary with the latitude, being greatest at the equator, and diminishing towards the poles in the following ratio: viz., 96 inches annually in the equatorial zone, 80 inches to latitude 20°, 29 inches to latitude 45°, and 17 inches to latitude 60°.

6. The quantity of rain thus varying, with some reference to the latitude, also to the position of the district in relation to the sea, and varying also from one year to another, is further affected by the season. Thus, the mean fall per month on an average of eight years in some districts of England has been recorded as fluctuating from 1·617 inch in March to 3·837 inches in November; the fall in each month being as follows :—

TABLE I.

		Inches.
January .	. . . . .	1·847
February	. . . .	1·971
March .	. . . .	1·617
April .	. . . .	1·456

Carried forward..... 6·891

	Inches.
Brought forward......	6·891
May	1·856
June ⌐	2·213
July	2·237
August	2·427
September	2·639
October	2·823
November	3·837
December	1·641
	26·614

7. This monthly quantity, being the mean of eight years, does not by any means indicate the monthly proportion for any one year, the variation being as great between the same months of different years, as it is from one year to another, or indeed from one latitude to another. Thus, during the eight years over which these observations extended, the quantity of rain falling in each month was as follows :—

TABLE II.

MONTH.	YEARS.							
	1836.	1837.	1838.	1839.	1840.	1841.	1842.	1843.
	Ins.	Ins.	Ins.	Ins.	Ins.	Ins.	Ins.	Ins.
January	2·40	2·40	0·31	1·40	3·95	1·50	1·36	1·46
February	2·04	2·85	2·65	1·45	1·32	1·02	2·02	2·42
March	3·65	0·75	1·55	1·92	0·34	1·65	2·20	0·88
April	2·57	1·32	1·35	1·65	0·34	1·85	0·47	2·10
May	0·70	0·94	0·84	1·22	2·62	1·68	1·85	5·00
June	1·80	1·86	2·85	3·31	1·33	3·00	2·00	1·56
July	2·29	1·30	2·35	4·36	1·68	2·80	1·93	2·09
August	2·24	3·00	0·95	3·65	1·90	3·62	1·40	2·66
September	2·60	1·38	2·47	3·22	2·31	4·00	4·50	0·63
October....	4·55	1·55	2·68	1·68	1·ʋ0	4·40	1·41	4·82
November.........	3·95	2·05	3·55	4·40	4·25	4·28	5·77	2·45
December.........	2·21	1·70	1·58	3·02	0·40	2·30	1·52	0·40
Quantity in } each year ... }	31·00	21·10	23·13	31·28	21·44	32·10	26·43	26·47

The greatest and least quantity falling in each month during this period is thus stated:—

TABLE III.

MONTH.	Maximum.	Minimum.	Difference.
	Inches.	Inches.	Inches.
January	3·95	0·31	3·64
February	2·85	1·02	1·83
March	3·65	0·34	3·31
April	2·57	0·34	2·23
May.................................	5·00	0·70	4·30
June	3·31	1·33	1·98
July.................................	4·36	1·30	3·06
August..............................	3·65	0·95	2·70
September	4·50	0·63	3·87
October	4·82	1·41	3·41
November...........................	5·77	2·05	3·72
December...........................	3·02	0·40	2·62

The third column shows the difference between the greatest and least fall in each month during the eight years, and thus represents the relative variableness of each month's rain. It thus appears that the fluctuation is least in February, and greatest in May.

8. As evidence of the great difference of quantity of rain which falls in similar latitudes, we may quote the following observations referring to the upland districts about Manchester, which we have compiled from Tables given by Mr. Homersham in his "Report on the Supply of Surplus Water to Manchester, &c."* These observations were made at eight stations during the four years 1844, 1845, 1846, and 1847; and at five other stations during the year 1847 only. The first column gives the name of the station at which the observations were made; the second shows its elevation above the mean level of the sea; the next five columns contain the depth of the rain in inches and decimal parts for each year, and mean depth of the four years; and the last column gives the names of the observers.

* Weale. 1848.

TABLE IV.

STATION.	Elevation in feet.	Depth of rain in inches during the years					OBSERVERS.
		1844.	1845.	1846.	1847.	Mean.	
Fairfield	220	26·35	38·90	30·20	40·75	34·05	Mr. J. Meadows.
Bolton	320	34·63	48·11	40·82	52·32	43·97	„ H. H. Watson.
Newton	350	..	..	..	34·69	34·69	„ J. Meadows.
Rochdale	500	34·41	51·64	42·04	51·72	47·45	„ J. Ecroyd.
Marple...............	531	29·40	38·80	32·35	43·70	36·06	„ J. Meadows.
Todd Brook Reservoir ..	620	..	..	..	38·39	38·39	„ Ditto.
Comb's Reservoir	720	42·70	51·10	38·10	51·30	45·80	„ Ditto.
Belmont, Sharples	820	50·00	55·00	49·80	61·40	54·05	„ J. Magnall.
Woodhead Tunnel	1000	..	..	..	33·12	33·12	„ J. Meadows.
Chapel-en-le-Frith	1121	33·00	43·80	38·80	44·00	49·90	„ Ditto.
Whiteholme Reservoir, Blackstone edge	1500	24·80	39·80	37·10	35·70	34·35	„ R. Mathews.
Brinks	1500	..	..	..	29·50	29·50	„ J. Meadows.
Comb's Ridge	1670	..	..	..	35·85	35·85	„ Ditto.
Mean of each of the four years at eight stations		34·41	45·89	38·65	47·61		
Mean of the one year at thirteen stations		..	..	..	42·49		

Among the many observations made upon this subject, it
must, however, be admitted that we have not yet the means
of instituting any very satisfactory comparison. To do this
we require careful observations carried on for a long series
of years, at stations selected for the purpose, and with apparatus of the same construction.

9. Any attempt to describe the several fluctuations which
are observed in the quantity of rain falling, or to explain the
causes of these fluctuations, beyond the few leading circumstances we have noticed, would involve an elementary
inquiry into the phenomena of rain far exceeding the limits
of these pages. But we may quote a few words from the
celebrated Dalton, which will be fully suggestive to the
studious mind in this interesting department of meteorological science. "The cause of rain, therefore, is now, I consider, no longer an object of doubt. If two masses of air
of unequal temperatures, by the ordinary currents of the
winds, are intermixed, when saturated with vapour, a precipitation ensues. If the masses are under saturation, then

less precipitation takes place, or none at all, according to the degree. Also, the warmer the air, the greater is the quantity of vapour precipitated in like circumstances." " Hence the reason why rains are heavier in summer than in winter, and in warm countries than in cold." *

10. The depth of rain which falls is ascertained by receiving it in a vessel of some form with a gauge connected, in which the depth may be accurately measured; but no instrument of the kind yet devised can be considered as entirely satisfactory in its action, or as giving results which will allow estimates of perfect correctness to be thence formed. The rain-gauge used by Dr. Dalton for a series of experiments extending from 1795 to 1819, or later, consisted of " a funnel of 10 inches diameter, and the top surrounded by a perpendicular rim of 3 inches high, to prevent any loss by the spray ; it was fixed in a proper frame with a bottle for the water, and stood above 2 feet above ground." Dr. Garnett, in 1795, suggested the addition to this simple form of gauge, of a cup inverted over the mouth of the bottle, and adapted to receive closely the neck of the funnel, so as to prevent the passage into the bottle of any water striking against or condensing upon the outer surface of the funnel. Gauges which have been subsequently used for many observations consist of a hollow cylinder of copper or other metal, 7 or 8 inches in diameter, and from 30 to 40 inches in length, with a perforated funnel or colander of the same diameter, fitted within the cylinder a few inches below the top. A float is placed within the cylinder and fitted with a staff which passes upward through a hole in the funnel, and, standing above the cylinder. serves to indicate the depth of rain accumulated within the cylinder. Experiments, with an apparatus adapted for the purpose, and called a *staff gauge*, have shown that the prolongation

---

* Memoirs of the Literary and Philosophical Society of Manchester; vol. iii., second series, 1819, p. 507. Several valuable papers, with detailed observations made during long series of years, by Dr. Dalton and others, are to be found in these Memoirs.

of the staff above the cylinder collects a great quantity of rain, and thus shows a greater depth than is due to the surface of the cylinder. This might be obviated by using a cylinder of glass inclosed in a suitable case, or a metal cylinder fitted with a glass panel, for observing the position of the float inside, and dispensing with the staff altogether. The apparatus must be partly sunk in the ground within a strong case, to prevent injury, and capable of being readily taken out when required.

11. The effectiveness of rain for all purposes of water-supply and drainage can be estimated only after determining the deduction due to the process of *evaporation*, by which the larger part of it is raised from the surface on which it has fallen, and, in the form of vapour, mingles with the atmosphere, to be again precipitated upon the earth and ocean. The proportion evaporated appears to be mainly dependent upon the temperature, heat promoting the process, and cold retarding it. The highest, lowest, and mean temperature in each month have been observed to be as follows :—

TABLE V.

MONTH.	THERMOMETER.		
	Highest.	Lowest.	Mean.
January	52·0	11·0	36·1
February	53·0	21·0	38·0
March	66·0	24·0	43·9
April	74·0	29·0	49·9
May...	70·0	33·0	54·0
June	90·0	37·0	58·7
July...	76·0	42·0	61·0
August.......................................	82·0	41·0	61·6
September	76·0	36·0	57·8
October......................................	68·0	27·0	48·9
November...................................	62·0	23·0	42·9
December...................................	55·0	17·0	39·3

And, accordingly, we find the proportion of rain evaporated corresponds with the temperature recorded thus :—being the mean evaporation of each month during the eight *years 1836 to 1843, and stated at per cent. upon the quantity of rain.

TABLE VI.

MONTH.	Evaporation per cent.	Remainder per cent.
January	29·3	70·7
February	21·6	78·4
March	33·4	66·6
April	79·0	21·0
May	94·2	5·8
June	98·3	1·7
July	98·2	1·8
August	98·6	1·4
September	80·1	13·9
October	50·5	49·5
November	15·1	84·9
December	00·0	100·0
Mean	57·6	42·4

The remainder stated in the second column shows the per centage upon the total quantity falling which is available for human purposes.

12. Besides the main condition of temperature, other minor circumstances affect the proportion of rain which passes from the surface in the state of vapour, and have to be considered in forming an estimate, from these records, of the available quantity of rain-water in any district. These minor conditions are chiefly the *structure* and the *state* of the supersoil and of the subsoil. Thus, if the structure be of an impermeable character, the water will lie upon the surface, while evaporation takes up more than its average quantity, being hindered only by the provision which may exist for passing the rain immediately to a more porous surface. On the other hand, a soil of excessive permeability will imbibe the water rapidly, and thus reduce the amount of evaporation. The *state* of the soil affected

by the frequency and extent of the showers will, moreover, determine in some degree the relative quantities of rain-water evaporated and retained. Thus, if the soil has acquired excessive hardness from long drought, or become super-saturated by excessive rain, evaporation will proceed more rapidly than percolation, and the effect of the fall be similarly diminished.

13. The average quantity remaining to filter through the soil, or to be made use of for the purposes of man, may be computed from the following Table, No.VII., which shows the average monthly fall during the same period of eight years as stated in Table No. I., the quantity evaporated, and the quantity remaining, in inches.

TABLE VII.

MONTH.	RAIN.		
	Total falling.	Evaporated.	Remaining.
	Inches.	Inches.	Inches.
January	1·847	0·540	1·307
February	1·971	0·424	1·547
March	1·617	0·540	1·077
April	1·456	1·150	0·306
May	1·856	1·748	0·108
June	2·213	2·174	0·039
July	2·287	2·245	0·024
August	2·427	2·391	0·036
September	2·639	2·270	0·369
October	2·823	1·423	1·400
November	3·837	0·579	3·258
December	1·641	0·164	1·805
	26·614	15·320	11·294

14. Of the quantity remaining and available, 11·294 inches per annum, it is desirable to notice the proportion due to the season. Thus, during the months of January, February, March, and October, the quantity is nearly uniform, varying only between 1·077 and 1·547. In the

month of December it rises to 1·805, while in November
it averages a depth of 3·258 inches. During the six con-
secutive months of April, May, June, July, August, and
September, the quantity remaining is comparatively small,
being always less than half an inch in depth. The following
Table, No. VIII., shows the monthly quantity in cubic feet
and weight of water remaining on each superficial acre, as
computed from the preceding Tables.

TABLE VIII.

MONTH.	Rain-water permanently deposited per acre.	
	Quantity in cubic feet.	Weight in tons.
January	4744	132·
February	5616	156·
March	3910	109·
April	1111	39·
May	392	11·
June	142	4·
July	87	2·42
August	131	3·61
September	1339	37·
October	5082	141·
November	11826	328·
December	6552	182·
	40932	1145·03

15. The following observations on evaporation and filtration,* for which we are indebted to the patient and care-
fully-conducted experiments of Mr. Charles Charnock, of
Holmfield House, near Ferry-Bridge (one of the Vice-
Presidents of the Meteorological Society of London), present some valuable facts for consideration.

* Quoted by J. H. Charnock, Esq., Assistant Commissioner under the
Drainage Acts, in a paper " On Suiting the Depth of Drainage to the Circumstances of the Soil," given in the Journal of the Royal Agricultural Society, vol. x. pt. ii. pp. 515 to 518.

TABLE IX.—An Account of Observations made, through a series of Five Years, at Holmfield House, near Ferrybridge, in the County of York, by Charles Charnock, Esq., with a view to determine the amount of Evaporation and Filtration under the several circumstances on the Magnesian Limestone Soil.

Months.	1842 Rain (1) On the Surface.	1842 Evaporation From Water (2) Exposed to both Sun and Wind.	1842 Evaporation From Water (3) Shaded from Sun, but exposed to Wind.	1842 Evaporation From Soil (4) When Drained.	1842 Evaporation From Soil (5) When Saturated.	1842 Filtration (6) Through the Soil from the Drain 3 ft. deep.	1843 Rain (1) On the Surface.	1843 Evaporation From Water (2) Exposed to both Sun and Wind.	1843 Evaporation From Water (3) Shaded from Sun, but exposed to Wind.	1843 Evaporation From Soil (4) When Drained.	1843 Evaporation From Soil (5) When Saturated.	1843 Filtration (6) Through the Soil from the Drain 3 ft. deep.
January	2·70	1·69	1·13	1·66	1·59	1·04	1·48	2·57	1·71	0·69	1·87	0·79
February	0·76	1·23	0·81	0·68	1·04	0·08	3·25	2·65	1·10	2·29	2·78	0·96
March	3·48	1·92	1·28	2·40	2·52	1·08	0·95	3·05	2·03	0·72	2·43	0·23
April	1·51	2·98	1·99	1·11	2·31	0·40	2·19	3·22	2·05	1·84	2·39	0·25
May	2·98	4·14	2·76	2·89	3·82	0·09	2·81	2·91	1·94	2·47	2·65	0·34
June	1·94	4·18	2·79	1·94	4·27	—	2·31	5·12	3·41	2·10	4·86	0·21
July	3·74	4·16	2·73	3·26	3·89	0·48	2·70	3·76	2·50	2·55	3·49	0·15
August	1·49	3·36	2·24	1·37	2·39	0·12	3·99	3·71	2·57	3·77	6·74	0·22
September	2·44	2·30	1·74	2·24	2·88	0·20	1·07	2·06	1·34	0·90	2·18	0·17
October	1·12	2·00	1·37	0·92	1·99	0·20	1·10	1·80	1·20	0·82	1·93	0·28
November	3·19	2·26	1·50	2·49	1·03	0·70	2·30	1·64	1·09	1·69	1·48	0·67
December	0·76	3·00	2·14	0·60	1·49	0·16	0·28	1·68	1·78	0·97	1·39	0·01
Totals	26·11	33·61	22·48	21·56	30·02	4·55	24·49	34·17	22·72	20·11	31·19	4·28

EXPLANATION.—Column 1.—Shows the Depth of Rain fallen, as registered by the ordinary Rain-Gauge.

Column 2.—Is the Amount of Evaporation from a Surface of Water fully exposed to both Sun and Wind.

Column 3.—Is the Evaporation from Water, shaded from the Sun, but exposed to the Wind.

Column 4.—Is the Evaporation from what represented drained or dry land.

Column 5.—Is the Evaporation from the same when saturated.

Column 6.—Is the Amount of Water which filtered through the soil.

Into a leaden vessel, of a foot square and three feet deep, was put two feet of gravel and calcareous sand, so as to represent the substratum of the farm, and the remainder filled up, to within an inch of the top, with an average quality of soil. At the bottom a pipe was inserted, which conveyed all the water which was filtered through into a bottle, which was regularly emptied and re-gistered. The vessel was inserted in the ground to within an inch of the surface, keeping the level of the soil, inside and out-side, alike, with an inch of the vessel above, to prevent any communication of water from without. The soil was kept free from weeds, and occasionally stirred, that it might not be more than ordinarily compact.

TABLE IX.—continued.

Months	1846 Rain (1) On the Surface	1846 Evaporation From Water (2) Exposed to both Sun and Wind	1846 Evaporation From Water (3) Shaded from Sun, but exposed to Wind	1846 Evaporation From Soil (4) When Drained	1846 Evaporation From Soil (5) When Saturated	1846 Filtration (6) Through the Soil from the Drain 3 ft. deep	1845 Rain (1) On the Surface	1845 Evaporation From Water (2) Exposed to both Sun and Wind	1845 Evaporation From Water (3) Shaded from Sun, but exposed to Wind	1845 Evaporation From Soil (4) When Drained	1845 Evaporation From Soil (5) When Saturated	1845 Filtration (6) Through the Soil from the Drain 3 ft. deep	1844 Rain (1) On the Surface	1844 Evaporation From Water (2) Exposed to both Sun and Wind	1844 Evaporation From Water (3) Shaded from Sun, but exposed to Wind	1844 Evaporation From Soil (4) When Drained	1844 Evaporation From Soil (5) When Saturated	1844 Filtration (6) Through the Soil from the Drain 3 ft. deep
January	2·18	2·07	1·58	1·22	1·94	0·96	1·74	1·53	1·02	1·28	1·49	0·46	1·31	1·61	1·08	0·85	1·50	0·46
February	0·47	2·59	1·69	0·14	2·09	0·03	0·73	0·71	0·47	0·43	0·94	0·30	2·22	1·31	1·08	1·66	1·11	0·56
March	0·93	2·56	1·55	0·86	2·16	0·07	1·88	2·91	1·94	1·33	2·89	0·55	2·27	2·13	1·42	1·58	1·50	0·69
April	5·97	1·91	1·27	2·98	1·49	2·99	1·54	4·79	3·19	1·45	4·06	0·39	0·27	3·83	2·55	0·27	3·42	—
May	0·82	4·52	3·02	1·58	3·89	0·07	2·24	2·84	1·99	1·97	3·26	0·27	0·42	5·77	3·85	0·42	4·86	0·04
June	1·63	4·08	3·25	2·74	4·73	0·07	3·18	3·10	2·06	2·93	2·98	0·25	1·24	5·31	3·58	1·20	4·95	0·33
July	2·90	4·44	2·96	2·46	4·39	0·16	3·49	2·86	1·99	8·30	2·79	0·19	2·76	4·17	2·78	2·43	4·28	0·41
August	2·65	3·00	2·05	1·00	3·28	0·19	4·61	2·56	1·70	4·24	2·43	0·37	2·85	4·70	3·14	2·44	4·63	0·30
September	1·07	2·99	1·99	2·40	3·14	0·07	1·36	2·79	1·86	0·45	2·29	0·41	1·92	4·91	3·28	1·62	3·96	0·24
October	4·09	2·23	1·49	0·83	2·76	1·69	3·36	2·77	1·84	2·69	2·82	0·67	1·41	2·79	1·86	1·17	2·93	0·51
November	1·15	1·63	1·09	2·40	1·74	0·07	1·01	2·13	1·41	0·73	2·20	0·28	1·98	2·84	1·80	1·47	2·78	0·06
December	1·36	1·79	1·10	1·09	1·67	0·27	3·04	3·57	2·38	2·36	2·87	0·60	0·35	0·79	0·53	0·29	0·93	—
Totals..	25·24	34·69	23·04	18·30	33·28	6·76	28·18	32·56	21·75	23·26	31·09	4·92	19·00	40·16	28·75	15·40	37·85	3·60

In these experiments, the evaporation from *saturated* soil was determined thus:—" A leaden vessel of 13 inches deep, and a foot square, was filled to within an inch of the top with soil, and placed in the ground, in the same manner as the previous vessel, with a pipe level with the surface of the soil to carry off the excess of top-water into a receiver. The same quantity of water was then daily supplied to this soil as the evaporating dish of column 2 showed was evaporated. The soil was stirred as in the former case, and thus represented wet and undrained land."

" In the first place, it is observable how much greater is the amount of evaporation from water than from land, and how near, as shown by columns 2 and 5 (Table IX., pp. 15, 16), the evaporation from wet land is to that from water itself— hence the wetter the land the greater the evaporation, and, as the well-known consequence, the greater its excess of coldness. We have a familiar illustration of nature's process in this particular, in the method often adopted to cool our wine on a hot summer's day, by wrapping a wet napkin round the bottle and exposing it to the full sun: as the moisture from the napkin is evaporated, the temperature of the wine declines to almost freezing point. The school-boy's experiment of producing ice before a fire, by incasing the vessel in wet flannel and adding a portion of salt to the water, is a similar example, with this additional lesson to the farmer—that to apply certain limes to wet land is only increasing the evil.

" You will then, in the second place, notice how much less the evaporation is in the shade than in the sun, and consequently that wet land must be the warmest when there is the least sun. From which cause, no doubt, arises that too vigorous growth of young wheat, so often observable on such land in the winter and spring months, which never fails to produce serious injury to the crop in all its subsequent stages. And thirdly, you will remark how comparatively small a proportion of the rain which falls is shown to be carried off by filtration. Taking the average of the five years' experiments, it will be seen that only 4·82 inches out of 24·6 inches of rain passed through the land to the depth of three feet. We might, therefore, be led at the first glance to infer that land in general stands less in need of drainage, or may be drained by a less perfect system, than is supposed to be requisite, did not daily experience oppose such a conclusion. We must, therefore, endeavour to reconcile this seeming incongruity, and deduce at the same time, from the facts disclosed, such data as may guide us in determining the essential requisites to ensure completeness of effect in drainage.

" Now, although there can be no reason to question the accuracy of the experiments on filtration made by Mr. Dickinson, and recorded in the Journal of the Royal Agricultural Society of England, Vol. V., Part I., yet there is a very considerable difference in the aggregate result, as shown by them and the account before us. ' The first important fact disclosed,' says the commentator, page 148, ' is, that of the whole annual rain, about 42½ per cent., or $11\frac{3}{10}$ inches out of $26\frac{6}{10}$,* have filtered through the soil :' whereas in the Holmfield House experiments there is only shown, as we have already said, 4·82 inches out of 24·6, or about $5\frac{1}{10}$ per cent. against 42½ per cent. This is certainly a very great and somewhat irreconcileable difference in the result of two experiments made professedly to ascertain the same fact. Now, on referring to the ' Memoirs of the Literary and Philosophical Society of Manchester,' Vol. V., Part II., you will find a paper on rain, evaporation, &c., from the pen of the celebrated Dr. John Dalton (the father of the science of Meteorology), wherein he explains a series of experiments made by himself and his friend, Mr. Thomas Hoyle, jun., to ascertain the amount of evaporation and filtration, and giving the following Table of results: †—

" ' Having got a cylindrical vessel of tinned iron,' says the Doctor, ' ten inches in diameter and three feet deep, there were inserted into it two pipes, turned downwards,

* See Table VII. p. 13.   † See Table X. p. 19.

TABLE X.

MONTHS.	Water through the Two Pipes.			Mean.	Mean Rain.	Mean Evaporation.
	1796.	1797.	1798.			
January	1·897	·680	1·774	1·450	2·458	1·008
February	1·778	·918	1·122	1·273	1·801	·528
March	·431	·070	·335	·279	·902 ·	·623
April	·220	·295	·180	·232	1·717	1·485
May...............	2·027	2·443	·010	1·493	4·177	2·684
June	·171	·726	...	·299	2·483	2·184
July...............	·153	·025	...	·059 ·	4·154	4·095
August............	...	...	·504	·168	3·554	3·386
September	...	·976	...	·325	3·279	2·954
October	...	·680	...	·227	2·899	2·672
November	...	1·044	1·594	·879	2·934	2·055
December.........	·200	3·077	1·878	1·718	3·202	1·484
	6·877	10·934	7·379	8·402	33·560	25·158
Rain	30·629	38·791	31·259			
Evaporation	23·725	27·857	23·862			

for the water to run off into bottles : the one pipe was near the bottom of the vessel, the other was an inch from the top. The vessel was filled up, for a few inches, with gravel and sand, and all the rest with good fresh soil. Things being thus circumstanced, a regular register has been kept of the quantity of rain-water that ran off from the surface of the earth through the upper pipe (whilst that took place), and also of the quantity of that which sank down through the three feet of earth, and ran out through the lower pipe. A rain-gauge of the same diameter was kept close by, to find the quantity of rain for any corresponding time.'

" You will notice that the general result of these experiments accords pretty nearly with that of the Holmfield account; and yet it may be readily conceived that circumstances of situation and stratification may often occasion as wide a difference in the amount of filtration as is shown between Mr. Dickinson's and Mr. Charnock's observations.

" On an examination of the *details* registered in the
account before us, it will be evident that the amount of
filtration is not exclusively dependent on the fall of rain;
but that a variety of other causes combine to affect its proportion. For instance, in March, April, May, June, and
July, of 1842, the fall of rain was 13·65 inches, and the
filtration for the same period was only 2·05 inches; whilst
in April, 1846, there was 5·97 of rain and 2·99 of filtration.
Similar instances are also noticeable in Mr. Dickinson's
details. From March to October, inclusive, of 1840, a fall
of 11·52 inches of rain is recorded, without any filtration;
but in November, 1842, the rain was 5·77, with 5 inches of
filtration. Dr. Dalton's table also shows the same variations. The lesson, therefore, derivable from these experiments, so far as regards filtration by drains, is one rather of
a speculative than of a definite character; for, although we
are assured filtration must be seeured, we are left with a
large and varying margin as to the proportion. We must
not, however, overlook the fact, that all the registered details show occasionally an amount of filtration nearly equal
to the rain that falls, and, therefore, in determining the
size of pipe to be used, the ready exit of this *maximum*
quantity must be provided for."

16. The precise *quantity* of water *required* for the agricultural purposes of any district depends upon the nature of
the soil and the crops, and the position of the district in
relation to the surrounding country. Thus, if a permeable
soil occupy an elevated site, the water deposited upon it
will pass rapidly, and perhaps before serving for the germination or the nutriment of the plant. If, on the other
hand, as is the far more common case in this country, the
soil be of a retentive character, and the site low in relation
to other districts, the water will be kept while the soil becomes saturated to so great an extent that the processes of
vegetable germination and growth are greatly impeded.
The soil exists in one of three conditions; 1st, in the form
of clay, being a dense mass consisting of finely comminuted
particles, but all of a highly tenacious kind; in a state of

slight moisture it becomes a clammy paste, and is never found so utterly devoid of moisture that its constituent particles are separable: it affords no passages for water, receiving it with difficulty, and retaining it in the same way. 2nd, in the form of sand or gravel, the particles of which are seldom or never united, and the soil is therefore full of passages or canals for water. Soil of this kind has no power either to oppose the admission or effect the retention of water poured upon it. And 3rd, existing in the form of a mixture of the aluminous, silicious, and calcareous elements, in endless variety of proportions, found as *clods*, and in this state affording two classes of passages for the ingress and permeation of water, viz., those remaining between the particles which are congelated in each clod, and those formed by the spaces between the clods. The former are sometimes called *pores*, and the latter *canals*. The power of admitting and retaining or discharging water exerted by these mixed soils, will exist in an endless variety of degrees, according to the mechanical formation of the constituent particles and clods. The state of soil which is most favourable for the germination and development of the plant is that of *moistness*, capable of being readily crumbled by the hand, and equally removed from the adhesive extreme of *mud* and the volatile one of *dust*. In this condition it will be found that the *pores* are filled with water, but the *canals* are not—these latter serving as passages for the air, which is one of the feeders of vegetable life; and we can, therefore, readily understand that, when water exists in such quantity that the soil is saturated with it, and all the pores or canals filled, its condition is unhealthy for the growth and development of plants.

The following extract, from an admirable lecture on Agricultural Science, by Dr. Madden, quoted by the General Board of Health in their " Minutes of Information," although of considerable length, claims a space here, for the valuable information it conveys on the fitness of soil for promoting vegetable germination.

" The first thing which occurs after the sowing of the

seed is, of course, *germination;* and before we examine how
this process may be influenced by the condition of the soil,
we must necessarily obtain some correct idea of the process
itself. The most careful examination has proved that the
process of germination consists essentially of various
chemical changes, which require for their development the
presence of air, moisture, and a certain degree of warmth.
Now it is obviously unnecessary for our present purpose
that we should have the least idea of the nature of these
processes : all we require to do, is to ascertain the conditions
under which they take place; having detected these, we
know at once what is required to make a seed grow. These,
we have seen, are air, moisture, and a certain degree of
warmth; and it consequently results, that wherever a seed
is placed in these circumstances, germination will take
place. Viewing matters in this light, it appears that soil
does not act *chemically* in the process of germination ; that
its sole action is confined to its being the vehicle by means
of which a supply of air and moisture and warmth can be
continually kept up. With this simple statement in view,
we are quite prepared to consider the various conditions of
soil, for the purpose of determining how far these will influence the future prospects of the crop, and we shall
accordingly at once proceed to examine carefully into the
mechanical relations of soil.

17. Soil, examined mechanically, is found to consist entirely
of particles of all shapes and sizes, from stones and pebbles
down to the finest powder; and on account of their extreme
irregularity of shape, they cannot lie so close to one another as to prevent there being passages between them, ·
owing to which circumstance soil in the mass is always
more or less *porous.* If, however, we proceed to examine
one of the smallest particles of which soil is made up, we
shall find that even this is not always solid, but is much
more frequently porous, like soil in the mass. A considerable proportion of this finely-divided part of soil, the *im-
palpable matter* as it is generally called, is found, by the aid
of the microscope, to consist of *broken-down vegetable tissue,*

so that when a small portion of the finest dust from a garden or field is placed under the microscope, we have exhibited to us particles of every variety of shape and structure, of which a certain part is evidently of vegetable origin.

18. On examining a *perfectly-dry* soil we perceive that there are two distinct classes of pores: 1st, the large ones, which exist *between* the particles of soil; and 2nd, the very minute ones, which occur in the particles themselves; and whereas all the larger pores—those between the particles of soil—communicate most freely with each other, so that they form canals, the small pores, however freely they may communicate with one another in the interior of the particle in which they occur, have no direct connection with the pores of the surrounding particles. Let us now, therefore, trace the effect of this arrangement. If the soil is *perfectly dry*, the canals communicating freely at the surface with the surrounding atmosphere, the whole of these canals and pores will, of course, be filled with air. If, in this condition, a seed be placed in the soil, you at once perceive that it is freely supplied with air, *but there is no moisture;* therefore, when soil is *perfectly dry*, a seed cannot grow.

19. Let us turn our attention now to that state of the soil in which water has taken the place of air, or, in other words, the soil is *very wet*. If we observe our seed now, we find it abundantly supplied with water, but *no air*. Here again, therefore, germination cannot take place. It may be well to state here, that this can never occur *exactly* in nature, because water having the power of dissolving air to a certain extent, the seed is in fact supplied with a *certain* amount of this necessary substance; and, owing to this, germination does take place, although by no means under such advantageous circumstances as it would were the soil in a better condition.

20. We pass on now to a different state of matters. Let us suppose the canals are open and freely supplied with air, while the pores are filled with water. While the seed now has. quite enough of air from the canals, it can never be

without moisture, as every particle of soil which touches it
is well supplied with this necessary ingredient. This, then,
is the proper condition of soil for germination, and, in fact,
for every period of the plant's development; and this condition occurs when soil is *moist* but not *wet*—that is to say,
when it has the colour and appearance of being well
watered, but when it is still capable of being crumbled to
pieces by the hands, without any of its particles adhering
together in the familiar form of mud.

21. Let us observe still another condition of soil; in this
instance, as far as *water* is concerned, the soil is in its
healthy condition—it is moist, but not wet, the pores alone
being filled with water. But where are the canals? We
see them in a few places, but in by far the greater part of
the soil none are to be perceived; this is owing to the particles of soil having adhered together, and thus so far obli-
terated 'the interstitial canals that they appear only like
pores. This is the state of matters in every *clod of earth*;
and you will at once perceive, on comparing it with a stone,
that it differs from it only in possessing a few pores, which
latter, while they may form a reservoir for moisture, can
never act as vehicles for the *food* of plants, as the roots are
not capable of extending their fibres into the interior of a
clod, but are at all times confined to the interstitial canals.

22. With these four conditions before us, let us endeavour
to apply them *practically* to ascertain when they occur in
our fields, and how those which are injurious may be
obviated.

The first of them, we perceive, is a state of too great
dryness, *a very rare* condition, in this climate at least; in
fact, the only case in which it is likely to occur is in very
coarse sands, where the soil, being chiefly made up of pure
sand and particles of flinty matter, contains comparatively
much fewer pores, and, from the large size of the individual
particles, assisted by their irregularity, the canals are wider,
the circulation of air freer, and, consequently, the whole is
much more easily dried. When this state of matters exists,
the best treatment is to leave all the stones which occur on

the surface of the field, as they cast shades, and thereby prevent or retard the evaporation of water.

23. We will not, however, make any further observations on this very rare case, but will rather proceed to a much more frequent, and, in every respect, more important condition of soil: an *excess of water*.

When water is added to perfectly dry soil, it of course, in the first instance, fills the interstitial canals, and from these enters the pores of each particle; and if the supply of water be not too great, the canals speedily become empty, so that the whole of the fluid is taken up by the pores: this, we have already seen, is the *healthy* condition of soil. If, however, the supply of water be too great, as is the case when a spring gains admission into the soil, or when the sinking of the fluid through the canals to a sufficient depth below the surface is prevented, it is clear that these also must get filled with water so soon as the pores have become saturated. This, then, is the condition of *undrained soil*.

24. Not only are the pores filled, but the interstitial canals are likewise full; and the consequence is, that the whole process of the germination and growth of vegetables is materially interfered with. We shall here, therefore, briefly state the injurious effects of an excess of water, for the purpose of impressing more strongly on your minds the necessity of thorough-draining, as the first and most essential step towards the improvement of your soil.

The *first* great effect of an excess of water is, that it produces a corresponding diminution of the amount of air beneath the surface, which air is of the greatest possible consequence in the nutrition of plants; in fact, if entirely excluded, germination could not take place, and the seed sown would, of course, either decay or lie dormant.

Secondly, an excess of water is most hurtful, by reducing considerably the *temperature* of the soil: this I find by careful experiment to be to the extent of $6\frac{1}{2}$ degrees Fahrenheit in summer, which amount is equivalent to an elevation above the level of the sea of 1950 feet. So that,

c

supposing two fields lying side by side, the one drained, the other undrained, and supposing them both equally well cultivated, there will be nearly as much difference in the amount and value of their respective crops, as if the drained one was situated at the level of the sea, and the other had an elevation as high as the most lofty of the Pentland Hills.* But, besides this, and what is nearly equally bad, the temperature is rendered unnaturally high during winter; whereas it has been proved that one great source of health and vigour in vegetation is the great difference which exists between the temperature of summer and winter, which difference amounts in dry soil to between thirty and forty degrees, while in soil very much injured by an excess of water, the whole range of the thermometer throughout the year will probably not exceed from six to ten degrees.

These are the two chief injuries of an excess of water in soil which affect the soil itself. There are very many others affecting the climate, &c.; but these are not so connected with the subject in hand as to call for an explanation here.

· 25. Of course all these injurious effects are at once overcome by thorough-draining, the result of which is to establish a direct communication between the interstitial canals and the drains, by which means it follows that no water can remain any length of time in these canals without, by its gravitation, finding its way into the drains.

26. Too much cannot be said in favour of pulverising the soil; even thorough-draining itself will not supersede the necessity of performing this most necessary operation. The whole valuable effects of ploughing, harrowing, grubbing, &c., may be reduced to this: and almost the whole superiority of *garden* over *field* produce is referable to the greater perfection to which this pulverising of the soil can be carried.

* Of course the field of high elevation must be *thoroughly* drained to equal even the *undrained* field at the level of the sea.

The celebrated Jethro Tull has the honour of having first directed the farmer's attention forcibly to this subject; and so deeply impressed was he with its infinite importance, that he believed the use of manure could be entirely superseded were this pulverising carried to a sufficient extent.

The whole success of the drill husbandry is owing, in a great measure, to its enabling you to stir up the soil well during the progress of your crop; which stirring up is of no value beyond its effect in more minutely pulverising the soil, increasing, as far as possible, the size and number of the interstitial canals.

27. Lest any one should suppose that the contents of these interstitial canals must be so minute that their whole amount can be of but little consequence, I may here notice the fact, that in moderately-well pulverised soil they amount to no less than one fourth of the whole bulk of the soil itself: for example, 100 cubic inches of *moist* soil (that is, of soil in which the pores are filled with water while the canals are filled with air) contain no less than 25 cubic inches of air. According to this calculation, in a field pulverised to the depth of 8 inches, a depth perfectly attainable on most soils by careful tillage, every imperial acre will retain beneath its surface no less than 12,545,280 cubic inches of air. A familiar illustration of the space occupied by the spaces between the particles of loosened soil is afforded by the fact that when soil is disturbed it more than fills the space it previously occupied.

28. Taking into calculation the weight of soil, we shall find that with every additional inch which you reduce to powder (by ploughing, for example, 9 inches in place of 8), you call into activity 235 tons of soil, and render it capable of retaining beneath its surface 1,568,160 additional cubic inches of air. And, to take one more element into the calculation, supposing the soil were not properly drained, the sufficient pulverising of an additional inch in depth would increase the escape of water from the surface by upwards of 100 gallons a day.

c 2

29. The great purpose of draining being, immediately, the improvement of the land, but, ultimately, the promotion and improvement of vegetable production, the preceding considerations as to the fitness of the soil for germination may be well followed by a brief enumeration of the rules for the application of water to plants, which, as laid down by De Candolle, refer

First, to the *quality* of the water used : that it should be well aërated ; the presence of atmospheric air is good, but of carbonic acid gas much better. The next qualities desirable are, that it should contain fertilising matters ; the water should be as little muddy as possible; the temperature of the water is of importance, especially for hot-house plants : the water used in hot-houses is allowed to stand for some time before it is employed, in order that it may have the temperature of the place; it is well that other water employed should stand for a time in the sun.

Second, to the *times* of the application:—In the winter time there should be little irrigation, because the plants are then dormant, and water is then superabundant. In spring time water is usually abundant. In summer it is wanting; and at that time the water should be given in the evening.

Third, to the *quantity* of the water to be applied, which should be varied according

a. To the *object* of the culture :—When for leaves, more water should be given than when for flowers ; less water should be given when for grains or fruits.

b. To the *depths* of the *roots:*—The application should be more frequent to the plants of which the roots are superficial; less frequent to deeper roots.

c. To the *structure* of the *foliage :*—Those which evaporate much (such as plants with large leaves), more frequently than perennial, or plants with thick leaves.

d. To the *consistence* of the *stalks*, and of the *roots:*—Roots with fleshy fibres do not thrive if too abundantly watered ; at the same time they are injured by dryness. Tuberculous or bulbous plants, or plants with fleshy leaves, can bear a long-continued dryness, and therefore infrequent, yet abundant, waterings suit them well.

e. To the *stage* of vegetation :—It is important to bear in mind that young germinating plants require light and frequent waterings; those that are in the height of growth abundant waterings; and when the fruit or seed is being matured the waterings should be infrequent. Those that have been transplanted require abundant watering.

f. To the *nature of the soil,* according to which these rules must be modified :—The lighter the soil the more frequent and plentiful must be the waterings. If it is a compact and clayey soil less watering will be required.

g. To the *state* of the *atmosphere :*—It will be readily conceived that the watering must be more frequent when the temperature is high, the sky clear, and the air dry, and during drought."

30. The *proper* serving of water for agricultural purposes, similarly with that for domestic and manufacturing uses, requires both adequate supply and discharge. That is, if the natural supply be deficient it becomes the business of the drainer to augment it; if excessive, to reduce it: but, in either or any case, his correlative object is to provide sufficiently for the discharge of the water as rapidly as vegetation has imbibed its nutriment from it, and the supply is replenished. The recognition of this essential principle founded the era of all modern improvements in the Art of Draining. The most skilful tenders of the soil were previously satisfied to drain the *surface* of the land. So long as they were enabled by superficial channels to get rid of the excess of water which appeared above ground, their work seemed to them complete, and the effects of subterranean reservoirs and aluminous sponges, made visible in stunted and unhealthy vegetation, were attributed to any causes rather than the true ones. In our third section, which will show in detail the methods of determining and

forming drains, these causes and the mode of treating them will be explained.

31. The facility or difficulty attending the artificial supply of water from rivers and subterranean sources depends upon the distance of the sources from the districts to be watered, their relative levels, and the geological structure of the soil. The map of the district will exhibit the first of these circumstances, and a corresponding section will show the second; the third may be inferred, with more or less exactness, from the superficial features of the country, but can be ascertained with certainty only by boring through the superincumbent strata until the spring or internal reservoir is arrived at.

32. The water of rivers is not generally available for the supply of the lands of districts of superior level, the expense of applying mechanical power for this purpose being too great to admit of extensive operations. For the supply of towns and buildings, however, this consideration is outweighed by the importance of the object. Pumps worked by water, steam, or other power, are applied to raise the water; and artificial channels, such as aqueducts or pipes, provided for its conveyance, with tanks or reservoirs for containing it, and submitting it to any desired operations of cleansing or purifying. If the district to be supplied lie on a level, near to that of the feeding river, the reservoirs are usually constructed contiguous to it, and receive and cleanse the water before its conveyance through pipes or other conduits. Thus, several of the companies who now supply water to London and its suburbs have reservoirs for these purposes on the banks of the river Thames, whence their supply is derived.

33. For the irrigation and draining of adjacent lands on accessible levels, the waters of rivers are conducted by artificial courses. The system of irrigation adopted in Lombardy is complete for this purpose, and the principle of it is illustrated in figures 1 and 2; *a a* is the feeder to

Fig.1.

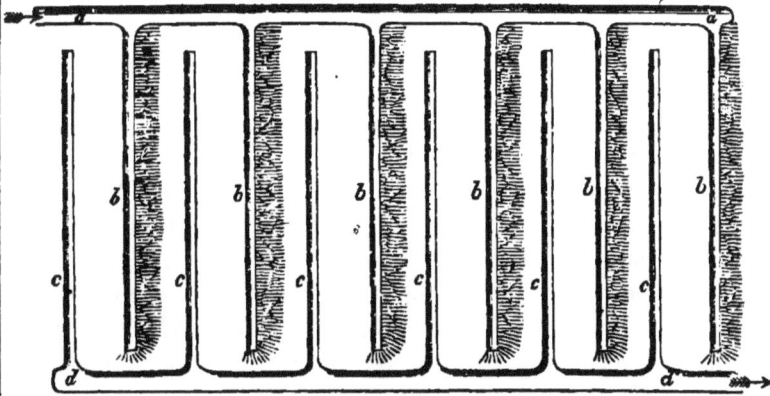

Fig.2.

Fig.3.

Fig.4.

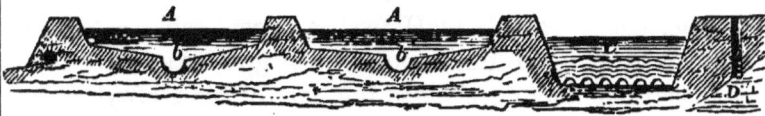

b b, the irrigating channels. The water flowing from these
spreads itself as a veil over the rectangular sections of land
between them, and thence passes into the draining channels *c c*, which are formed at a lower level than the supply
channels, and is received in the common drain *d d*, through
which it passes, and becomes the means of irrigation and
draining to other similar districts in succession. In Lombardy, the width of land between the channels *b* and *c* is
usually about 22 feet, and the difference of level between
the irrigating and the draining channels about 6 inches.
These water-meadows, or *marcite*, are thus irrigated in
summer during several hours about once in each week;
and from the end of September to the end of March the
process proceeds continuously, the water being turned off
only while the grass is being cut.

34. Instances of irrigation by *submersion*, by *filtration*,
and by *regurgitation*, or subterranean irrigation, are mentioned by Count Manetti, the Superintendent of the Royal
Gardens at Monza, near Milan, in these words:—" The
irrigation by submersion is, in Lombardy, limited to rice
fields. Elsewhere, as, for instance, in Tuscany, it is employed to improve the soil by the 'deposit of earthy matter
from the water, whilst in France and Germany it is employed
both for arable lands and meadows, leaving them under
water till a scum appears, which indicates that the crust of
the soil begins to decay. The irrigation adopted in Lombardy for arable and pasture lands, as well as for meadows,
is by filtration, for one could scarcely call submersion that
very thin veil of moving water so skilfully spread over the
land by our irrigators, who, in this point, are the best
agriculturists in the world. The irrigation by regurgitation
(more properly subterranean irrigation) is not in use in
Lombardy; but in Switzerland, in the neighbourhood of
Berne, and especially at Hofwyl, a considerable extent of
land is irrigated in this manner with great success. The
famous Fellenberg reclaimed the bogs of his Hofwyl estate

by the application of subterraneous drains, so contrived that by stopping their mouths when the surface of the soil is too dry, he compels the water to swell back to the roots of the grass. This mode of irrigation is not only adapted to grassy lands after they have been drained, but to every other description of light soils, especially in hot climates. It was common in Persia long ago."

35. A method of irrigation by water-meadows, similar to that described (par. 33), has been adopted in Wiltshire and the neighbouring counties, the soil being thrown up into beds, the water running along the crown of each bed, and discharging on each side.

36. Another method adopted chiefly, until lately, in Devonshire, is that known by the name *Catch-water Meadow* or *Catch-meadow*. In this arrangement the gutters are drawn along natural slopes, and the water falls from the upper one to that immediately below it, which spreads it anew equally over the surface lower down. This system, originating in an almost mountainous country, has been of late years transferred to land quite as level as any on which the ridge-water meadows were made in Wiltshire.

37. Experiments have been recently made upon the economical effect of substituting an apparatus of pipes or flexible hose for irrigating lands, for the channels and gutters previously described. The chief advantages apparent from the use of pipes either of iron or earthenware, or the flexible hose with lateral openings, as water-carriers or conductors over the water-meadows, have been thus recited: —that the surface now occupied by fixed carriers for water-meadows would be saved; that the flexible hose may be carried across depressions or over undulations of surface, with closed lengths, without the expense of permanent works; that with these tubular carriers less water will suffice, and, therefore, less waste from evaporation; that the apparatus may be at once removed for the adaptation of the land to arable or varied cultivation. The hose may

be carried over hedges, ditches, or even small streams.
With a slight covering, it may be carried temporarily over a
road, or under it, through a road-drain.

The distribution by the hose and jet admits of various
modes of appliance with steam-power, from the heaviest
fall of a thunder-shower within the range of the jet, to a
shedding in the shape of a mist by a skilful operator, or
the shedding in various full streams upon the ground.
Horticulturists deem various niceties in watering essential
to good production. In the practice of the new distribution
on a large scale these points of skill appear to have been
little, if at all, attended to; and although crops now
deemed extraordinary were got without them, it is probable
that, with them, further increase of production will be
attained.

38. When the several specified advantages of the pipe-
distribution are considered, they appear to the engineering
inspectors who have examined the different works, to pre-
ponderate greatly, even for common agricultural purposes,
over the cheapest methods of distribution by catch-water
meadows, and to induce a far higher and more profitable
order of cultivation. When, indeed, the question relates
to land which is valuable from its contiguity to a town, and
to a cultivation yielding 10l. or 20l. or more per acre, the
annual cost of the land occupied by the water-carriers
alone exceeds the annual cost of a complete system of iron
pipes. Thus, in the newly-formed water-meadows at Edinburgh, the open gutters occupy about one thirty-seventh of
the area irrigated, and in many places a greater proportionate space is occupied for the purpose. As the rent of
the land is there 20l. per acre, the annual value of the space
devoted to gutters is 10s 9d. per acre, or about equal to
the average annual working expenses, including interest on
capital, of the pipe distribution; and it would be clearly a
saving to the owners of those meadows, and of several
others, to fill up the gutters, to abandon the practice of
distribution by submersion, and to adopt that of pipe dis-

tribution. Besides saving land, filling up the gutters would be removing impediments to the passage of carts, and to numerous other agricultural operations. Where a sufficient fall is obtained for pipe distribution by gravitation, it is cheaper than the common catch-water meadows.

39. The power derivable from the prompt application of plain water to arable cultivation may be said to be unknown to the agriculture of this country, and very little known in that which has heretofore been distinguished as " market garden cultivation ; " and it is only imperfectly practised in horticulture. Wheat has, under some circumstances, been watered with great success. In the market-garden cultivation at Naples and Paris, effects are stated to be produced by skilful watering which are unknown in this country. At Naples the water is distributed by regular channels of irrigation. At the market gardens of Paris, it is skilfully distributed by hand labour, by the use of the scoop, at great expense indeed, but for which the extra produce compensates. Some of the more eminent market-gardeners would, however, appreciate the advantages of any appliances for the cheap distribution of water. A quantity of water equal to the fall from a heavy thunder-shower may be distributed by engine power at an expense not exceeding a few pence per acre.

40. The cheap power of distributing water may often be of importance to the agriculturist, to facilitate the working of the land at those times when it is hardened by drought, and when, for ploughing or other work, extra labour, often more than double the ordinary amount, is necessary. On such occasions, labourers wet the ground to facilitate the working with the spade. When water is available, and when the ground may be thoroughly wetted at a trifling expense, the farmer may, by such an application, work with two horses where otherwise four would be required.

41. Districts considerably above river level, or so situated that no water can be conveyed to them from that source, may be artificially supplied from wells, or by making com-

munication with the lower and saturated strata of the soil.
Thus, the ridges of land lying above the reach of the Nile
were perforated with wells by the ancient Egyptians, as a
substitute for the inundation by which the lower banks
were fertilized. The Chinese also resort to wells for the
purpose of irrigation. In many parts of the East, where
the natural supply by rain is deficient, works on a very
large scale have been constructed for obtaining water sufficient for irrigating. In Hindostan, Japan, China, Java,
Tartary, &c., the supply is, to a great extent, drawn from
wells; and in Bengal, Ceylon, the Carnatic, &c., immense
tanks have been for ages constructed to contain the valuable
liquid.

42. The principal rivers noted for the periodical rising
of their waters, are the Nile, the Ganges, the Euphrates,
and the Mississippi. Of these, the Nile, which flows from
the Jibel Kumri Mountains, begins to rise in June, and by
the middle of August attains an elevation of 24 to 28 feet,
the inundation flooding the valley of Egypt for a width of
12 miles. The Ganges, flowing from the Himalayas, rises
32 feet from April to August, and creates a flood of 100
miles in width. The Euphrates, from Mount Ararat, rises
12 feet between March and June, and covers the Babylonian
plains. The Mississippi, flowing from the Stony Mountains, rises with the melting of the snows from March till
June, forming a vast belt of watered surface. At the distance of 1000 miles from the ocean it is said to rise 50 feet,
while nearer the sea its rise is considerably reduced by the
vast tract which it covers.

43. The periodical rise of river waters gives facilities for
their systematic distribution over the higher districts to a
most important extent by the construction of canals, of
which the ancient Egyptians largely availed themselves.
Their canals, branching in various directions, are said to
have amounted to 80 in number, and to have extended to
60 or 100 miles each in length. Similarly, the great
cavities, called the Lakes of Mœris, Behira, and Mareotis,

are considered to have been reservoirs artificially formed for collecting vast stores of water to be afterwards distributed for irrigation.

44. Among the means of artificial supply, the construction of wells has always been resorted to as a certain method of obtaining water in cases where no other was practicable. Failing rain and rivers, subterranean sources have, from the earliest times, been sought for, the formation of wells being one of the most ancient engineering expedients. The primitive wells of Greece are described as being surmounted with massive marble cylinders; those of Thrace consisted of arched excavations in the sides of rocks, where the water was directly obtainable from springs; the springs of Turkey were converted into fountains, and " the castle of Cairo contains a curious well, sunk in the rock to the depth of 280 feet, and having a circumference of 42 feet. The water of this well filters through the sand from the Nile, and, being impregnated in its passage by salt and nitre, it has a brackish taste."

45. As a modern discovery, or, properly, revival, that of *Artesian Wells* is a valuable source of the artificial supply of water. The *theory* of these wells is simply this:—that water, descending through the permeable strata of the earth, and reaching either a cavity or a bed of spongy materials, will accumulate there if its egress is prevented by an impervious surrounding stratum; then, if an artificial opening or well be made into this water-bearing bed, the water will rise upward in it to a height, and with a force, due to the superior elevation and fertility of the source. Having been long adopted in Artois, these wells have received the modern name of Artesian wells. This contrivance appears to have been introduced into this country from France and Italy about the year 1790. At Mortlake, near the Thames, a well was driven through the clay and sand into a bed of soft chalk to a depth of 375 feet, and a good supply obtained through a bore of 3½ ins. in diameter. The strata penetrated were as follow:—gravel, 20 ft.

London clay, 250 ft.; plastic sands and clays, 55 ft.; hard chalk with flints, 35 ft.; soft chalk, 15 ft.

46. The celebrated Artesian Well at Grenelle, in France; formed under the direction of M. Mulot, is an instance of the difficulties and success of these works. This work occupied eight years and a half of anxious exertion, and the

STRATA.	Total Depth from Surface.	Depth of each Stratum.		Formations of London and Paris Basins.
	Feet.	Feet.		
Gravel and sand	13	13		
Cockle shells ....⎫				
Quartzose sand, with fine particles of sulphuret of iron...............				
Fine sand.................			⎱	Plastic clay.
Argillaceous sand..................... ⎬	180	167	viz. :	
Mottled				
Clay.........................				
Sand and clay, with nodules of limestone ....⎭				
White chalk, with layers of flints ...	328	148	formations,	White chalk.
White chalk, alternating with strata of dolomite and small pieces of silex ....⎬	886	558	London	
Gray chalk with particles of silex ...	1049	163	the	
Gray chalk compact without silex ...	1453	404	with	Gray chalk.
Green chalk and particles of the⎫ silicate of iron			Corresponding	
Blue argillaceous chalk.............. Blue argillaceous and sandy chalk, with particles of mica, and veins of green chalk...............⎬	1634	181		Chalk marl.
Clay with iron pyrites, nodules of⎫ the phosphate of lime and fossil débris				
Green sand .... Clay and greenish sand, with grains of quartz .... Argillaceous sand..................... Green and white sand ....⎭	1794	160		

water first rose on the 26th of February, 1841. The strata
bored through are as shown in the table page 38 — the
depth being measured from the surface :—the second column
shows the depth of each stratum, and the third column ex-
hibits the resemblance between the formations of the basins
of London and Paris.

The sand, in which the water is obtained, continues below
this depth. The boring was commenced at a diameter of
20 inches, and diminished as the tubes descended, so that
at the depth of 576 feet, employing 4 columns of tubing,
the diameter was 12 inches. The fifth column of tubing
reaches 1148 feet, with a diameter of 10 inches. The sixth
reaches 1345 feet, and has a diameter of $8\frac{1}{4}$ inches. The
seventh and last tube reaches 1771 feet, with a diameter
of $6\frac{3}{4}$ inches. The lower 23 feet in the clay are not tubed.
During the progress of the work several accidents of a discouraging nature occurred : the rods and chisels sometimes
became detached, and fell to the bottom. The chisel also,
when in the chalk, sank at one stroke 85 feet, and became
so firmly fixed, that M. Mulot found it necessary to enlarge
the hole on all sides. All difficulties were at length, however, surmounted, and on the day mentioned the rods suddenly sunk several feet; the workmen found that all
resistance had ceased, and that the water-bearing stratum
was attained. After a few hours a column of water rose to
a height of nearly 2000 feet. The subsequent operation of
lining the bore was a work of great importance, in order to
prevent the sides of the hole falling in through any of the
less compact strata, and at the same time prevent the possibility of the water escaping, or the pressure being lost by
any fissures that exist, or may be formed in the strata
through which the water passes. The arrangement of
these tubes requiring a regular diminution of diameter in
the manner of the tubes of a telescope, it is essential that
the relative dimensions be calculated with great exactness,
otherwise the lower tubes are found to become too small to
admit the water, and it is then necessary to remove the

whole of the tubes deposited, and enlarge the bore accordingly.   At Grenelle it was five times necessary to remove
the whole of the placed lining, and enlarge the bore of the
well.   Wrought iron had been employed for lining on previous occasions, but had failed, one remarkable instance of
which may be mentioned.   The water of an Artesian Wel[1]
at St. Cyr, near Tours, rises from the sand beneath the
chalk, and the tubular lining to the well was of iron.   The
supply, however, diminished in every succeeding year, and
M. Bretonneau caused the tubing to be drawn up, which
was $\frac{3}{10}$ths of an inch in thickness, and found well preserved;
but at the joints of the pipes several circular holes were
discovered, two or three centimes in diameter.   This effect
has been accounted for by an assumed electro-chemical
action, but, however caused, it led to the rejection of iron
as a material for the tubing.   Copper tubes $\frac{1}{10}$th of an inch
in thickness have been applied at Grenelle.

47. The supply from the Puits de Grenelle was reported
in 1841 to exceed 880,000 gallons every 24 hours, and the
cost of the work was about £10,000.   Some of the Artesian
Wells sunk at Tours were found to yield less than when
opened.   A greater number, however, have produced an
augmented quantity, and the probability is that the defi-
ciencies have arisen from imperfections in the lining of the
bores.   In the province of Artois, where Artesian Wells
have existed upwards of 300 years, no diminution has ever
been observed.   The subterranean sheet of water which
supplies these, extends over a space of several hundred
square leagues, in comparison with which the outlets to
these wells are almost inappreciable.

48 The deficiency of supply by which Artesian Wells
are rendered inoperative, usually becomes evident before
any very great depth is reached, although, if the water-
bearing stratum happens to crop out at any points however
distant from the boring, the supply is liable to deficiency,
and the pressure necessary to force the water upwards is
also perhaps lost.   Thus, previous to the operations at

Grenelle, just described, a boring was executed at the Jardin des Plantes; but the water never reached the surface, although it rose to within a few feet of it. This fact was afterwards accounted for by the discovery that the sheet of water which supplied this boring, being the same that feeds the fountains of St. Ouen and St. Denis, crops out on the banks of the river Seine, between Chaillot and Saint Cloud. From this sheet, that which supplies the wells at Tours and Elbœuf is separated by the entire chalk formation. M. Champoiseau communicated to the Academy of Sciences, in 1840, the result of experiments he had made at Tours, to ascertain if any connection existed between his Artesian Well and the neighbouring rivers. These experiments were conducted during the months of March, April, and May, while the waters of the rivers were fluctuating, but no corresponding change was found in the well-waters, which did not show any variation either in their quantity or clearness. The temperature of the water of Grenelle was found to be 81°·7 Fahr., and its quality far more pure than that of the Seine, or of Arcueil. From an analysis by M. Pelouze, it appears that 100 cubic inches of the Grenelle water gave only 3·5 grains of extraneous matter, whilst a similar quantity of water from Arcueil or the Seine yielded 4·3 grains mechanically suspended, and 11·6 grains of chemical impurities.

49. Borings similar to those for Artesian Wells have been executed for the purpose of getting rid of superfluous water and liquid matters. An embankment near Val de Fleury, for the left bank Versailles Railway, was drained in this manner by means of *absorbing wells*. A stratum of clay and sand soaked with the rains of the previous year forced the bank from its position, and destroyed the works. Borings were made, the first of which reached 20 yards in depth, where it arrived at the upper part of the chalk, full of fissures, and which speedily absorbed the water. The subsequent borings were carried to 35 and 40 yards, in order to reach the chalky fissures which communicate with the

Seine, and feed the neighbouring wells. Absorbing wells
have also been used in France to dispose of the refuse of
the lay-stalls. M. Mulot, who superintended the Grenelle
Artesian Well, executed a boring for this purpose at Baudy.
Through a perforation 244 feet in depth, two absorbing
strata were obtained, one at a depth from 133 to 155 feet,
in chalk mixed with silex, and the other from 210 to 244
feet, in argillaceous sand, and green and gray sands containing lignites and pulverized shells. By the first 70, and
the latter 140 cubic yards of refuse liquid were absorbed.

50. The question of relative levels as affecting the practicability and expense of draining operations in the raising
and removal of water and even of soils, has been rendered
far less important by the application of steam-power. The
expense of raising 43,000 gallons of water a hundred feet
high by a Cornish engine of 25-horse power is only a shil-
ling; and with an engine of 180-horse power, 80,000 gallons
are lifted for that sum, coal being 12s. per ton. In the
potteries, what is called " slip," that is, clay mixed with
powdered flint and granite, with about one ton and a half
of water to one ton of solid matter, is pumped and distributed;* and there is no doubt that where water is available, and where the operation required is on a sufficiently
large scale, lands might be " clayed" and earths distributed
much more effectually and cheaply by this than by any
other method. The greater weight of the " slip" referred
to, as compared with that of water, increases the labour of
pumping about one-third.

51. The power of water in carrying matters in suspension
is much neglected in agricultural as well as in engineering
operations. Earths may, when properly diluted, be distributed, by the pump worked by steam-power, through a
hose with open apertures, not only at a cheaper rate than
by any other method, but in a far superior mode, being
finely comminuted and evenly spread. In Germany, where

* General Board of Health.—" Minutes of Information."

water can be obtained at a high level, and gravitation can
be used, improvements are effected by the distribution of
earths on an extensive scale, the principle of the mechanical
distribution by hydraulic power being the same as warping.
In Tuscany the large work of the " *bonificamento* " of the
Maremma is a work by means of water-power so applied by
which upwards of two feet in thickness of solid earth has
been spread over forty square miles of country; a mass of
earthwork equal to nearly $82\frac{1}{2}$ million cubic yards, regularly
deposited as if rammed. On an estimate for some work
on a large scale in this country it appeared that the working
expense of spreading clay by means of a hose would be
little more than 2s. 6d. per inch of depth per acre, equal to
134 cubic yards, the expense of carrying and spreading
which, by man and horse-power, would have been very
considerable.

52. The following example of the comparative expense
of removal of earth by cartage and in suspension in water
is given in the sanitary report:—" A contract was about to
be entered into by the West Middlesex Water Company for
hauling out from their reservoir at Kensington the deposit
of eight or ten years' silt, which had accumulated to the
depth of three or four feet. The contractor offered to
remove this quantity, which covered nearly an acre of surface, for the sum of 400l., in three or four weeks. The
reservoir was emptied (of the water) in order to be inspected
by the engineer and directors before the contract was
accepted. It occurred to one of the officers that the cleansing might be accomplished more readily by merely stirring
up the silt to mix it with water; and then, if a cut or outlet
were made in the main pipe used for conveying the water
to London, that it might be washed out. He accordingly
got thirty or forty men to work in stirring up the deposit,
and accomplished the work at the cost of 40l. or 50l., and
three or four days' labour, instead of so many weeks.
When the directors went to see the basin, to decide upon

the contract, the reservoir was as free from any deposit as a house-floor."*

53. On the peculiar qualities of water depends its fitness for agricultural, manufacturing, and domestic purposes. Chemical researches have put us in possession of much valuable knowledge upon this subject, which it behoves the land-worker and the engineer equally to avail themselves of. Pure water, as proved by the early experiments of Priestley and Cavendish, about the year 1780, consists of the two gases, oxygen and hydrogen, in the proportion of 85 parts, by weight, of oxygen, to 15 of hydrogen. This pure liquid can be obtained only by distilling water as it is found in the several states of rain-water, river-water, sea-water, and spring-water. The water obtained from each of these sources contains foreign matters of some kind, the nature and effects of which, as ingredients in the water we have to employ, are well deserving our best attention.

54. Liebig has proved, by experiments made in the laboratory at Giessen, that rain-water contains ammonia. All the rain-water used for these experiments was collected at a distance of 600 paces (south-west) from the town, and while the wind was blowing towards it. Several hundred pounds of the water were distilled in a copper still, and, upon evaporating some of it with muriatic acid, an evident crystallization of sal-ammoniac was observed. The same eminent chemist has fully satisfied himself of the presence of ammonia in snow-water. By evaporating the snow with muriatic acid, crystals of sal-ammoniac were obtained; and from these crystals the ammonia was liberated by adding hydrate of lime. In these experiments Liebig observed that the inferior strata of snow always contained a larger proportion of ammonia than that lying upon the surface.

* Under the old practice, sewers were cleansed from deposit by buckets, and the deposit removed by cartage, at an expense of 10s. per load, by contract.ˈ By means of flushing, or by water, the cleansing and removal were effected at a cost of from 3d. to 8d. per load.

The origin of this ingredient and its utility in the vegetable economy are details of a most interesting study, but we cannot afford space to pursue the inquiry.

55. Sea-water contains, besides carbonic acid, ammonia, &c., the following salts :—according to Marcet,

Chloride of sodium	26·660
Chloride of magnesium . . .	5·152
Sulphate of soda	4·660
Sulphate of lime	1·500
Chloride of potassium . . .	1·232
	39·204

making a total of 39·204 parts of salts in 1000 parts of sea-water. An analysis of the water of the North Sea, made by Clemm, differs slightly from this. Clemm's is as follows :—

Chloride of sodium	24·84
Chloride of magnesium . . .	2·42
Sulphate of magnesia	2·06
Chloride of potassium	1·31
Sulphate of lime	1·20
	31·83

showing a total of 31·83 parts of salts in 1000 parts of sea-water. These salts are, by the constant evaporation from the surface of the sea, floated over the earth and carried down by the rain, thus replenishing vegetation with the salts essential to its growth and existence.

56. The waters obtained from rivers, springs, and wells, are all impregnated, in a greater or less degree, with foreign substances, and also hold others in a state of mechanical suspension. These impurities are of four kinds, viz. :—

1st. The Mechanical.
2nd. The Animal.
3rd. The Vegetable.
4th. The Mineral or Saline.

Although the purification of water from these matters may belong peculiarly to our Second Division, it will be well to consider it under the First, in order to establish correct notions of the qualities of water, whether applied to agricultural, manufacturing, or domestic uses. The process of filtration separates only the first of these. The saline matter contained in water may be distinguished as *neutral* and *alkaline*. The neutral salts are gypsum, common salt, &c. The alkaline portion consists of earthy bicarbonates, such as those of lime and magnesia, and alkaline bicarbonates, such as those of potash and soda. The principal cause of that quality of water, termed "hardness," arises from the presence of the earthy salts mentioned, and sometimes iron-salts; and the same property is evinced if the water contains an excess of carbonic acid. Exposure to the air will diminish the hardness of water, as far as that quality is occasioned by the excess of carbonic acid; and it will have a similar effect, but in a much diminished degree, upon waters which owe their hardness to the presence of the earthy bicarbonates.

57. The economical results dependent upon the qualities of the water supplied to towns are of extreme importance, and therefore deserve attention. Thus, the bicarbonates of lime, &c., affect, to a great degree, the value of water in its application to many manufacturing purposes, and to the production of steam and the heating of pipes for artificial warming. The incrustation of boilers is a well-known theme of consideration in the economy of steam-power, and, moreover, frequently becomes operative as the ultimate cause of accidents, in the case of explosions. In its domestic applications, the properties of water are equally important. The quality of hardness occasions a necessity for a great additional consumption of soap in all the processes of washing and cleansing. And this resistance to the cleansing action becomes, as is universally known, the cause of increased mechanical effort on the part of the operator, and a corresponding increase of wear and injury to the clothes

acted upon. Dr. T. Clark, who has given much attention
to this subject, and is the patentee of a process for testing
the hardness of water, conceives that a very considerable
expenditure arises from these causes in a large town supplied with hard water.

58. For the purpose of comparison, Dr. Clark adopts
the effect of the presence of one grain of chalk in one
gallon of water as a standard, or *one degree of hardness;* and
he gives the results of some of his analyses as follows :—
The hardness of the waters supplied through pipes in
London varies from 11° to 16°, or equal to the effect of 11
to 16 grains of chalk per gallon. The pipe-water of Manchester has 12° of hardness. The water of Glasgow 4°·5.
Of Edinburgh about 5°. Newcastle-upon-Tyne Company's
water, nearly 5°. Thames water near Mortlake had 14°·2
hardness, while the average of many trials upon Thames
water, after conveyance through pipes, gave only 11°·8. The
inference, therefore, is, that it had lost 2°·4 of its original
hardness during its passage and exposure.

59. The outline of Dr. Clark's process may be gathered
from the following abridged extract from the specification
of his patent :—" Chalk forms the bulk of the chemical
impurity that the process will separate from water, and is
the material whence the ingredient for effecting the separation will be obtained. In water, chalk is almost or altogether insoluble, but it may be rendered soluble by either
of two processes of a very opposite kind. When burned,
as in a kiln, chalk loses weight. If dry and pure, only
9 oz. will remain out of 16 oz.; these nine will be soluble
in water, but require 40 gallons for entire solution. Burnt
chalk is called quicklime, and water holding quicklime in
solution is called lime-water, and is clear and colourless.
The 7 oz. lost by burning the 16, consist of carbonic acid
gas, which, dissolved under compression by water, forms
soda-water. The other mode of rendering chalk soluble in
water is nearly the reverse. In the former mode, one
pound of pure chalk becomes dissolved in water, in conse-

quence of losing 7 oz. of carbonic acid. To dissolve in the
second mode, not only must the pound of chalk not lose
the 7 oz. of carbonic acid, but it must combine with 7 additional ounces of that acid. In such a state of combination, chalk exists in the waters of London, dissolved, invi-
sible, and colourless like salt in water. . . . . A pound of
chalk dissolved in 560 gallons of water by 7 ounces of carbonic acid, would form a solution not sensibly different, in
ordinary use, from the filtered water of the Thames, in the
average state of that river. Chalk, which chemists call
carbonate of lime, becomes bicarbonate of lime when dissolved in water by carbonic acid. Any lime-water may be
mixed with another, and any solution of bicarbonate of
lime with another, without any change being produced.
But, if lime-water be mixed with a solution of bicarbonate
of lime, the mixture acquires the appearance of whitewash,
and chalk is precipitated, leaving the water above perfectly
clear. This operation will be understood by supposing
1 lb. of chalk, after being burned to 9 oz. of quicklime, to
be dissolved, and form 40 gallons of lime-water; that
another pound is dissolved by 7 oz. of extra carbonic acid,
so as to form 560 gallons of a solution of bicarbonate of
lime, and that the two solutions are mixed, making up 600
gallons. The 9 oz. of quicklime from the 1 lb. of chalk
unite with the 7 extra ounces of carbonic acid that hold
the other pound of chalk in solution. These 9 ounces of
quicklime and 7 ounces of carbonic acid form 16 oz., or
1 lb. of chalk, which, being insoluble in water, becomes
visible at the same time that the other pound of chalk,
being deprived of the extra 7 oz. of carbonic acid that kept
it in solution, reappears. Both pounds of chalk will be
found at the bottom of the subsidence. The 600 gallons
of water will remain above, clear and colourless, without
holding in solution any sensible quantity either of quicklime
or bicarbonate of lime."

60. All the methods of mere mechanical clearing of
water are one or other of two processes, viz., *settling*, or

subsidence, and *filtration*. The first of these processes is of a negative character, consisting simply in letting the water remain for a considerable period in an undisturbed condition. It is well known that, if a quantity of water, having particles of any foreign matters of greater specific gravity than water floating or diffused within it, be allowed to continue in a quiescent state for a sufficient length of time, these particles will subside to the bottom of the water, which is thus left comparatively clear and limpid. In order to accomplish this purpose on a great scale, reservoirs are constructed, in which the water is accumulated and permitted to remain, and from which it is delivered as required. Such reservoirs are termed *subsiding* or *settling* reservoirs.

61. The East London Water Company, which draws the water from the River Lea, near Lea Bridge, and supplies the eastern part of the metropolis and suburbs, has 20 acres of settling reservoirs. The arrangement is this:—The water is introduced through a canal, two miles long, into a wide canal, or small reservoir, at the end of which there are two sets of gates, each of which opens a communication with a separate reservoir. The water is admitted into both of these reservoirs, but drawn from only one of them at a time, the other remaining closed. Thus the water remains for one day in each reservoir alternately, while, in time of floods, it may be shut off altogether from these reservoirs for four or five days.

62. The value of all merely settling reservoirs can be derived only by drawing the water from the *upper* part of them. It is evident that, while the subsidence is going on, the whole bulk of the water is clarified only in proportion to its distance from the bed; and thus, the lower down that the point of exit is situated, the less clear must be the water that passes away.

63. To make the principle of subsidence fully effective, it is likewise necessary that the water should remain for some period, probably 24 hours at least, *entirely undisturbed.*

D

If any motion is permitted, the subsidence is interrupted, if not arrested. The reservoir should therefore be filled, and then totally closed both to ingress and egress. At the expiration of 24 hours, the upper part of the water should be *gently* drawn off. If the extent of supply will admit, the lower portion of the water may afterwards be let off for manufacturing or inferior purposes, or allowed to mingle with another fresh portion. If both the supply and the discharge be conducted at a sufficiently slow rate, and enough time be allowed for the quiet completion of the subsidence, the bulk of the water will always maintain a high degree of mechanical clearness, and the intermixture of the water remaining after each drawing off with the incoming water, will not involve any material loss of time in the process.

64. The process of *filtration* is effected by providing a bed of easily permeable materials, in which the water deposits the solid particles which it held in suspension, and finds its way to the lower bed in a comparatively clear state. The filtering materials employed in large filters are, sand and gravel of various degrees of fineness, pebbles, and shells. These latter, by their calcareous properties, act chemically on the water to a trifling extent, or while they retain free carbonic acid; the other materials admit the passage of the water, but prevent that of such solid particles as are larger than the interstices between the particles of the materials. The filtered water is collected in brick tunnels, constructed in the lower filtering stratum, and having apertures in the joints to admit the water. Fig. 3 is a section of a filtering reservoir as constructed for the Chelsea Water Company. In this reservoir the water comes in contact first with a bed of fine sand, *a*, which arrests the mechanical impurities. It thence passes through the strata, *b*, of coarse sand, *c*, of pebbles and shells, and *d*, of fine gravel, into the lowest bed, *e*, which consists of large gravel, lying upon a firm foundation of clay, 18 inches thick, and having the brick culverts, *fff*, built within it. The

clay bottom must, of course, be rendered sufficiently compact to resist the passage of the water; and, if no clay be
found, it will be necessary to form an artificial bed for the
purpose. The collecting tunnels are here constructed of
blocks of brickwork in cement, and partly open-jointed.
They are three feet in diameter, and two half-bricks in
thickness. The water is admitted to the filtering-bed at
nine places, the ends of the supply-pipes being fitted with
curved boards to diffuse the water, and prevent any disturbance of the upper stratum of sand. The quantity
filtered in this bed, which is 240 feet long, and 180 feet
wide, is 72 gallons per superficial foot of the filtering-bed
daily, according to the demand. The undulating surface of
the bed allows parts of it to be drained when necessary,
without removing the water from the adjacent hollows. It
is found that the sediment penetrates only from six to nine
inches in depth, and the removal of one inch in thickness
of the fine sand, every fortnight, is found sufficient to
secure the proper action of the apparatus. Air-drains are
provided to admit the escape of the condensed air which
may collect in the tunnels. It has been found that it is
necessary, in all cases, to remove the old sand before introducing fresh sand ; otherwise a film is formed on the original sand which will resist the passage of the water.

65. The first and current expense of this system of filtration is estimated by Mr. James Simpson, the engineer of
the Chelsea Waterworks, to be as follows :—

First cost of filtering-bed, exclusive of land . . £11,700

Annual expense of raising water in filtering-bed .   800
   (From the River Thames, close adjoining, ·
        raised by steam engine.)
Annual expense of cleansing and renewal . . .   800
Five per cent. interest on outlay of capital . .   585

Total annual cost, exclusive of land . . £2,185

The quantity filtered being 3,136,320 gallons daily, or 1,144,756,800 gallons annually, or at the rate of about 2183 gallons for one penny.

66. The system of cleansing adopted by the Southwark Water Company embraces settling reservoirs, as well as filtering-beds or reservoirs, and some peculiarities in the formation of the former deserve notice. The section, fig. 4, will clearly show the construction. A A are the settling reservoirs, having an area of between four and five acres, and being 13 ft. 6 in. deep, and faced with gravel. The bed was found to be springy in some places, and there lime was mixed with the gravel, forming an impermeable concrete. The beds are formed with a slight inclination from the sides towards the middle, along which an inverted arch, b, is formed of brickwork in cement, 6 ft. wide, and 3 ft. 6 in. deep. This invert is an essential improvement, and, with the inclined bed, gives great facilities for cleansing, by sweeping the deposits into the invert, and flushing it away with a current of water from an upper reservoir. The filters are constructed similarly to those of the Chelsea Works just described. The series of filtering substances consists of coarse gravel, 1 ft. deep; rough screened gravel, 9 in. deep; fine screened gravel, 6 in. deep; hoggin, or fine gravel, 9 in. deep; and fine wash gray river sand, 3 ft. 6 in. deep. The water is gradually drawn from the settling reservoirs, A A, on to the surface of the sand, on the filter-bed, c, and is permitted to percolate through brick culverts, formed with open joints in cement. The filtered water passes from these into close brick tunnels, by which it is conducted into the well of the pumping engine, D.

67. The expense of filtering by this, the Battersea filter, is stated by Mr. Joseph Quick, the engineer for the Works, not to exceed 350l. per annum, the quantity filtered being 2,160,000 gallons per diem, or 66 gallons per superficial foot. At this rate the annual quantity filtered will be 788,400,000 gallons, and the cost about one penny per

9386 gallons. At the Bleaching Works at Dukinfield, 500,000 gallons are filtered daily, at a cost of 156*l.* per annum, or at the rate of 4874 gallons for one penny.

68. Water which has been subjected to the process of subsidence only still usually contains finely-comminuted particles of solid matter, from which the subsequent process of filtration is necessary to cleanse it. The settling reservoirs having answered the double purpose of depositing the grosser solid particles, and of effecting all the chemical *softening* of the water which can be effected by mere exposure to the atmosphere, the filtering reservoir completes the process of depositing, and sends the water forward in a tolerably pellucid condition. But beyond these processes, and altogether irrespective of any chemical improvement of its constitution, it is found that water which has remained in an exposed reservoir, and subject to the action of light —made so much more effective by the transparency of the filtered water—does, in some states and temperatures of the atmosphere, betray unequivocal symptoms of vegetable formation within it, and, if the action proceeds, animal life, in the form of minute animalculæ, rapidly succeeds. It has therefore been suggested, that the filtering process could be still further improved, if the water were submitted to a subsequent passage through some filtering medium calculated to detain any such vegetable or insect productions as might be formed on the surface of the filtering-bed, and by chance find their way with the water into the tunnels beneath.

69. When water is drawn from a river having a sandy or gravelly bed in its vicinity, it is comparatively easy and inexpensive to form a natural and highly-effective filter. Thus, at Nottingham, the Reservoir, which is formed on the banks of the River Trent, about a mile from the town, is excavated in a stratum of clean gravel and sand, through which the water slowly percolates to a distance of 150 feet from the river. The deposited solid matter thus remains on the bed of the river, from which it is removed by the

natural action of the current. The reservoir being exposed
to the solar influence, vegetation is sometimes produced,
and which is removed at intervals of three weeks in summer, and six weeks in winter, by pumping out the water
and sweeping. Besides the reservoir, there is a tunnel
filter, which passes through a similar stratum for a considerable distance up the adjacent lands. This tunnel is 4 ft.
in diameter, and half-brick thick, laid without mortar or
cement, costing about 10s. a foot, including excavation to a
depth of 12 feet.

70. An arrangement, somewhat similar to the last
described, has been successfully carried out on the River
Clyde, a few miles above Glasgow. At the selected spot
there is an extensive round bank of sand. A tunnel was
constructed in this bank parallel to the edge of the river,
and also to the surface of the water and below the level of
the water. This tunnel being constructed of bricks set in
mortar below, but bricks without mortar above, received the
water, which afterwards percolated into wells, from which it
was pumped up for use. Similar natural filters were
attempted at other points contiguous to the Clyde, but most
of them failed, from the interception of springs of water of
a harder quality than that from the river. In some cases,
also, the natural springs intruded water containing iron,
and injurious to the purposes for which the supply was
required. Natural filters must, therefore, also be considered
with reference to their liability of interruption by natural
springs of a different or inferior quality. Beyond this, they
should always be designed with a reference to the *lowest*
level to which the river water may fall at any season, or
under any circumstances; and this necessity sometimes
involves a depth for the pipes, or other constructive difficulties, which altogether mar the economy and advisability
of the arrangement. If this last precaution be not adopted,
it will happen at the driest season of the year, when the
maximum supply is required, that the reduced level of the
river will be *below* the fixed level of the filtering tunnel,

which thus becomes dry and inactive. The only alternative which then remains is, to draw the water directly from the river, and thus the filter remains useless at the season when it is most desirable that it should be performing its highest duty. The work of cleaning the tunnels is, moreover, by no means an easy one; and, considering all the circumstances and liabilities of these expedients, it would appear that they are of very limited application.

71. Let us recapitulate the heads of the subjects over which we have already passed.

The sources whence the land is supplied with water, without artificial aid, are rain and tidal rivers, as in the Nile, Euphrates, Ganges, Mississippi, &c., besides such springs as rise spontaneously to the surface, either upward, by the pressure of internal reservoirs at a higher level, or at the outcrop of the strata of the earth. The artificial sources are the ocean, rivers, streams, and wells. The quantity of rain which falls over the earth appears to vary with the latitude, the distance from the ocean, the season, and other circumstances, the nature and influence of which we do not yet understand. The effect of the rain, as a source of water, and a cause of the necessity for drainage, is limited by the quantity which passes off from the surface of the earth in a state of vapour. The quantity so raised depends upon the temperature which prevails while the process of evaporation is going on. Efficient drainage requires the supply and the discharge of water to be duly regulated, the supply to be sufficient, and not in excess, and the discharge to proceed correlatively. The quantity required for the complete irrigation of a district is determinable by reference to the nature of the soil and the crops, and the position of the district in relation to surrounding tracts of country. The state of soil most favourable to vegetable growth is that of moistness, having water between the particles, but none between the clods or masses of earthy matter. Among the artificial sources of water, that yielded by the ocean requires chemical changes in order to

fit it for domestic purposes, and its applicability for those
of agriculture is necessarily limited by remoteness. River-
supply is attainable only for the adjacent lands of low level,
unless it be forced up to the higher districts by mechanical
means, which are afforded by steam-pumping at a comparatively small cost. The water of streams which are tributary to rivers is applicable for superior levels, and may, by
judicious diversion and extension through artificial channels,
be made widely useful. Wells are generally available by
mechanical agency, and in some cases without it, provided
a subterranean reservoir exists, and is subject to sufficient
pressure from a higher source. All water at our command
for practical use is more or less impure. Thus, rain-water
contains ammonia, and sea-water a variety of salts. The
water from rivers, springs, &c., contains several kinds of
impurities. These impurities are dispelled only by a compound process, or rather series of processes, by which such
matters as are mechanically suspended in the water are
allowed to subside, or are arrested by filtering media, and
the chemical impurities are absorbed and withdrawn by
suitable agents. A brief notice of several varieties of filtering apparatus concludes this section of our first Division.

72. The filters already described, which, acting by the
spontaneous percolations of the water through the apparatus,
may be termed *self-acting*, have been further improved by
adding means for their *self-cleansing*. The arrangements
introduced for this purpose at Paisley, and other places, by
Mr. Robert Thom, are illustrated, in principle, by the
adjoining figures 5, 6, and 7. Fig. 5 is a plan, and figs. 6
and 7 sections taken at right angles to each other through
the filter. In these filters, which are provided with layers
of gravel and sand, the foul water is admitted at the top,
and descends through these strata to undergo filtration;
but the construction also admits of an occasional forced
ascent of the water through these media, by which the foul
particles are raised, deposited on the upper surface of the

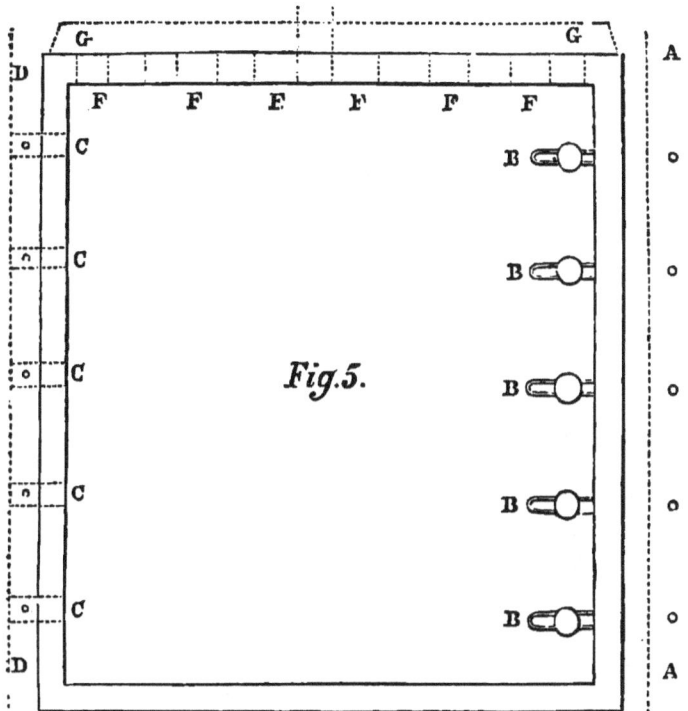

Fig.5.

Fig. 6.

Fig. 7.

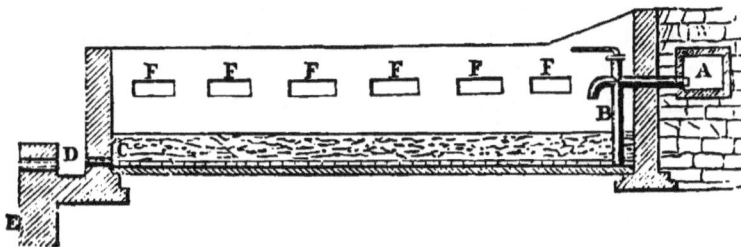

sand, and eventually carried away through a foul-water
drain. The Paisley filter is 100 feet long and 60 feet wide,
arranged in three compartments, each of which may be
used separately while the others are cleansing. They are
excavated, on a level site, to a depth of 6 or 8 feet, surrounded with retaining walls built in cement, and puddled
behind. The bed is puddled 1 foot thick, and cemented
pavement is laid upon it. It is then covered with fire-
bricks laid on edge, and with spaces ¼ inch wide between
them. These are covered with flat tiles, perforated with
holes ⅕th inch diameter. Over the drains thus formed,
six layers of gravel, each 1 inch deep, and of finer particles
than the one below, are evenly spread, and overlaid with
2 feet depth of clean, sharp, fine sand, the upper 6 inches
of which are mixed with ground animal charcoal. The
water is admitted through a stone pipe, A, and vertical iron
pipes, B B, each having an upper and lower outlet to the
filter. These pipes are fitted with valves, by which either
of these communications is opened and the other closed.
The clean water passes from the bottom of the filter
through openings c c, fitted with stop-cocks, into a drain, D,
and thence into the clean water basin, E. When the
cleansing is going on, these connections are shut off, and
access is given to the foul matter through holes, F F, to a
drain, G. The cost of this filter was less than 600l., and
the quantity of clean water produced every 24 hours, on an
average, is 106,632 cubic feet. Trap rock, from the hills
above Greenock, has since been substituted by Mr. Thom
for the charcoal with perfect success, and considerable
economy: one part of the charcoal was mixed with eight or
ten parts of the sand. The charcoal is sometimes laid in
deep layers, without mixture, and is then worth reburning
for a second use.

73. In the use of charcoal as a filtering agent, an attempt
is made to effect something more than the mere mechanical
clearing of the water by absorbing some of the gases with
which it is chemically adulterated. How far this expedient

is valuable is, however, very questionable. The power of charcoal to act in this manner is well known to depend upon its being thoroughly and *recently burnt* and *dry.* Moisture diminishes this absorbing power, and in a short time the chemical action of the charcoal ceases. Some difference, doubtless, exists, in this respect, between animal and vegetable charcoal, but neither of them can be admitted as an effective chemical agent in the purification of water, without requiring a costly rapidity of renewal quite impracticable upon an extended scale.

74. With a view to promote the mechanical action of filters, many arrangements of internal partitions have been suggested. One of the best of these is exhibited in fig. 8,

Fig. 8.

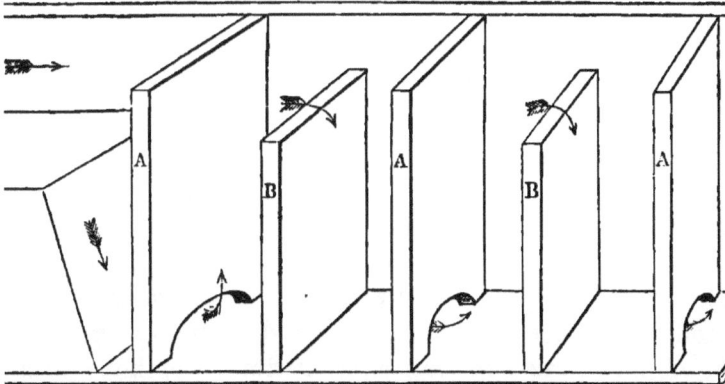

and was successfully applied in Switzerland, by Sir Henry Englefield, upwards of forty years ago. This filter is divided into chambers by parallel partitions, A B, which admit the passage of the water alternately above and below them. The intermediate spaces may be filled with filtering materials of uniform quality. The course of the water must evidently be in the direction of the arrows; and the effect of this arrangement is, that the floating impurities are retained on the surface, while the heavier particles sink to the lower level.

75. An apparatus for close filtering, within an iron water-tight box, has been introduced by M. Maurras; and its principal novelty consists in interposing the strata of fine sand between flat iron cases, perforated with holes and filled with sand of particles larger than the holes in the cases, with an arrangement of sluice-cocks, &c.; the process of cleaning was effected by sudden and violent currents of water. A machine of this kind, 5 ft. 6 in. by 5 ft. 6 in., working under a head of water of 12 ft. 6 in., is said to have filtered an average quantity of 150,000 imperial gallons in 24 hours. A filter of this kind was tried for four months at the works of the New River Water Company, but the experiment does not appear to have been prosecuted, or the invention adopted.

76. In the year 1841, the Council of Health of Paris reported upon several processes which had been tried for filtering the waters of the Seine. The two principal plans noticed were those known as "Fonveille's" and "Souchon's." The apparatus used in the first of these consists of several layers of sponge, sand, and charcoal, disposed alternately in a close vessel. The filtration is accelerated by a considerable pressure upon the water. This arrangement was found to produce the most clear and least impure water; but, although this superiority was attributed to the charcoal, it was admitted that this effect required a very frequent washing, drying, and renewal of it. Souchon's process, which is most extensively used, consists in passing the water through layers of woollen tissue, formed of clippings of wool laid on the frames which form the bottom of the filter. The water filters through five of these layers, of which the two lowest are the thickest, and are changed at intervals of about five days. The upper layers are changed twice or thrice a day. The water thus filtered is stated to have been inferior to the other, but the quantity passed through was greater, being as 162 to 110.

SECTION II.

Upper and Lower Districts.—River-watered and Sea-coast Districts.—
Reclamation of Land.—Modes of Draining, Pumping, &c.—Water-wheel as
applied for Draining and supplying Upland Districts.

77. The principal division of districts and lands, as subjects for watering and draining, is derived from their relative
levels. The sources at command and methods of proceeding for high and low tracts are perfectly dissimilar, and
hence the natural and necessary distinction which is adopted
as the head of this section. And as the plains and valleys
are far more extensive in themselves than the hill tops and
uplands, and equally superior in importance as recipients
of the drainer's care, it is proper to turn our attention to
them in the first instance.

78. In this first class, the Lower Districts, we propose to
include the following varieties of surface, viz.:—1st, the low
lands forming the margins of seas and rivers; and, 2nd,
generally, the valleys in which natural watercourses have
been formed, such as rivers, streams, &c.; 3rd, valleys in
which lakes, or similar expanses of water, do or might exist,
and which, with that adaptation, have a continuously-curved
or basin-like contour; 4th, and plains which, although they
may have a superior elevation to adjacent districts on one
side, are correspondingly low in relation to the hills on the
other. The sections sketched in the foreground of the figs.
Nos. 9 to 12 will illustrate the relationship of levels referred
to in each of these four varieties.

79. The watering and drainage of districts belonging to
the first of these varieties (fig. 9) are frequently reduced to
the sufficiently heavy task of getting rid of a large surplus
of water which collects from the adjacent estuaries of large
streams, or is detained in the form of evaporation from the
surface of the sea, and condensed by low temperature. If
the level of the district is above that of the sea and river-
mouths, surface drainage, of properly-determined depth and
extent, with ample main conducting channels, will suffice

Fig. 9.

Fig. 10.

to keep the land in a tolerably dry condition. An opportunity very seldom exists in such districts for tapping, or getting rid of the excess by opening a communication with a lower and permeable stratum. Rock in some·cases, and bog in others, usually form the inferior deposits. If the former, surface draining is certain of success, although the construction will probably be expensive; but if the sub-

Fig. 11.

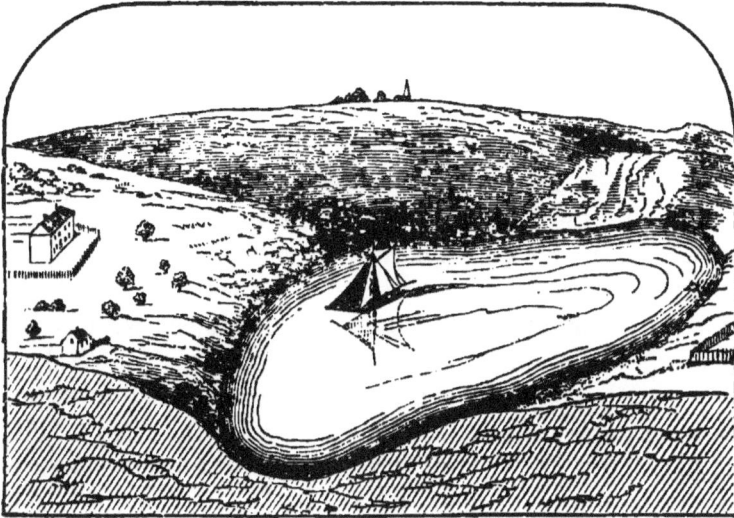

Fig. 12.

stratum be bog, and its bed below the river or sea level, boring to lower strata is presented as the only chance of success.

80. If, however, the level of the district be below that of
the contiguous waters, it will be manifestly impossible to
dry the land without embanking. And it will be necessary
either that this work be sufficiently substantial to prevent
the ingress of the water, or that the surface of the land be
simultaneously raised artificially until it has a superior
level, or that mechanical means be constantly employed to
pump out the surplus water. Our own island has been
preserved in its borders, nay, extended, by works of this
class, which we shall now have to notice.

81. In pursuing this branch of our subject, we have the
great gratification, through the kindness of Mr. Weale, of
referring to one of the earliest and most interesting records
of engineering art in this country—the celebrated account
of the Fens, by Sir William Dugdale, the earliest edition
of which was published in the year 1652, and the second
edition, in folio, which we have consulted, in the year 1772,
under the care of Charles Nelson Cole, Register to the
Corporation of Bedford Level. Dugdale had been in the
employment of the Corporation since the year 1643, and
"published this History of Imbanking and Draining, at
the request of Lord Gorges, who at that time had the principal direction of their works, and was, after their incor-
poration, for many years their Surveyor-General." The
first sixteen pages of Sir William's book are occupied in
brief notices of foreign works of this kind, beginning with
Egypt, and thence passing to Babylon, Greece, and Rome,
quoting his authorities from Herodotus, Strabo, Pliny, and
others, and ending with Holland and the Netherlands.
We cannot forego one short extract in reference to the
troublesome Tiber, whose later tricks, we all remember,
plunged so many of the poor Romans in ruin. Our quo-
tation shows that, even in the time of Tiberius, public
improvements were scandalously thwarted, as they are in
our own day, by the petty jealousies of cities and corpora-
tions. "To restrain the exorbitant overflowing of this
stream (the river Tiber), which was not a little choaked with

dung and several old buildings that had fallen into it, I find that Augustus Cæsar bestowed some cost in the clearing and scouring of it; and that after this, through abundance of rain, the low grounds about the city suffering much by great inundations thereof, the remedy in preventing the like for the future was, by the Emperor Tiberius, committed to the care of Ateius Capito and L. Aruntius. Whereupon it was by them discussed in the senate, whether, for the moderating the floods of this river, the streams and lakes, whereby it increased, should be turned another way; but to that proposal there were several objections made from sundry cities and colonies; the Florentines desiring that the Glanis might not be put out of its accustomed channel, and turned into the river Arnus, in regard much prejudice would thereby befall them. In like manner did the inhabitants of Terano argue; affirming that, if the river Nar should be cut into smaller streams, the overflowings thereof would surround the most fruitful grounds of Italy. Neither were those of Reate (a city in Umbria) silent, who refused to stop the passage of the lake Uelinus (now called Lago de Terni), into the said river Nar. *The business, therefore, finding this opposition, was let alone.* (!) After which, Nerva or Trajan attempted likewise, by a trench, to prevent the fatal inundations of this river; but without success."

82. The earlier works of the Dutch were well followed up by the contemporaries of Dugdale. He describes their works " within the space of these last fifty years" to have included the " draining of sundry lakes, whereof sixteen were most considerable, by certain windmills, devised and erected for that purpose. The chiefest of which lakes, called the Beemster (containing above eighteen hundred acres), is made dry by the help of LXX of those mills, and walled about with a bank of great strength and substance." " The other lakes, so drained, as I have said, do lie about the cities of Alcmare, Horne, and Purmerende; and are vulgarly called de Schermere, de Waert, de Purmer,

and de Wormer." "Neither have the attempts of these
people, by the like commendable enterprises, in South
Holland, about the cities of Leyden, Dort, and Amsterdam,
had less success, there having been divers thousands of
acres, formerly overwhelmed with water, made good and
firm land, within these few years, by the help of these
engines." We shall have more to say by-and-by of the
Dutch draining, as further extended within our own times,
but meantime pass on to a notice of our own works in a
similar department.

83. Dugdale shows, by "circumstantial testimonies," that
the Romsey marshes were reclaimed by the Romans, and
then quotes the ordinances of Henry III., "that all the
lands in the said marsh be kept and maintained against
the violence of the sea, and the floods of the fresh waters,
with banks and sewers." The execution of these ordinances
appears to have provoked much litigation, and Edward I.
found it necessary to issue letters patent for the repairing
of the banks and ditches. Further disputes followed, and
led to new letters patent two years afterwards. Edward II.,
Edward III., Richard II., and several succeeding sovereigns,
repeated their patents for the like purpose. Similar Royal
commissions were instituted for preserving the lands in
East Kent, "for the digging of a certain trench, over the
lands, lying between Gestlinge, and Stonflete, and from
Stonflete to the town of Sandwich; to the intent that the
passage of the water called Northbroke, which was at
Gestlinge, should be diverted; so that it might run to
Sandwich." Also "for the repair and safeguard of the
banks and ditches, from the overflowing of the tide, betwixt
Dertford, Flete, and Grenewich," and thence to London
Bridge. The banks, &c., in Surrey, "betwixt Lambehethe
and Grenewiche;" of Middlesex, "betwixt the hospital of
S. Kathrine's, near the Tower of London, and the town of
Chadewelle;" some parts "within the precincts of West-
minster;" "betwixt a place called the Neyt and Temple
Bar, in London, then broken and in decay by the force of

the tides," were also to be repaired by Royal letters patent; besides the marshes in the suburbs of London, in Essex, and in Sussex. On the coasts of Somersetshire, Gloucestershire, and Yorkshire, &c., works of repair were also provided for under the care of Commissioners, severally appointed by letters patent from the kings. Of Somersetshire, Dugdale observes, "that the overflowings, both of the sea and fresh rivers in some parts of this country, were heretofore likewise exceeding great. I need not seek far for testimony; the rich and spacious marshes below Wells and Glastonbury (since, by much industry, drained and reduced to profit) sufficiently manifesting no less. For, considering the flatness of those parts, at least twelve miles eastward from the sea, which gave way to the tides to flow up very high; as also that the filth and sand, thereby continually brought up, did not a little obstruct the out-falls of those fresh waters which descend from Bruton, Shepton Malet, and several other places of this shire, all that great level about Glastonbury and below it (now for the most part called Brentmarshe) was, in time past, no other than a very fen; and that place, being naturally higher than the rest, accounted an island, by reason of its situation in the bosom of such vast waters."

84. The history of the works of embanking and draining in the counties of Lincoln and Cambridge affords evidence of the skill and labour which had then been applied to these objects. The good abbots appear to have acted as conservators of the low lands in Lincolnshire. Thus, in the isle of Axholme, "one Geffrey Gaddesby, late abbot of Selby, did cause a strong sluice of wood to be made upon the river of Trent, at the head of a certain sewer, called the Maredyke, of a sufficient height and breadth for the defence of the tides coming from the sea; and, likewise, against the fresh waters descending from the west part of the before-specified sluice to the said sewer, into the same river of Trent; and thence into Humbre:" "John de Shireburne," Geffrey's successor, pulled down this timber

sluice, and ' did new make the same sluices of stone, sufficient (as he thought) for defence of the sea tides, and likewise of the said fresh waters;" but jurors, appointed under
patent of Henry V., reported that these stone sluices were
"not strong enough for that purpose, being both too high and
too broad; and that it would be expedient, if the then abbot
would, in the place where those sluices of stone were made,
cause certain sluices of strong timber to be set up, consisting of two flood-gates, each flood-gate containing in itself
four foot in breadth, and six foot in height." They also
recommended "one demmyng" to be made, "without the
said sluice, towards the river of Trent."

85. It was upon districts such as those we are now considering that the art of draining was first practised. Here
the matter was one of obvious necessity. In wet fields and
moist pastures, our ancestors found no positive demand for
improvement; the evil was seen and recognised in its full
extent, but the only tangible effect was to depreciate the
value of the land, and induce a preference for districts
where nature provided a more sufficient drainage. But on
the sea-coast, and especially in the neighbourhood of the
outfalls of rivers, the evil of neglect was too apparent to be
disregarded; the ocean spread over its common bounds,
and the waters of the river, choked up with silt, passed
their limits, the pasture fields became swamps,—in some
cases the land disappeared by degrees, and the inheritance
of ages became merged in the boundless waters.

86. The first work was to cut channels at intervals
through the threatened district (selecting the lowest levels
for them, where a choice was afforded), in which the excess
of water might be collected and conducted to a main drain
cut parallel to, or at an angle with, the coast or river, the
transfer of the water from one to the other, and from the
main to the sea or river, being, when necessary, regulated
by sluices. The earth removed from these collecting and
main drains, being cast up on either side of them, at once
increased their available depth, formed boundaries to the

passing water, and raised causeways for the passage of men and animals. Thus arose the combined arts of draining and embanking.

87. The maps of the fens of Cambridgeshire and Lincolnshire exhibit a multitude of illustrations of the works here referred to ; but we may select those executed in one district as examples of the whole. This district consists of the lowland or level about the river Ancholme in Lincolnshire, and is situated on the south side of the river Humber, about ten miles below its junction with the river Trent, containing about 50,000 acres of land. It is bounded on the east by a ridge of chalk hills, which extend from the Humber nearly 24 miles N. and S. From this ridge the Ancholme receives the drainage of about 100,000 acres. A lower ridge of oolite and sandy limestone divides it on the W. from the Trent Valley, and contributes to the Ancholme the drainage of some 50,000 acres more, and on the S. a low diluvial ridge divides the district from the Witham Valley. The Ancholme thus receives the drainage of a total of 200,000 acres. The valley varies from one to three miles in width, and the total bulk of waters daily poured through the river is estimated at 140 millions of cubic feet, being sufficient to cover the entire level to a depth of $2\frac{1}{4}$ inches. The principal portion of the district lies below the level of high water spring tides in the Humber, being in some places as much as 9 feet below that level. From a map of the valley published in Dugdale, and bearing date in 1640, it appears that the course of the Ancholme was originally very tortuous, being probably enfeebled and choked up by the alluvial deposits from the overflowing of the Humber. At that time, however, a straight channel had been cut, extending from the Humber to Glentham Bridge (a distance of 18 miles), and several drains formed, leading to the new Channel. Figs. 13 and 14, which are reduced sketches of the plan and section given by Dugdale, show the general direction of the old and new channels, and the drains as they existed in 1640. In the previous

year Sir John Munson became the undertaker for improving the draining works of this district, having a period
of six years allotted for their execution, and a part of the
lands, extending to 5827 acres, assigned to him, free of all
commons, titles, charges, interest, and demand, of all or
any persons whatsoever.

88. In the year 1801, the late Mr. Rennie reported upon
the best means of completing the drainage and navigation
of the level; and recommended that the drainage of the
high lands should be separated from that of the low lands
by main drains, commonly called *catch-water drains*, formed
at a higher level than the others, and arranged with separate
sluices for discharging into the Humber. This recommendation was well founded on the observation that the
greater force and rapidity with which the waters from the
upper districts reached the river than those from the lower,
had the effect of driving the latter over the level, the sluices
being inadequate to discharge the entire bulk of water
during the periods while the river-tide permitted the sluice
doors to remain open. Another and highly-important purpose which the catch-water drains fulfil, is that of providing
a reserve supply of water which, during dry seasons, may be
applied to the lower lands, thus promoting the objects
which in those districts are usually associated with drainage, viz. irrigation and navigation. Mr. Rennie had already
adopted a similar system of drainage on a more extensive
district, that of the East, West, and Wildmore fens, near
Boston; but his Report upon the Ancholme level was not
then adopted. Twenty-four years later, however, an Act was
obtained, viz. in 1825, for effecting improvements recommended by Sir John Rennie, and comprising the formation
of the catch-water drains, as proposed by his late father in
1801. Sir J. Rennie advised that the river Ancholme should
be straightened, widened, and deepened, so as to double its
capacity; that a new sluice be formed at Ferraby, having
its cill 6 ft. lower than the old one; together with a new
lock, 20 ft. wide, so as to serve the double purpose of ad-

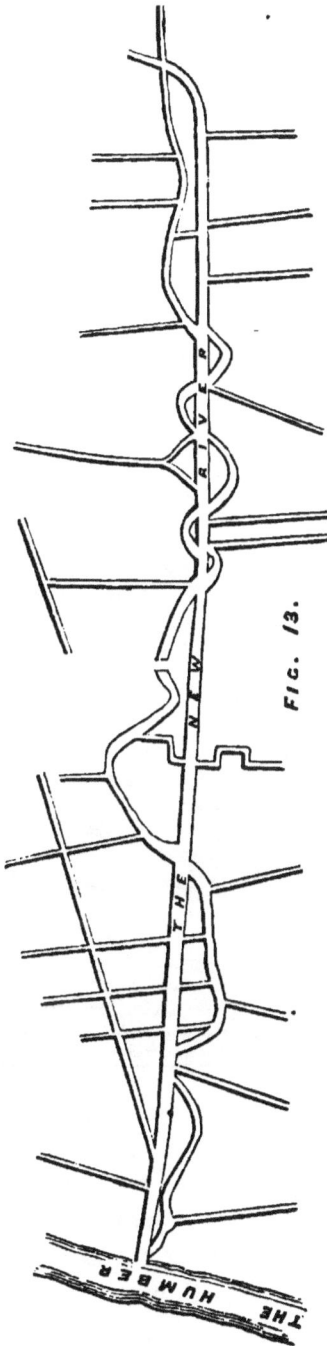

FIG. 13.

THE NEW RIVER

THE HUMBER

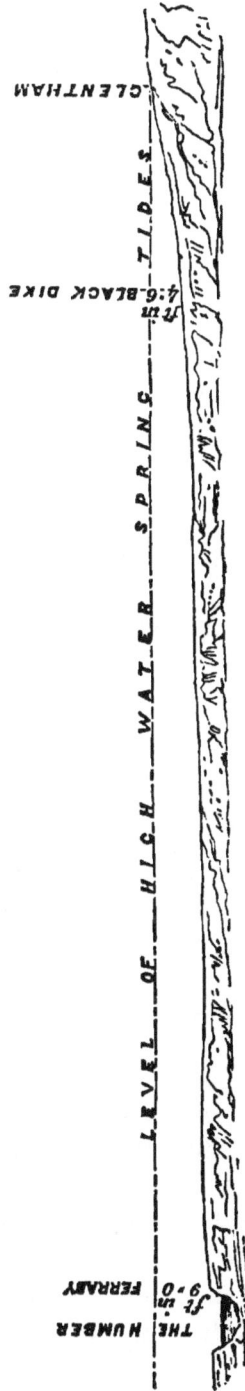

FIG. 14.

GLENTHAM

4:6.-BLACK DIKE
ft in

TIDES

SPRING

LEVEL OF HIGH WATER

FERRABY
9·0
ft in

THE HUMBER

mitting larger vessels, and affording a greater discharge for
the drainage waters during floods : that all old bridges
which obstructed the flow should be removed, and a new
lock be formed 18 miles above Ferraby sluice. These
several works were executed accordingly, and the entire
level of Ancholme has been converted into a rich arable
district, capable of producing superior crops of every kind.
Sir John Rennie also recommended the formation of reservoirs, with overfalls and weirs to receive the sand and
mud brought down from the upper part of the country, and
thus prevent its accumulation in the river.

89. Fig. 15 will give a general idea of an arrangement of

Fig. 15.

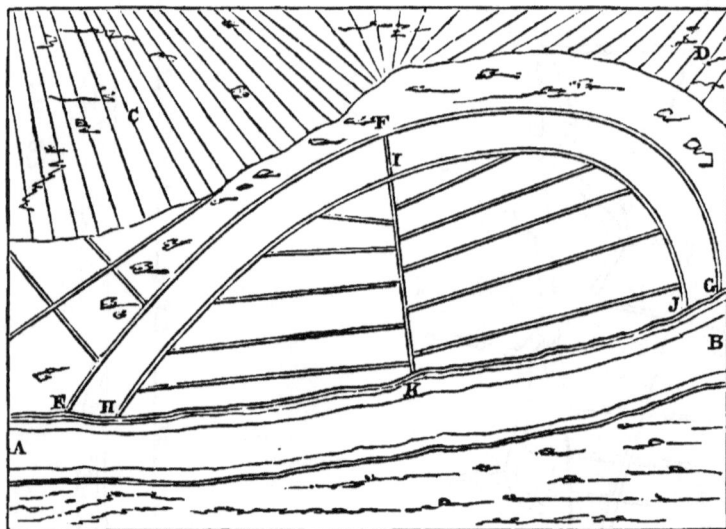

drains, which will be suitable for a level district with high
land behind it.   In this fig. A B is the river, and C D the
high land.   E F G represent a catch-water drain for receiving
the waters from the high land ; H I J, a parallel main
drain for the level, with another main drain I K.   Between
the main drains the level is intersected with minor drains,
which have a fall either way towards the mains.   The

catch-water drain is adapted to discharge directly into the river ; or by closed sluices at E G, and an open one at F, its contents may be directed into the main level drain at I, and made to assist the irrigation of the level in dry seasons. Sluices will be required at E, F, G, H, I, J, and K, by the regulation of which the water may be collected and disposed of in any manner required for the preservation and improvement of the district.

90. Figs. 16, 17, and 18 represent sections of drains of

Fig. 16.

Fig. 17.

Fig. 18.

large size, adapted for works of the kind here referred to. Drains of these sections, formed with a·fall of 18 in. per mile, will discharge as follows:—fig. 16, 10-ft. drain, will discharge 1193·4 cubic feet per minute; fig. 17, 15-ft. drain, 2880 cubic feet per minute; and fig. 18, 18-feet drain, will discharge 4642 cubic feet per minute. A good section for an embankment against the sea for these works is shown in fig. 19, in which A represents the embankment of earth; B, a solid wall or dyke of puddle; C, a facing wall

E

Fig. 19.

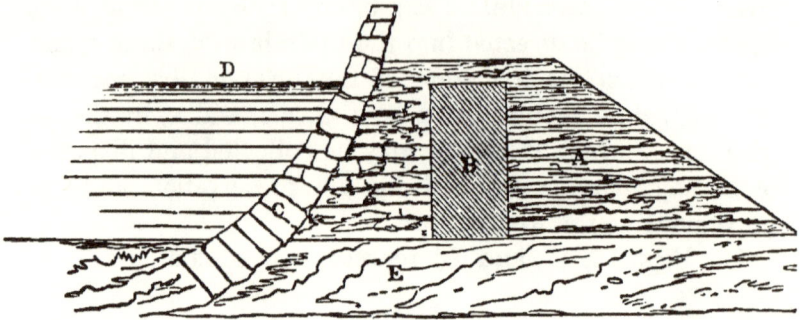

of masonry; D, the high-water level of the sea or bay; and
E, the natural bed.   The form of the front wall must
be adapted to resist the action of the waves, and the embankment must have an internal slope, according to the
nature of the materials of which it is composed : for ordinary materials, a base of 1·5 to a perpendicular height of 1
will insure the necessary stability and firmness.

91. If the entire embankment be formed of loose stones,
with occasional facing only of laid masonry, as in the case
of the celebrated breakwater at Plymouth, a form of less
steepness must be adopted for the river front of the embankment.   By way of illustration we may refer to fig. 20,

Fig. 20.

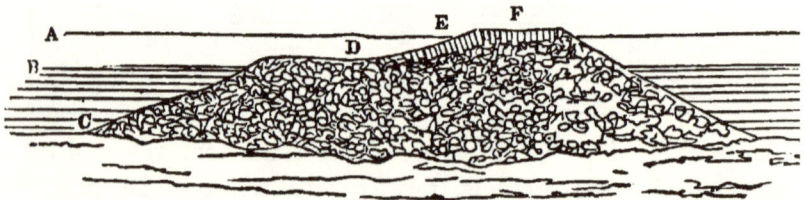

which shows a section of the Plymouth breakwater.   The
line A A shows the level of high-water spring tides; B B, low
water spring tides; c c, original bottom, varying from 40 to
45 feet below low-water mark; D, the fore shore; E, sea
slope; F, top, 45 ft. wide.   The mass of the work is composed of limestone, from the Overton quarries, distant four
miles from the spot.   The stone is raised in blocks varying

from one quarter to ten tons and upward in weight, which are promiscuously thrown into the sea, care being taken that the greater number of the large blocks are thrown upon the outer or sea slope, and that the whole are so mixed together as to render the mass as solid as possible, the rubbish of the quarry and screenings of lime being flung in occasionally to assist the consolidation of the materials. The form of the outer slope, below low-water line, has been effected by the action of the sea, and is ascertained to be at from 3 to 4 feet of base to 1 of perpendicular altitude. From low water upward the work has been set artificially and inclined at 5 to 1. The inner slope next the land is nearly 2 ft. base to 1 altitude. The foreshore shown at D, which is from 30 to 70 ft. wide at different parts of the work, rises from the toe of the slope, to a height of 5 ft. above low water at its outer extremity, and serves to break the waves before they reach the main work; thus diminishing their force, and, at the same time, preventing the recoil of the wave from undermining the base of the slope.

92. The several sluices, gates, &c., constructed for the Ancholme drainage, being of the best description, may be briefly described as applicable for similar works in future. The *sluice* at Ferraby consists of three openings, each 18 ft. wide, with cills 8 ft. below that of the old sluice, and from 2 to 3 ft. below low water of spring tides in the Humber. The *lock* is 20 ft. wide in the clear, and 80 ft. long between the gates, giving a clear water-way of 74 ft., with an additional fall of 8 ft. The masonry is of best Yorkshire stone; and the foundations, which are in alluvial silt and clay, are upon piles 12 in. diameter, of beech, elm, and fir, from 24 to 28 ft. in length, and fitted with wrought iron hoops and shoes. When the piles were driven and the heads levelled, the earth was excavated to a depth of 2 ft. below them, and the spaces filled with blocks of chalk rammed soundly in, and grouted with lime and sand. Cap-cills of Memel fir, elm, or beech, 12 in. square, were fitted on the pile-heads

and firmly spiked down, the intermediate spaces being afterwards filled with solid brickwork, set and grouted with best Roman cement. The whole was then covered with a 3-in. flooring of Baltic fir-plank, bedded in lime, pozzolana, and sand. Inverted arches of solid stonework, 18 in. deep at the crown, are built upon this platform, and the work carried upon them. Two *sluice gates* were provided for each opening in the sluice, with *draw-doors* fitted in a water-tight groove by means of pinions, of wrought iron, which work in screws connected with vertical rods. These draw-doors are for regulating the navigation level (which is 13 ft. 8 in. above the cill), and to preserve a depth of 8 ft. 9 in. at Brigg, which is 9 miles distant, and 6 ft. 6 in. at Haarlem Hill lock, 18 miles distant. The *gates* are self-acting, being shut by the tide, and opened by the head of fresh water as soon as the tide falls below the level of the inside water. Four pairs of *lock-gates* were provided for the lock; two pairs pointing to the sea, and of sufficient height to exclude the highest tides: the other two pairs, pointing to the land, are high enough to control the navigation of the level. These gates were wholly constructed of the best English oak, well fitted together with wrought iron straps and bolts. The lock is filled and emptied through side culverts in the masonry, provided with cast-iron sluices, sliding upon brass faces, and worked with pinions and screws of wrought iron. The works also included several bridges of various spans and forms of construction.

93. In the application of *catch-water drains* it is preferable to discharge their contents at a higher point of the river, or main receiving channel, than that at which the low land drains are emptied. This principle was very successfully adopted by the late Mr. Rennie, in the drainage of the East, West, and Wildmore Fens, bordering on the river Witham, and comprising about 75,000 acres. The drain for the high land waters was made to discharge into the Witham, at a distance of three miles above the discharge of the low land waters.

94. The drainage of a low fenny district being arranged as far as the judicious selection of separate channels for the high and low lands, and provision made, with sluices, &c., for their communication with each other and with the river at pleasure, it remains to consider the state in which this river must·be maintained in order to give efficiency to the internal system of drains by which the district is traversed. For this purpose it is evidently necessary that the channel be adequate in dimensions and suitable in form to maintain an active and sufficient current through it, and these conditions require a direct course and proper fall for the channel. If the direction be tortuous, the projecting banks will be washed into the bed and impede the flow of the current, and if the bed be on a dead level, or have an inadequate inclination, the flow will be sluggish, and lend no assistance to the discharge. Besides these conditions, it is necessary that the outfall of the river into the sea be of ample dimensions and unencumbered with shoals, bars, or other solid accumulations. These arise from the depositions of alluvial matter, which is liable to be brought in by the tides from the neighbouring coast, and also brought down with the drain-water from the interior country. This matter remains suspended in the water until the velocity is diminished, which generally occurs at the entrance to the river, owing jointly to the reduced inclination of the river bed near the sea, and the resistance suffered from the wind and waves, and it is then deposited, and by continual augmentation forms a fatal obstruction to the efficiency of the current. To determine the precise fall or inclination required for the bed of the channel, many experiments have been tried, but it will evidently be, to a considerable extent, controlled by the obstructions which may exist to the discharge of the waters. If the outfall be unimpeded, 4 or 5 in. per mile will be sufficient fall, but if obstructions exist, in the form of old bridges, sinuosities, &c., from 12 to 18 in. per mile will be found requisite.

95. Among the notable works of this kind which have been executed in this country, we may mention those for

improving the rivers Ouse and Nene. The chief defect in the former existed above the town of Lynn, where the river turned almost at right angles to its general course, and in a length of 5½ miles formed a semicircle of only 2¾ miles in diameter. The channel was, moreover, so irregular in width and encumbered with shifting sands, that the tidal and drainage waters were unable to force a passage, and disastrous inundations were the results. In the year 1724 this evil was understood, and a proposition made by Bridgeman for improving the river by making a direct cut which should intercept the bend here described. Succeeding engineers concurred in this recommendation; but it was not until the year 1817 that an Act was obtained for executing this important work, which was named the Eau Brink Cut, and confided to the late Mr. Rennie. The works were finished on the 19th July, 1821, and have proved highly successful, lowering the low-water line in the river several feet, and completing the drainage of more than 300,000 acres of land.* A work of similar character was executed in the year 1829, by Telford and Rennie, at the outfall of the river Nene, which commences about five miles below Wisbeach, and terminates after a length of five miles in the great estuary of the Wash. The benefits of this improvement have been very great; the low-water mark has been lowered 10 ft. 6 in., and a district of more than 100,000 acres, formerly a stagnant marsh, has been brought into cultivation.

96. Closely allied with the drainage of low lands are the operations by which their boundaries are extended, and large districts actually *reclaimed* from the action of the sea. This is effected by judiciously controlling the deposit of the alluvial materials which are washed down with the drainage waters and thrown back by the tide. This requires the formation of embankments or opposing barriers, by which the removal of those materials is prevented. A

---

* The Great Level of the Fens contains about 680,000 acres, formerly of little value, but now rich in corn and cattle.

similar artificial mode of depositing the solid matters contained in the water is practised in the interior districts by surrounding them with embankments, and admitting and discharging the water by means of sluices and canals. This method has for many years been adopted with great success in the rivers Trent, Ouse, and Humber.

97. Districts lying below the level of the adjacent river, or so little above it that drains of adequate capacity must have their beds below the water line, necessarily require artificial means of discharging the drainage waters into the receiving channel or river. In the low lands of Holland this is commonly the case, and accordingly we find the Dutch were early adopters of contrivances for this purpose. Fig. 21 shows the relative conditions of the drain and of

Fig. 21.

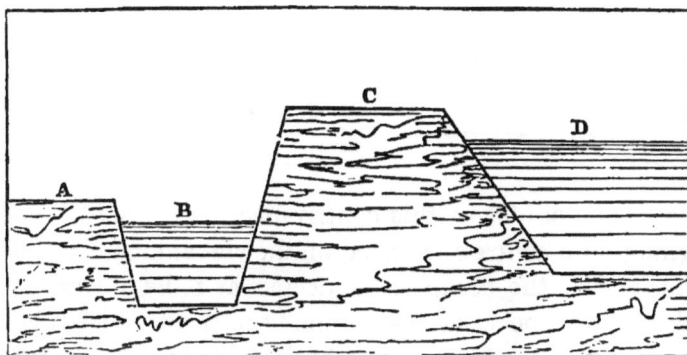

the river into which its contents are required to be discharged. A represents the general level of the district; B, that of the water in the drain to be discharged; C, the top of embankment; and D, the high-water level outside. To transfer the contents of the drain B into the main channel D, it is, evidently, only necessary to erect upon the embankment, pumps, buckets, or scoops, which shall bring the water up on the one side and discharge it on the other. Among the earlier machines employed by the Dutch were *scoop wheels*, which they worked by means of windmills, and continued to use for many ages.

98. A new form of scoop or alternating trough has been
designed by Mr. W. Fairbairn, and adapted to be worked
by the single-acting Cornish engine.   Fig. 22 will serve to

Fig. 22.

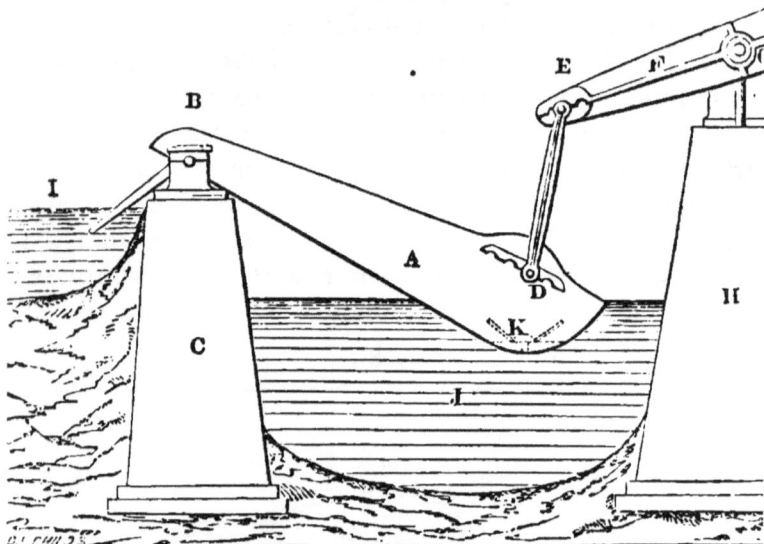

give a general idea of this contrivance.   A is the bail scoop,
turning on a centre at B, fixed on the embankment c.   The
other end of the scoop is connected at D by a connecting
rod with the end E, of the engine-beam F, of which G is the
centre, and erected upon suitable foundations, H.   I represents the level of water in the river, and J, the drain from
which the water is to be discharged.   The action of the
apparatus will be evident from inspection of the figure.
The engine employed is of the reciprocating kind, and by
raising a weight suspended at the other end of the engine-
beam F, the bailing scoop A descends, and becomes filled
with the drainage water through the opening valves at K.
The weight having been raised to the height of the stroke,
descends by its own gravity, and raising the end, D, of the
scoop, discharges its contents into the river at I.   This apparatus is well adapted to be worked by the single-acting

Cornish engine, and while the length of stroke in the cylinder always remains the same, the dip is regulated as required by shifting the connecting rod at the ends D and E. The scoop is made of iron boiler plates, and is 25 ft. long and 30 ft. wide, with two partitions across it to strengthen the sides and afford bearings for the valves at K. The machine is adapted to raise 17 tons of water at each stroke, and, with an engine of 60-horse power, will do a duty equal to 3 lbs. of coal, per horse power, per hour.

99. The greatest improvement, however, effected in mechanical draining is by the employment of the steam engine for this purpose. In the year 1820, Rennie applied one of Watt's engines to the working of a large scoop wheel for draining Bottisham Fen, near Ely. Since that time large districts have been efficiently drained by steam power; and of them we may enumerate the following:—

	Containing	Drained by	
	Acres.	Engines.	Horse power.
Deeping Fen, near Spalding, Lincolnshire	25,000	2	80 and 60
Marsh West Fen, Cambridgeshire	3600	1	40 ,,
Misserton Moss, with Everton and Graingeley Carrs	6000	1	40 ,,
Littleport Fen, near Ely	28,000	2	30 and 40
(75 wind engines were employed in this district before steam was used.)			
Middle Fen, Cambridgeshire	7000	1	60 ,,
Waterbeach Level, between Ely and Cambridgeshire	5000	1	60 ,,
Magdalen Fen, near Lynn, Norfolk	4000	1	40 ,,
March Fen district, Cambridge	2700	1	30 ,,
Feltwell Fen, near Brandon	2400	1	20 ,,
Soham Mere, Cambridgeshire	1600	1	40 ,,
(Formerly a lake: the lift is here very great.)			

100. If the drainage from the high lands be discharged through catch-water drains, that from the low levels will consist of the rain water only, and as this, in the fen districts on the eastern side of England, seldom exceeds the

E 3

average of 26 in. in depth per annum, of which a large
quantity is carried off by evaporations and absorption,
2 in. in depth or 1½ cubic ft. of water on every square yard
of surface is the ordinary maximum quantity to be lifted
per month. Adopting the admitted standard of horse-
power, viz. 33,000 lbs., raised one foot per minute, and the
weight of a cubic foot of water to equal 62¼ lbs., or 10 lbs.
per gallon, a horse's power will raise 300 gallons, or 52·8
cubic ft. of water 10 ft. high per minute. The total quantity to be raised per acre per month, viz. 7260 cubic ft.,
may thus be raised a height of 10 ft., and discharged in
about 2 hours and 10 minutes. Upon this calculation,
which Mr. Glynn (a high and practical authority in these
matters) has found to be supported in practice, it appears
that a steam engine of 10-horse power will raise and throw
off the drainage water due to a bistrict of 1000 acres of
fens, in each month, in 232 hours. or less than 20 days,
working 12 hours a day. The scoop-wheels used for raising
the water resemble an undershot water-wheel, but, instead
of being moved by the force of the water, they are adapted
for forcing the water upward, deriving their motion from the
steam engine. The float boards or ladle boards are of wood,
and fitted to work within a track or trough of masonry :
they are usually about 5 ft. long, that is, they are immersed
in the water to that extent, the width or horizontal dimen-
sion of them being varied, according to the power of the
engine and the head of water to be provided for, from
20 in. to 5 ft. The lower end of the wheel track communicates with the main drain, and the higher end with the
river, the water of which is excluded by a pair of doors,
pointing like the gates of a canal lock, and closed when the
engine ceases to work. The wheels are of cast iron, and
fitted in parts. The float boards are attached to the wheel
by oak starts, stepped into sockets cast in the periphery of
the wheel for that purpose. The wheel is fitted with cast-
iron toothed segments, working into a pinion upon the
crank shaft of the engine. If the level of water in the

delivering drain and in the river does not vary much, one speed for the wheel is sufficient; but if the tide rises to any great extent, it is found desirable to have two speeds of wheel work, one to be used at low water, and the more powerful combination to act against the rising tide. It is usually not necessary to raise the water more than 3 or 4 ft. above the surface to be drained, and that only when the river is filled by long-continued rains or floods from the upland. If the main drains be $7\frac{1}{2}$ ft. deep, and the floats dip 5 ft. below the surface of the water, 1 ft. in depth will be left below them to admit the passage of weeds or other matters, and the water will yet be kept 18 in. below the surface of the land. If the wheel dips 5 ft. below the drain-water level, and the level of the water in the river is 5 ft. above that in the drain, the wheel will be said to have " 10 ft. head and dip," and should be 28 or 30 ft. in diameter. For a dip of 5 ft. and head of 10 ft., that is, " a head and dip of 15 ft.," Mr. Glynn used wheels of 35 ft. to 40 ft. in diameter. A wheel of 40 ft. diameter, and situated on the ten-mile bank near Littleport in the Isle of Ely, is driven by an engine of 80-horse power. The largest quantity of water discharged by one engine is from Deeping Fen, near Spalding. This fen comprises 25,000 acres, drained by two engines of 80 and 60-horse power. The 80-horse power engine works a wheel 28 ft. diameter, with float boards $5\frac{1}{2}$ ft. by 5 ft., and moving with a mean velocity of 6 ft. per second. When the engine has its full dip, the section of the stream is $27\frac{1}{4}$ ft., and the quantity discharged per second is 165 cubic feet, equal to more than $4\frac{1}{2}$ tons. These two engines were erected in 1825, before which time the district had been kept in a half cultivated condition (being sometimes wholly under water) by 44 windmills.

The land now grows excellent wheat, producing (in 1848) from four to six quarters to the acre. In many districts land was purchased by persons who foresaw the consequences of these improvements, which they could now sell

at from 50l. to 70l. per acre. This increase in value has
arisen not only from the land being cleared from the injurious effects of the water upon it, but from the improved
system of cultivation it has enabled the farmers to adopt.
The fenlands in Cambridgeshire and great part of the
neighbouring counties are formed of a rich black earth,
consisting of decomposed vegetable matter, generally from
6 to 10 ft. thick, although in some places much thicker,
resting upon a bed of blue gault containing clay, lime, and
sand. When steam-drainage was first introduced, it was
usual to part the land and burn it, then to sow rape-seed,
and to feed sheep upon the green crop, after which wheat
was sown. The wheat grown upon this land had a long
weak straw, easily bent and broken, carrying ears of corn of
small size, and having but a weak and uncertain hold by its
root in the black soil. Latterly, however, chemistry having
thrown greater light upon the operations of agriculture, it
has been the practice to sink pits at regular distances
through the black earth, and to bring up the blue gault,
which is spread upon the surface as manure. The straw,
by this means, taking up an additional quantity of silex,
becomes firm, strong, and not so tall as formerly, carrying
larger and heavier corn, and the mixture of clay gives a
better hold to the root, rendering the crops less liable to be
laid by the wind and rain, whilst the produce is most luxuriant and abundant. Mr. Glynn has applied steam-power
to the drainage of land in fifteen districts, all in England,
and chiefly in the counties of Cambridge, Lincoln, and
Norfolk, to the extent of more than 125,000 acres, the
engines employed being seventeen in number, of sizes
varying from 20 to 80 horses, and having an aggregate
power of 870 horses. The same engineer has also drained,
by steam-power, the Hammerbrook district, near Hamburg,
and designed the works for draining a level near Rotterdam,
which have been carried out by the Chevalier Conrad.* In

---

* Abstract of a Report on the Application of Steam Power to the

British Guiana the steam engine has been made to answer the double purpose of drainage and irrigation. Some of the sugar plantations of Demerara are drained of the super-fluous water during the rainy season and watered during the dry season.

101. Recurring to fig. 10, p. 62, the districts there illus-trated will require methods of drainage determined by the inclination of the surface. If this be comparatively level, the drains may be generally cut with beds parallel, or nearly so, to the surface, and arranged to deliver into one or more main drains having lower beds, but still above the low-water level of the river or receiving channel, and from which the water can be let off when the tide is down by providing

Fig. 23.

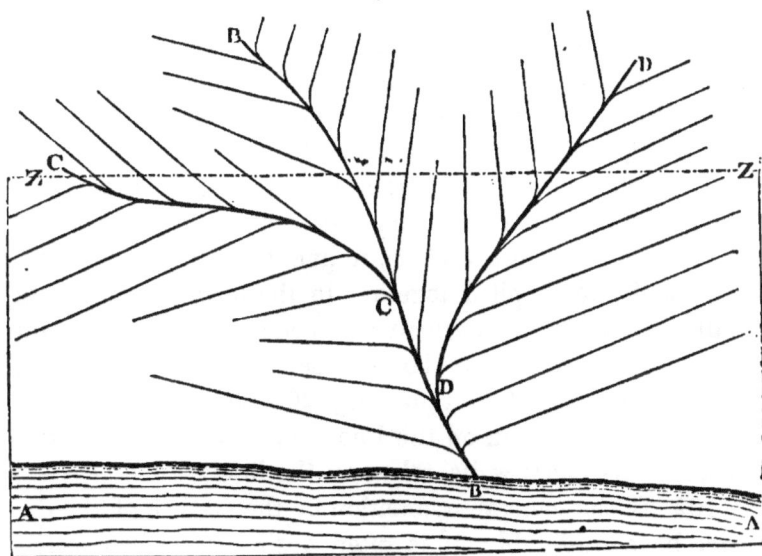

sluices suitable for the purpose. If the surface undulate, the main drains must be laid in the hollows, and the feeders be distributed over the higher parts, and made to commu-nicate with the mains. Small sluices fixed at intervals,

Drainage of Marshes and Fen Lands, to the British Association for the Ad-vancement of Science, 1848, by Joseph Glynn, F.R.S., M. Inst. C.E.

both in the main and minor drains, will, by intercepting the water, permit an accumulation when desired for flood-

Fig. 24.

ing or irrigating the higher lands. Figs. 23 and 24 show a plan and section of a district of this character. A A is the river or receiving channel; B B the principal main drain; and C C and D D two other main drains delivering into it; each of the mains receiving the drainage from the feeders or minor drains. Fig. 24 is a section supposed to be taken on the line z z on the plan. Two imperative rules require to be observed in these arrangements, viz. that all the junctions shall be curved, and that no two feeders shall enter the main drain at opposite points. If these rules are neglected, the currents will be interrupted at these points, and mischief may arise from flooding when the drains become filled in wet seasons. It is also advisable, if the ground be of a loose texture, to guard the junctions with a few rough stones piled together in the form of a retaining wall; or, for greater permanence, concreted with lime and gravel, as shown in the plan and sections, figs. 25, 26, and 27, of which fig. 25 is a plan, fig. 26 a section through the ordinary drain taken on the line Y Y; and fig. 27 a section through the guard-walls, taken on the line X X.

102. If the general inclination of the surface of the district be considerable, it is often desirable to form catchwater drains, or series of drains at different elevations, communicating with each lower one successively by falls. By this method great facilities are obtained for regulating the management of the waters, so that any required quantity can be retained to compensate for seasons of drought; while, moreover, the falls are applicable as water-power, and may be used for a variety of purposes. Fig. 28 is a

Fig. 25.

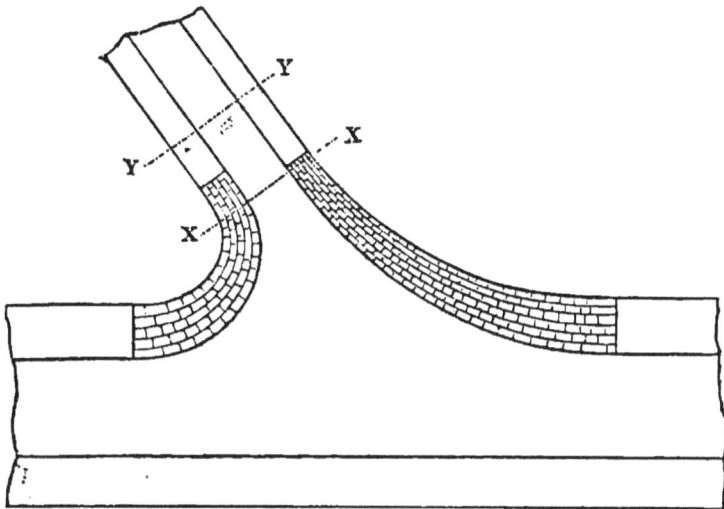

Fig. 26.

Fig. 27.

plan, and fig. 29 a section of a district drained in this
manner. A A, B B, and c c, are the main or catch-water
drains, each of which receives the drainage from the minor
drains or feeders connected with it, and delivers it to the
next lower main, through the channels *a a*, *b b*, and *c c*,
each of which has sluices fitted to it, while the water forms
a series of falls at the points marked *y*. Or the water from
the superior levels may be received in reservoirs constructed
for the purpose and in the places of the catch-water drains,
and there disposed of for agricultural, manufacturing, or
domestic purposes.

103. In fig. 11, p. 63, we have sketched an inland body

Fig. 28.

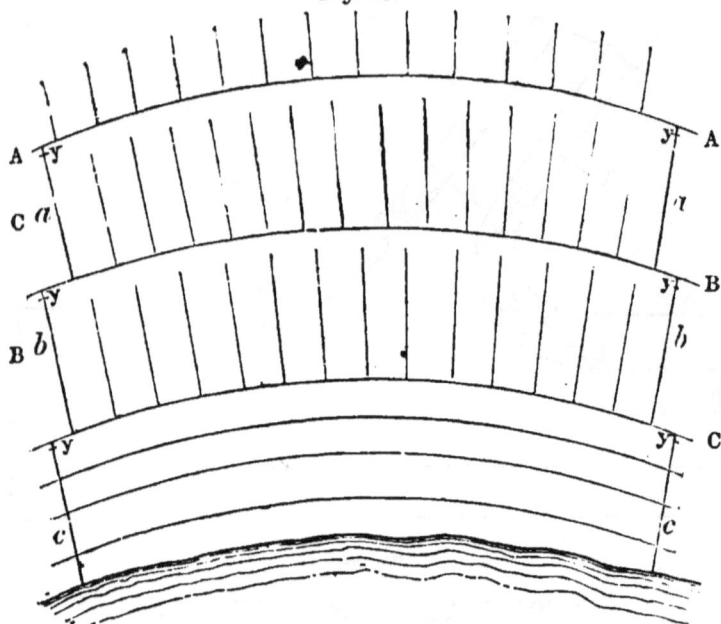

Fig. 29.

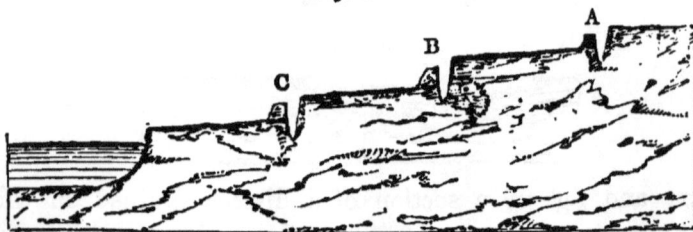

of water, or lake, which receives the drainage of the adjacent districts, and to these, thus situated, the same methods
of draining as those just described are generally applicable.
The formation of lakes upon the surface of our globe appears to have resulted from three causes, viz. the outcrop-
ping of internal springs or sources of water; subterranean
communication with seas; or, the flowing down and accumulation of the surface waters from the surrounding and
more elevated districts. Lakes formed by the first of these
causes being constantly fed and replenished, may be re-

garded as permanent reservoirs; and those formed by the second are dependant upon the preservation of their inlet from the ocean; but those which receive their supply from the drainage of the lands around, appear destined to extinction by the constant deposit within them of the solid matters brought down by the water. Thus, the Black Sea, the Caspian, and Arral, are fairly supposed to have originally formed one vast lake, the ridges in which have now become elevated, so as to form permanent boundaries between them. The Caspian, also, has evidently become reduced in extent, as proved by the marine matters now found at a distance from its present shores.

104. Fresh-water lakes, of considerable extent and little depth, are sometimes worthy of being entirely drained for the sake of cultivating the site they occupy. One of the most recent examples of this class of works is the drainage of the Lake of Haarlem in Holland. This lake is situated between Leyden and Amsterdam, and communicates with the Zuyderzee. The bottom of it consists of a rich alluvial deposit well fitted for agriculture. A Dutch engineer, popularly known by the name of " Leeghwater," or " drier up of water," formed a project for draining this lake in 1623, and another proposal for the same object was brought forward at the end of the last century, when steam was first employed in draining; similar works having been already executed in the Beilm and Diem. The area of the Lake of Haarlem is equal to 45,230 acres, and its average depth about 14 ft., the cubic contents being equal to 800,000,000 of tons of water. One part of the lake is 13 ft. under the level of the tide. The longest side of it is parallel to the sea, and separated from it only by a very narrow strip of land. Observations, continued during a period of 91 years, show that the maximum quantity of rain which falls upon the lake amounts to 36,000,000 tons of water monthly. The Dutch Government having appointed a commission of engineers to report upon the best means of draining the lake, many proposals were sub-

mitted and examined, and it was ultimately determined to adopt the plan recommended by Messrs. Gibbs and Dean. These gentlemen employed three engines for the purpose of draining the lake, each being of great power, whereby the total current cost was much less than would be incurred by using a greater number of smaller engines. These three engines are named the " Leeghwater," the " Cruquius," and the " Van Lynden," after three celebrated men of these names, who had interested themselves in the draining of the lake.

105. Of these three engines the " Leeghwater" was first erected, with suitable houses and pumping machinery. The first step in this work was to construct an earthen dam of a semicircular form, inclosing about $1\frac{1}{2}$ acre of the area of the lake, and adjoining its bank. The space inclosed by this dam was then cleared of water by a small steam engine, and the foundations for the houses and machinery commenced. These foundations consisted first of 1400 piles, which were driven to the depth of 40 ft., into a stratum of hard sand. Upon these piles, and at the depth of 21 ft. below the surface of the lake, a strong platform was laid, and upon this a wall, pierced with arches, was constructed, at the distance of 22 ft. from the intended position of the engine-house. Upon this wall a thick flooring of oak was laid, between the wall and the engine-house. The pumps rest upon the platform, beneath and opposite to the arches, and their heads pass through the floor just described, standing about 3 ft. above its level. Into the space left between the engine-house and outer wall, the water raised by the pumps was received and discharged from it on either side of the boiler-house, through sluice gates, into the canals conducting to the sea sluices. The general arrangement of the engine, boilers, pumps, and sluices, will be understood from Fig. 30, in which A represents the engine; B, the boiler-house; C C, the pumps; and, D D, the sluices through which the water was dis charged. The *engine* has two steam cylinders, one within

Fig. 30.

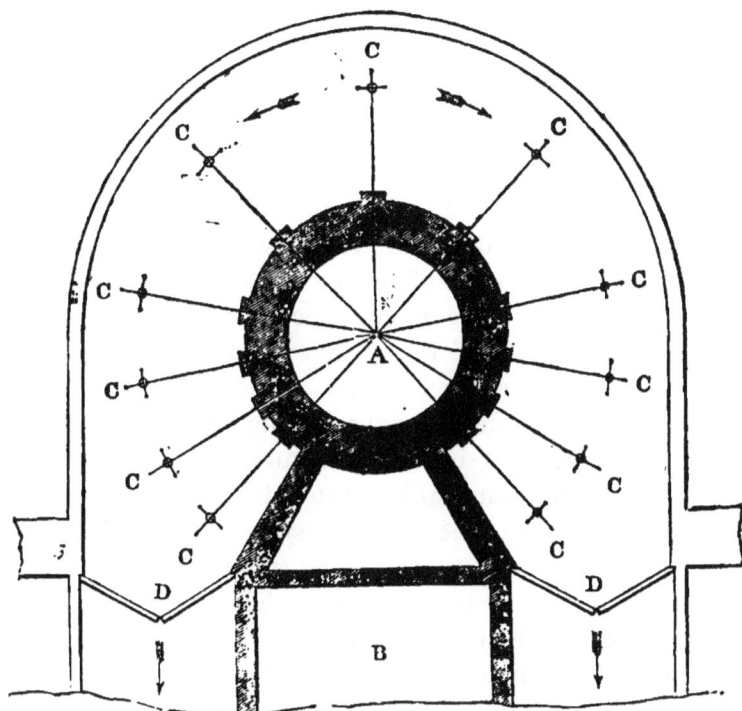

the other, united at the bottom, but with a clear space of
1½ in. between them at the top under the cover, which is
common to both. The large cylinder is 12 ft., and the
small one 7 ft. in diameter. The small cylinder is fitted
with a piston, and the large cylinder with an annular piston.
These pistons are connected by one main piston rod (of
the internal cylinder) 12 in. diameter, and four small rods
(of the annular piston) 4½ in. diameter each, with a great
cap or cross-head, having a circular body 9 ft. 6 in. diameter, and formed to receive the ends of the balance beams of
the pumps. The *pumps* are eleven in number, and each of
them 63 in. diameter, with a cast-iron balance beam turning
upon a centre in the wall of the engine-house, one end of
which is connected with the great cap of the engine, the
other to the pump rod. Each pump rod is of wrought

iron, 3 in. diameter, and 16 ft. long, with an additional
length of 14 ft. of patent chain cable attached to the pump
piston.   The steam and pump pistons have a *stroke* of 10
ft. in length ; each pump is calculated to deliver 6·02 tons
of water per stroke, or 68·22 tons for the eleven pumps.
The quantity actually raised was found to be about 63 tons.
The *action* of the engine is as follows :—The steam being
admitted, the piston and great cap are thereby raised, and
the pump pistons make their down stroke.   At the top of
the steam stroke a pause of one or two seconds is made, to
enable the valves of the pump pistons to fall out, so that,
on the down stroke of the steam piston, they may take
their load of water without shock.   In order to sustain the
great cap and its dead weight during this interval, an
hydraulic apparatus is brought into use, which consists of
vertical cylinders, into which water is admitted, forcing
upward two plunger poles which sustain the cap, the water
being prevented from returning by spherical valves fitted
at the lower part of the cylinders.   The arrangement of
the *two* steam cylinders is adopted in order to bring the
load under immediate command, the varying character of
which would otherwise require occasional alteration of the
dead weight to overcome it, which would involve great
delays and inconvenience. By the use of the two cylinders,
the dead weight raised by the small piston did not usually
exceed 85 tons, the extra power required being derived
from the pressure of the return steam at the down stroke
upon the annular piston.   A skilful regulation of the expansion and pressure of steam in the small cylinder thus
enables the engine-man to provide for all cases of difference
of resistance without the delay of altering the dead weight.
Respecting the *power* of the "Leeghwater," it appeared,
from experiments conducted by a sub-committee of the
Commission, that the engine would do a duty equal to
raising 75,000,000 lbs., one foot high, by the consumption of 94 lbs. of good Welsh coal, and exerting a net
effective force of 350-horse power.   The lift being 13 ft.,

the engine worked the eleven pumps simultaneously; the net weight of water lifted being 81·7 tons, and the discharge 63 tons per stroke. When the site of the lake is cultivated, the surface of the water in the drains will be kept at 18 in. below the general level of the bed; but during floods the waters of the upper level of the country will be raised above their usual height, and the lift and head will be increased to 17 ft. To test the power of the engine to meet these cases, the eleven pumps were worked simultaneously, without regard to economy of fuel, and 109 tons net of water were raised, per stroke, to the height of 10 ft. The *boilers* of the Leeghwater engine are five in number, cylindrical, and each 30 ft. long, and 6 ft. in diameter, with a central fire tube 4 ft. in diameter. Under the boilers a return flue passes to the front, and then divides along the sides. Over the boilers, and communicating with all of them, is a steam chamber, 42 ft. in length, and 4 ft. 6 in. in diameter; from which a steam pipe, 2 ft. in diameter, conveys the steam to the engine. The *consumption of fuel* is 2½ lbs. of coals per horse power per hour, when working with a net effect equal to the power of 350 horses. The *cost* of the " Leeghwater" and machinery was 21,000*l.*, and of the buildings and contingencies, 15,000*l.* It was calculated that the entire cost of the works for draining the lake would be 100,000*l.* less than would have been incurred by adopting the ordinary system of steam engines and hydraulic machinery, and 170,000*l.* less than the expense of applying the system of windmills hitherto prevailing in Dutch drainage. The *annual* cost of the three methods was thus estimated:—by three engines, such as the Leeghwater, 4500*l.*; by windmills, 6100*l.*; and by ordinary steam-engines, 10,000*l.*

106. The several methods of draining, as already explained in reference to figs. 9, 10, and 11, are also more or less applicable for districts of the kind sketched in fig. 12, and also for the second class or Upper Districts. Thus, the drainage from •the high lands has to be received and

collected in catch-water drains at the base of the hills, and means taken for combining these waters with those from the level district, or for keeping them separate, as may be required. Or reservoirs may be formed in connection with the catch-water drains, so that irrigation may not be necessarily suspended in cases of drought or deficiency of rain water.

107. Upland districts are liable (even with all the aid that can be rendered by economy of the natural supply) to suffer from an inadequate command of water. Thus, if, as shown in fig. 31, the surface of the district A A have a

Fig. 31.

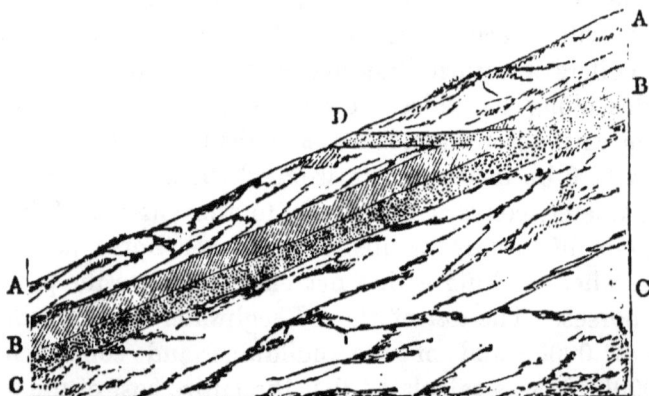

stratum of clay or other impervious material, B B, immediately beneath it, the outer stratum will remain always comparatively dry, the rain and drainage waters eagerly flowing downward, while the clay resists their passage into the subsoil. Beneath the resisting layer, however, a permeable and saturated soil, as c c, is often situated, and in these cases an adit drain at D, or other convenient point, will bring the internal water to the surface, and probably aid the supply of the district with the drainage waters from a higher and overcharged level. Internal springs are also, in some cases, available for this purpose, and may be brought into use by simple and inexpensive means. If these resources fail, it may become desirable to apply mechanical power for raising the necessary quantity of water from a river or other reservoir at a lower level.

108. Various forms of apparatus have been devised and applied for the purpose of raising water, some of which are actuated by the accumulated force of small streams from superior levels ; but these admit of very limited application for draining purposes. Pumping engines, worked by steam power, form the only class of machines at present available, by which the required accession of water can be, under all circumstances, brought up from the lower source. If the lower source, however, be a tidal river, the pumps may be worked by an undershot water-wheel placed upon it, and the water be delivered above into an artificial channel or aqueduct, and thence conducted to the higher levels.

109. A very extended and valuable experience of the powers of steam-pumping engines has been obtained in the mines of Cornwall, from the records of which a few facts may be usefully gleaned in this place as authentic data for application in many draining operations. The number of engines employed in these mines was, in 1822, 52, doing an average *duty* of 28,900,000. In the year 1843, 36 engines are reported, but their average *duty* had risen to 60,000,000. The *duty* is measured by the number of pounds weight of water raised 1 ft. high by the combustion of one bushel of coal. Thus, while in 1822, each bushel of coal raised less than 29,000,000 lbs. of water 1 ft. high, the same fuel was able, by improvements in the details of the engines, to raise 60,000,000 lbs. in 1843. The *best* engine in 1822 was a double cylinder one by Woolf, the *highest* duty of which was 47,200,000. The best engine in 1842 was a single cylinder (85 inches) engine, by Hocking and Loam, the highest duty of which was 107,500,000. This engine was erected in 1840, and was especially intended to work more expansively than had hitherto been practised. The boilers were made smaller in diameter than usual, and of stronger plate, so as to stand a

higher pressure of steam, the working elasticity being fixed at 40 lbs. per square inch above the atmosphere. Also an extra number of boilers was provided, in order to give an increased proportion of heating surface, and the strength of the working parts of the engine and machinery was augmented to withstand the strain caused by the great force of the steam on the piston at the commencement of the stroke. The progress of the application of the expansion principle has been intimately connected with the deepening of the shafts of mines. In order to render this principle effective in practice, to any great extent, it is necessary that a considerable load be moved by the engine-stroke. As the mines were deepened, the weight of the pump rods and balancing machinery necessary for draining them was of necessity increased ; thus furnishing the load required, and affording at once occasion and opportunity for gradually extending the improvement derivable from the principle of expansion.

110. Motive power may frequently be obtained from streams of drainage water collected or received from superior levels, and economically applicable to pumping and the actuating of mills and other agricultural machinery.*

* As an example of the adaptation of water power derivable from drainage to agricultural purposes, the arrangement adopted upon the estate of Lord Hatherton, in Staffordshire, may be aptly adduced. " His lordship has there had collected very cleverly the drainage water of the higher lands of his estate ; he has erected several ponds for storing it, and he has it carried to his farm-yard, where it drives a powerful water-wheel, which does all the thrashing, milling, chopping, &c., and drives a saw-mill besides. From the mill the water is carried in canals of gentle fall to lower meadow ground, where it is used in extensive and profitable irrigation. Drain-water always contains more or less of the manure and soluble parts of the soil in suspension ; and the fertilising properties of the drain-water on this estate are particularly marked by the very luxuriant growth of grass it produces on the meadows. This experiment forms a noble example of an economy in agriculture worthy of imitation, and is one which can be carried out to a greater or less extent on all farms having surfaces at different altitudes."— *Answer by the late James Smith, Esq., of Deanston, to Query No. 13, issued by the Metropolitan Sanitary Commissioners.*

Machines which derive their motive force from water are constituted mainly of a wheel or revolving lever apparatus, actuated either from the circumference or from the centre. In the former case, the wheel is usually made to revolve vertically upon an horizontal axis, and receives the impulse afforded by the weight and motion of the water at a level above the periphery of the wheel, or just below the axis, or identical with the lowest position which the periphery assumes in the course of its rotation. The wheels are distinguished in each of these arrangements respectively, as *overshot, breast,* and *undershot.* Water-wheels actuated from the centre derive their motion from the resistance offered by arms or vanes to the *centrifugal* disposition of the water, which thus *reacts* and produces a rotatory motion in the opposite direction. They are thus commonly known as "*reaction water-wheels,*" and in France have received the name of "*turbine,* or *horizontal water-wheel,*" from the position of the wheel, the axis being *vertical.* The celebrated French experimenters, Poncelet and Morin, have ascertained that overshot wheels and turbines produce an effect equal to from 60 to 80 per cent. of the power exerted; that breast-wheels produce from 45 to 50 per cent; and that undershot-wheels produce only from 27 to 30 per cent., being thus the least effective of all.

111. As the inventor of turbines, M. Fourneyron has attained considerable celebrity in France, and is reported to have realised an useful effect equal to 87 per cent. of the power expended. The proportion of effect to power is, however, not the only criterion of the usefulness or adaptability of these machines. Many circumstances are usually present which will dictate, or enable us to arrive at, a selection of such an apparatus as will be practically found to yield ample useful effect. Thus M. Fourneyron has produced a good average effect from a simple apparatus, with a *fall of water of only nine inches.* There are many places, especially in hilly districts, where high falls of water are found, and where the nature of the ground affords facilities

F

for making reservoirs, so as to insure a constant supply, where the height of the column of water may compensate for the smallness of its volume. And there are other situations where a great volume of water rolls with a very trifling fall. In either of these cases the turbine may be applied with great advantage. It, moreover, occupies a very small space in comparison with a water-wheel of the same power; its speed is high, and the expense of its construction greatly below that of any other effectual mechanism for deriving a rotatory motion from a head of water.

112. The turbine of M. Fourneyron consists of a horizontal water-wheel, in the centre of which the water enters; diverging from the centre in every direction, it enters all the buckets at once, and escapes at the circumference or external periphery of the wheel. The water acts on the buckets of the revolving wheel with a pressure in proportion to the vertical column or height of the fall; and it is led or directed into these buckets by stationary guide curves, placed upon and secured to a fixed platform within the circle of the revolving part of the machine. The efflux of the water is regulated by a hollow cylindrical sluice, to which a number of stops, acting simultaneously between the guide curves, are fixed. With this short cylinder, or hoop, they are all raised or lowered together by means of screws communicating with a regulator or governor, so that the opening of the sluice and stops may be increased or diminished in proportion as the velocity of the wheel may require to be accelerated or retarded. This cylindrical sluice alone might serve to regulate the efflux of the water, but the stops serve to steady and support the guide curves and prevent tremor.

113. One of these machines, erected by M. Fourneyron for M. Caron, was of 50-horse power, the fall of water being 4 ft. 3 in., and the useful effect varied with the head and the immersion of the turbine from 65 to 80 per cent. Another erected at Inval, near Gisors, for a fall of 6 ft. 6 in., the power being nearly 40 horses, expended 35 cubic

feet of water per second, and produced an useful effect of 71 per cent. of the force employed. One with a fall of 63 ft. gave 75 per cent.; and when it had the full height of column for which it was constructed, viz. 79 feet, its useful effect is said to have reached 87 per cent. of the power expended. Another, with 126 ft. fall, gave 81 per cent., and one with 144 ft. gave 80 per cent. In 1837, M. Fourneyron erected a turbine at St. Blasier, in the Black Forest of Baden, for a fall of 72 ft. The wheel is made of cast iron, with wrought-iron buckets; it is about 20 in. in diameter, and weighs about 105 lbs.; it is said to be equal to 56-horse power, and to give an useful effect equal to 70 or 75 per cent. of the water power expended.

114. The turbine is adapted, when applied to tidal waters, to work with one flow only; and to improve on this arrangement, and produce a continuous movement both with the rise and fall of the tide, is the object aimed at in Mr. Gwynne's "Double-Acting Balanced Pressure Wheel," which is said by the inventor to effect a saving of from 33 to 50 per cent. on the first cost (as compared, it is presumed, with the ordinary water-wheels), to produce an useful result equal to 85 per cent. of the power employed, and to maintain a perfect operation irrespective of floods or large accumulations of back water. This contrivance consists mainly of a flat cylindrical casing, with a vertical spindle passing through its centre, and carrying the internal wheel or arrangement of buckets which receive the impulse of the water entering at the periphery, the peculiar feature of the invention consisting in the shape of the partitions or buckets, which are adapted to present a direct surface to the action of the water in its passage through, whether it passes in one direction or the reverse of it.

115. Dr. Barker's mill, which was formerly neglected as being useless for practical purposes, is now recognised as involving important principles of action. It consists of a vertical tube, terminating in an open funnel at top, but closed at the lower end, from which project, at right angles'

two horizontal tubes in opposite directions, in communication with the vertical tube, and having closed outer ends. Each of these horizontal arms, however, has a round hole on one side of it (the two holes being opposite to each other), and the vertical tube being mounted on a spindle or axis is kept full of water flowing into the top. The issue of the water from the holes on opposite sides of the horizontal arms causes the machine to revolve rapidly on its axis, with a velocity nearly equal to that of the effluent water, and with a force proportionate to the hydrostatic pressure given by the vertical column, and to the area of the apertures; for there is no solid surface at the apertures to receive the lateral pressure, which acts with full force on the opposite side of the arm. According to the celebrated Dr. Robison, this unbalanced pressure is equal to the weight of a column, having the orifice for its base, and twice the depth under the surafce of the water in the trunk for its height. Desaguliers, Euler, John Bernoulli, and M. Mathen de la Cour, have treated of this machine, and the last-named author proposed (in 1775) an arrangement by which any fall or column of water, however great its height, may be rendered available. This proposition was to bring down a large pipe from an elevated reservoir, to bend the lower part of it upwards, and to introduce into it a short pipe with two arms, like Dr. Barker's mill reversed, and revolving on an upright spindle in the same manner; the joint of the two pipes being so contrived as to admit of a free circular motion without much loss of water.

116. In the year 1841, Mr. Whitelaw essayed an improvement of this machine, and obtained a patent for it This contrivance appears to consist mainly in the modifications suggested by Dr. Robison and M. Mathon de la Cour, and in the bending of the two horizontal arms into a form resembling that of the letter S. In this machine, the water is discharged from the ends of the arm in the direction of the circle described by their revolution, or in that of a tangent to it, the capacity of the arms increasing as they

approach the centre of rotation, so as to contain a quantity of water at every section of the arm inversely proportionate to its velocity at that section, with the view of economising the centrifugal force. The transverse sections of the arms are everywhere parallelograms of equal depth, but of width increasing from the jet at the outer extremity of the arm to the central vertical pipe. In a model of this form, with a fall of 10 ft., the diameter of the circle described by the ends of the arms being 15 in., and the aperture of each jet 2·4 in. in depth, by ·6 in. in width (the area of each orifice being thus 1·44 in.), the water expended was 38 cubic feet, the velocity 387 revolutions per minute, and the effect equal to 73·6 per cent. of the power employed.

117. Mr. J. S. Gwynne publicly exhibited, at the *Passaic Copper Mine*, U.S., in January, 1849, his "direct acting balanced pressure centrifugal pump," and obtained patents for the invention in the United States, 1850, and in England in March, 1851. The "balanced centrifugal pump," as described by the patentee, has a rotatory action, by which a centrifugal movement is given to the inclosed water, which it discharges in radial lines coincident with the direction of the centrifugal force, into a flattened spheroidal chamber, constituting the body of the pump, and having but one exit pipe, placed at a tangent with-its circumference. The water, as it is thrown off from the open periphery of the revolving piston, is forced up the discharge-pipe in quantities, and at a rate, proportioned to the speed at which the piston is driven. The piston is formed of two concave discs placed parallel, with their concave surfaces towards each other. Between these discs is a single arm, or impeller, radiating from a boss, or hollow axis, mounted on a shaft, which may work horizontally, vertically, or obliquely. The impeller varies in breadth; its narrowest part is at the outer edge of the piston, and it becomes gradually broader, until its edge intersects the inner surface of the opening in the suction side of the piston, from which line, to its extremity at the boss, its

edges continue parallel to each other, and at right angles
to the axis of the shaft. In working the pump, the water
is poured into the piston, at its centre, through a circular
opening in one of its sides and concentric with it. The
piston is inclosed in a case placed parallel and concen-
trically with the discs, and which acts as a receiver. From
the circumference of this case, and at a tangent to it, the
discharge-pipe rises perpendicularly. To prevent the water
rotating in the case, and to give it a direction upward to
the discharge-pipe, a stop or plate is provided. The joint
between the suction pipe and piston is carefully made, and
so situated, that no sand, gravel, or other gritty matter can
lodge in or near it. Mr. Gwynne also describes a "balancing nut," and claims that or any other contrivance for
"equalising the lateral pressure on the piston, which would
give rise to very serious inconveniences in the use of the
pump, when great elevations of water were to be obtained;
for, in raising it to great heights, the pressure would be excessive, amounting to many tons." As applicable to works
of drainage and irrigation, the patentee announces the
sizes, powers, and prices of his pumps as follows:—

Size of Pipes.		Gallons of Water raised 30 ft.	Equal to Horses' Power.	No. of revolutions per minute of Piston required to raise Water.				Price.
Dis-charge.	Suc-tion.			10 ft.	20 ft.	30 ft.	60 ft.	
in.	in.							£
6	7	1320	15	500	700	800	1200	40
9	10	3000	35	375	525	600	900	85
12	13	5310	60	250	350	400	600	200
18	20	12000	136	171½	240	275	412½	437
24	26	21000	240	125	175	200	300	750

118. In November, 1848, Mr. Appold exhibited a model of a
rotatory pump as a convenient one for draining purposes,
and made experiments on it with 6, 24, and 48 arms or vanes.
A pump of this description was shown at the Great Exhibition of 1851, and experimented upon by the jury. In
this pump the fan revolving vertically was 1 ft. diameter,
and 3 in. wide, having an opening one-half the total

diameter in the centre of each side for the admission of the water, and a central division plate extending to the circumference, to give a direction to the two streams of water, and convenient for fixing on the shaft; the 6 arms curved backwards, terminating nearly tangential to the circumference. The revolving fan was fixed on the end of the horizontal driving shaft, passing through a stuffing-box in the side of the casing, and it worked between two circular cheeks, running close without actually touching, by which the outer revolving surfaces were shielded from the water, but a free ingress allowed for the water, and a large space left all round the periphery of the fan, facilitated the discharge of the water. In the experiments instituted by the jury, the power employed was measured with great care by means of Morin's dynamometer, and the following results obtained :—

Appold's Centrifugal Pump, with Curved Arms.

Height of Lift.	Discharge per Minute.	Revolutions per Minute.	Velocity of circumference.	Percentage of effect to Power.
ft.	gallons.		ft. per minute.	
8·2	2100	828	2601	59
9·0	1664	620	1948	65
18·8	1164	792	2988	65
19·4	1236	788	2476	68
27·6	681	876	2751	46

With Straight Inclined Arms.

18·0	736	690	2168	43

With Straight Radial Arms.

18·0	474	720	2262	24

A large pump constructed on this plan erected at Whittlesea Mere, for the purpose of draining, was reported, in July, 1852, to have been then working for nearly a year with complete success. This pump is 4½ ft. diameter, with an average velocity of 90 revolutions, or 1250 ft. per minute, and is driven by a double cylinder steam engine, with steam 40 lbs. per inch, and vacuum 13¼ lbs. per inch; it raises about 15,000 gallons of water per minute, an average height of 4 or 5 ft. The cost of the engine and pump was about 1600*l*. The following experiments were tried to ascertain the percentage of effect obtained from the pump; the power employed being measured by taking indicator figures from the engine, deducting in each case the power that was indicated when the engine was working at the same speed without the pump, which was found to take 10·6-horse power. The quantity of water discharged was measured by calculating the overflow from an opening 6 ft. wide in each case.

	Experiment First.	Experiment Second.	Experiment Third	Experiment Fourth.
Velocity of circumference of pump, in feet, per minute...	1159	1357	1301	1329
Height of lift, in feet and inches	3 0	4 1	5 0	5 11
Depth of water at points of overflow, in feet and inches, A.	1 4	1 5½	1 3¾	1·2
Ditto, at 17 ft. distance, B. ...	1 7	1 8½	1 6¾	1·5
Gallons discharged per minute, according to the depth, A.	12,429	14,223	11,706	95,45
Ditto, ditto B.	16,104	18,023	15,288	13,606
Theoretical discharge	17,400	21,587	15,768	12,803
Horse power effective in raising the quantity, A.	11·34	16·88	17·79	17·17
Ditto ditto B.	14·70	22·38	23·24	24·49
Horse power employed in working the pump.	23·00	40·90	29·90	39·80
Percentage of effect to power employed, by calculation, A.	49	41	60	43
Ditto ditto B.	64	55	78	61

The centrifugal pump has been found more advantageous

for lifts below 20 ft. than for higher lifts, and its most advantageous application is as a tidal pump, where the height of lift is continually varying, because the lower the lift, the greater is the discharge, the speed of the pump remaining the same. This form of pump has also been applied with advantage as an assistant or feeder to a water-wheel, to keep the latter going constantly in the summer time, when short of water.

SECTION III.

Means of Conveying, Distributing, and Discharging Water.—Drains and Watercourses.—Forms, Sizes, and Methods of Construction.—Implements employed.—Shallow and Deep Draining.—Stone, Tile, and Earthenware Drains, &c.

119. Having in the preceding sections shown the general principles of draining, as applicable according to the general profile of the district, we have now to direct our attention to the details of the system; to show the methods to be selected with reference to the nature of the soil and the position of the substrata, and to consider the arrangement, form, size, and construction of the drains which it may be necessary to provide in order to promote the objects of agriculture.

120. The *nature* of the several soils which we have to deal with will be best understood by regarding the manner in which they have been formed, and the several materials of which they are constituted. The formation of all soils may be very clearly traced to the disintegration, by mechanical and chemical agencies, of rocks and minerals which contain alkalies and alkaline earths. In the mountainous districts of perpetual snow, the most refractory rocks are crumbled into fragments, which, being rounded by the action of glaciers, or pulverised into dust, are borne down by the rivers and streams, and deposited upon the plains and valleys below. Some of the most remarkable proofs of the influence of the air, water, and carbonic acid upon

F 3

the constituents of rocks, are exhibited in parts of South America, where the elements of the silver ores are gradually dissolved and dissipated by the action of these agents in the winds and rains, and the metal, resisting the destruction, is left exposed in sharp angular projections from the surface of the rock.*

121. The yellow clay which occurs so frequently in Denmark is considered by Forchammer to have been formed from granite, the felspar of which has undergone change, while the mica has not; the quartz forming the sand of the clay. The blue clays, having no mica, appear to have been formed from sienite and greenstone. The great stratum of clay which occurs at Halle has resulted from the disintegration of porphyry. Most of the sandstones contain silicates with alkaline bases, and in the sandstone of the Holy Mountain, near Heidelberg, fragments of felspar are observed partly changed into clay, and visible at white points in the sandstone. Felspar is unable to resist the solvent action of water when saturated with carbonic acid. Clays formed by the disintegration of felspars containing potash are free from lime; those formed from Labrador spar, which is the principal component of lava and basalt, contain lime and soda. Most rocks, as felspar, basalt, clay, slate, porphyry, and the numerous varieties of the limestone formation, consist of compounds of silica with alumina, lime, potash, soda, iron, and protoxide of manganese; and from the fact that most of these ingredients are susceptible of uniting with oxygen, the cause of the disintegration of the rocks which they constitute may be readily and fairly inferred. Of these constituents, the protoxide of iron has a great disposition to absorb oxygen from the atmosphere; thus forming the higher oxide or peroxide of iron. This property is indicated by the reddish brown colour of the rich ferruginous soils, while the black colour of the subsoil shows the presence of the protoxide. In the process of subsoil-ploughing, this protoxide frequently be-

* Darwin, Liebig, &c.

comes exposed, and the consequence is, that the fertility is impaired until the protoxide is converted into peroxide, and the red colour becomes apparent upon the surface. Struve has proved by experiments, that water which is impregnated with carbonic acid decomposes all rocks containing alkalies, and then dissolves a portion of the alkaline carbonates.

122. Soils being thus *formed* by the disintegration of rocks, their *properties* are evidently dependent, first, upon the nature of the several components of these rocks; and, secondly, upon the effects produced upon these components by the action of the air and water to which they are subsequently exposed. Of the various kinds of soil, the principal constituents are—1. *sand*, 2. *lime*, and 3. *clay*. The first two of these, containing no other inorganic substances except siliceous earth and carbonate or silicate of lime, afford no nutriment whatever for vegetation. The clay or argillaceous earth constitutes the fructifying element in all soils, and is produced by the disintegration of aluminous minerals, among which are the felspars, mica, &c. The fertilising properties of argillaceous earths appear to arise from their containing alkalies and alkaline earths, with sulphates and phosphates, ingredients which are never absent from these earths. This valuable property of the argillaceous earths is also aided by the peculiarity of their texture, which affords great facilities for retaining moisture. Vegetable life, however, requires, besides the nourishing properties found in the argillaceous earths, to be supplied with air and moisture, and while alumina gives no aid to the passage of these essentials, chalk and sand do give it by their mechanical formation. Hence, " land of the greatest fertility contains argillaceous earth and other disintegrated minerals, with chalk and sand in such a proportion as to give free access to air and moisture."* The *clays* are therefore to be regarded by the drainer as *impermeable* and *retentive* materials, and the *sands* and *limes* as *porous* ma-

* Liebig.

terials; and the infinitely varied proportions in which these
matters are found combined in soils, determine the degree
in which each soil will facilitate or impede the passage
of water through it.

123. With this knowledge of soils, we may proceed to
the *arrangement* of the drains required for regulating the
supply of water to the lands of a district. This arrangement will be varied according to the contour of the surface,
and the position of the substrata. If this be level, and the
texture of the soil uniform, the drains may be at once
planned, with mains at certain intervals, and minor drains
or feeders at right angles to the mains, and parallel among
themselves, as shown in fig. 32. The inclination in the

Fig. 32.

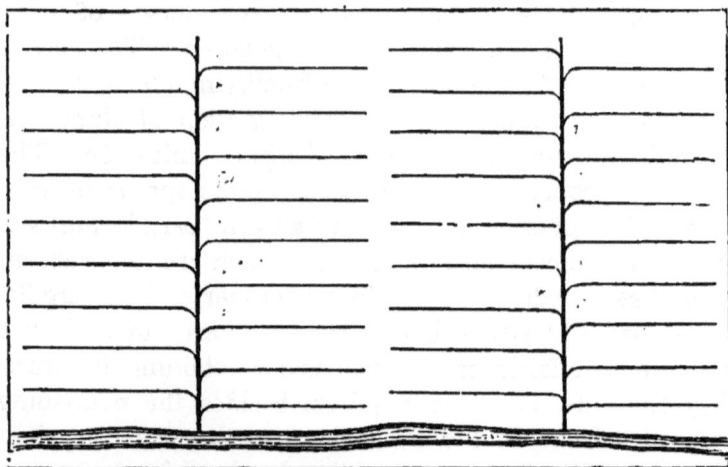

bed of the drains, which is necessary to assist the discharge
of their contents, must be obtained by cutting them deeper
towards the receiving channel, as shown in fig. 33, which
is a longitudinal section of one of the main drains, with
the feeders discharging into it. This increase of depth
also provides the additional capacity wanted in the drains
as the water accumulates in them. An undulating surface
will require the main drains to be arranged at the lowest

Fig. 33.

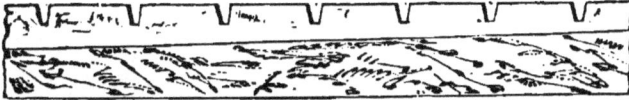

levels, and the minor channels conducted into them with
due reference to the capacity of the mains to discharge
their united contents.   Fig. 34 is a plan of a surface of

Fig. 34.

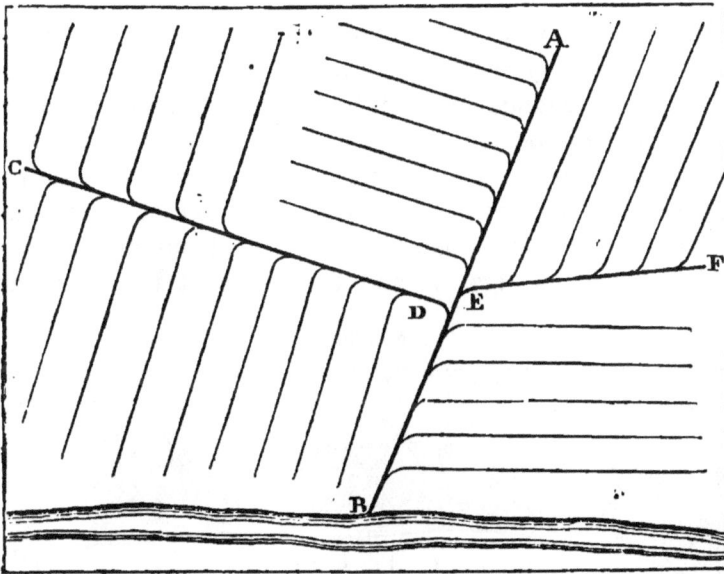

this kind, the lines A B, C D, and E F, showing the position
of the hollows in which the main drains are to be laid, the
minor ones being arranged so as to divide the total discharge among the mains as nearly equally as the nature of
the surface will conveniently admit.   The capacity of the
mains, as also of the feeders, must, of course, be determined according to the quantity of water for which passage
is required, and modified also by the steepness of the fall.
If the fall is considerable, a drain of smaller dimensions
will suffice than will be necessary if but little inclination

can be obtained. That part of the main drain from E to B will, in any case, require enlarged capacity, as it receives the entire drainage. Fig. 35 is a section of part of the

Fig. 35.

mains, in which it will be seen that, as the surface inclines, the bed of the drain will have sufficient fall if laid at equal depth from the surface of the ground throughout. The drains are arranged parallel to each other, which is evidently a good rule where it can be observed, the surface being thus divided into equal spaces, and the drainage made at once perfect and simple.

124. The arrangements of drains here described suppose the texture of the soil to be the same throughout the surface to be drained; but if soils of different retentive power appear upon the surface, it becomes essential to arrange the drains with reference to the line of junction of the soils. Thus, let figs. 36 and 37 represent the plan and

Fig. 36.

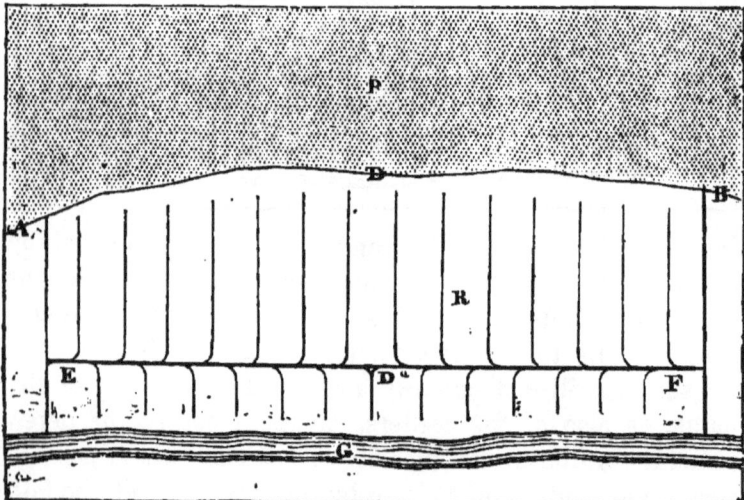

Fig. 37.

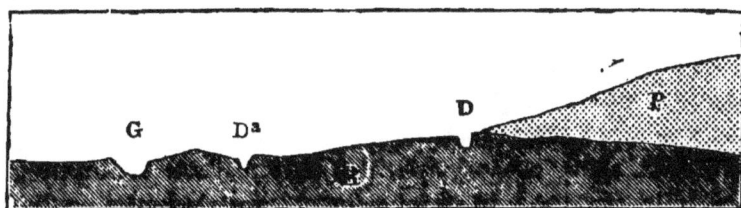

section of a district, whereof the higher ground is of porous materials, P overlying the clay, or other retentive soil (marked R), as far as the line A, D, B, where the clay first appears upon the surface. The water, percolating through the bed P, and prevented from descending by the clay, will accumulate along the line A, D, B, and form a swamp unless got rid of. In this case, therefore, a drain must be laid along this line, while a main drain for the remainder of the ground, which slopes towards the brook at G, must be formed on the line E, Dᵃ, F. This latter drain receives the contents of the minor drains, which are to be laid parallel, as shown in fig. 36. The upper drain, A, D, B, discharges into the brook G by the cross ducts A E and B F. The two main drains must, if the surface permit it, be laid with their beds dipping either way from the middle at D and Dᵃ, so as to insure the free passage of the drainage water.

125. The general arrangement of the drains being, as already stated, controlled by the superficial contour and texture of the soil, cannot be properly determined without a careful reference to the sectional strata of the district. If these consist of materials of various degrees of porosity, their relative positions, not only on the surface, wherever the substrata may outcrop, but also in the section, must be regarded. In this kind of consideration consists one great field in which so much improvement *has been* already effected, and so much more *may be*, in the practical art of land-draining. The history of the art, indeed, informs us that the Romans consulted the nature of the soil in selecting the form of their drains; that they provided for drain-

ing from springs and subterranean sources as well as from
the surface; and that they were thoroughly conversant with
the superiority of covered and deep drains in certain circumstances. Without attempting to pursue this history,
however, which is abundantly interesting, but far beyond
our space, we may recall to the minds of our readers the
fact that, some century ago, the only method of draining
generally practised in this country consisted of forming
the surface into rude ridges and furrows, and cutting open
trenches by the hedges to carry off some of the superabundant moisture. But, more than this, we are called
upon to attest the fearful truth, that our own little island
still contains thousands upon thousands of acres of land in
this same disgraceful condition, which are commonly con-
demned as *bad* lands, and regarded as evidences of the
misfortunes, instead of the ignorance, of their cultivators.
The modern art is not yet a century old. It appears to
date principally from the methods instituted in the year
1764 by one Mr. Joseph Elkington, a Warwickshire farmer,
who happening to drive an auger through the bed of a
trench, discovered the existence of a water-bearing stratum
beneath, by drawing the water from which, the surface and
supersoil became thoroughly drained. The late Mr. Smith,
of Deanston, and others, subsequently extended the principle of consulting the texture of the subsoils, and have
adapted the depth, capacity, and construction of drains to
these varieties of texture.*

* As early as the time of the Commonwealth, *deep draining* was advocated by Captain Walter Blith, who says—" Wherever you see drayning
and trenching, you shall rarely finde few or none of them wrought to the
bottome. But for these common and many trenches, ofttimes crooked too,
that men usually make in their boggy grounds, some one foot, some two, I
say away with them as a great piece of folly, lost labour and spoyle. And
for thy drayning trench it must be made so deepe that it go to the bottome
of the cold spewing moyst water that feeds the flagg and rush—a yard or 4
feet deepe if ever thou wilt drain to any purpose." And " to the bottome
where the spewing spring lyeth thou must go, and one spade's graft beneath,
how deepe soever it be, if thou wilt drain thy land to purpose."

126. Proceeding to examine the several varieties of the structure of soils which are met with, we propose to consider the strata under three leading characters, as exhibited in the following diagram, viz. the *porous*, or readily permeable; the *retentive*, or comparatively impervious; and the *semi-porous*, or mixed.

Characters of Strata.

Porous or Pervious, marked P in Sections.		Sand, Gravel, &c.
Retentive or Impervious, marked R in Sections.		Clays, Marl, Dense Rocks, &c.
Mixed or partly Pervious, marked M in Sections.		Loam, Soft Chalk, and Surface Soils, of mixed ingredients.

This diagram, therefore, will serve as an index to the several sequent illustrations, numbered from 38 to 51 inclusive.

127. In fig. 38 we have a common arrangement of strata,

Fig. 38.

the surface-soil (which for its thinness need not be regarded), on a porous stratum, and this lying upon a retentive one. The method to be 'adopted, in this case, will

partly depend upon the thickness of the several strata.
The water falling upon the surface will saturate the super-
soil, and, being impeded by the lower stratum, will not pass
away except by frequent drains, arranged with •regard to
the inclinations of the surface. If the clay beneath be of
considerable thickness, so that the average depth of its bed
from the surface exceeds 5 ft., it will be economical to limit
the depth of the drains to the upper stratum, and thus
avoid interference with the clay. The land, however, will
be brought into a drier condition by penetrating the clay
with the main drains at least. But, by doing so, if springs
exist below the clay, they may thus become exposed, and
the work of draining thereby augmented. Borings should,
therefore, be made along the lines of lowest surface, and
especially at any points where wet appears to gather, and,
if any springs are thus detected immediately below the
clay, they should be *tapped*, that is, have communications
opened to the drain, so that their contents may pass away
through the proper channels.

128. When the porous bed is beneath the clay or retentive bed, as shown in fig. 39, it will be advisable to cut the

Fig. 39.

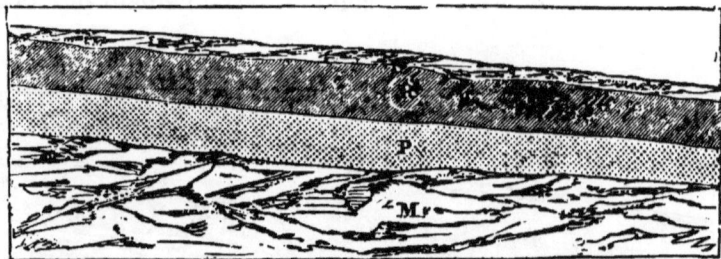

drains through the clay into the lower stratum, provided
the latter is of sufficient depth to bear the water, and assist
the drainage of the district. Supposing the entire depth of
the lower bed from the surface does not exceed 6 ft., and it
be desirable to economise the water for subsequent irrigation, or other purposes, it will be well to cut the drains

completely through both strata, and thus clear off the whole
of the subwater, as well as that from the immediate surface. This arrangement of strata requires the drains to be
laid at small intervals, and of ample capacity, as the dense
super-soil will keep all the water which falls upon it, and
also that which reaches it from superior levels. Some beds
of clay are of such thickness that no practicable drains can
be made to communicate with the substrata. When this
is the case, the general system of drainage should consist
of small drains laid very closely, so that the worked mass
of clay may become thoroughly freed from an excess of
water.

129. Fig. 40 represents a similar succession of strata to

Fig. 40.

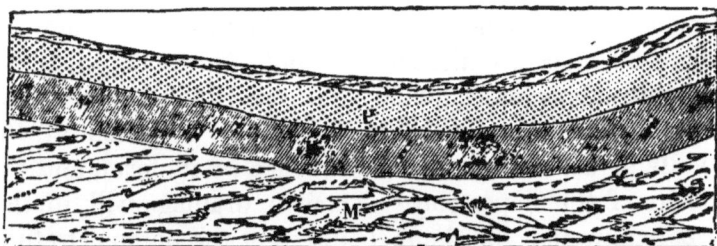

those shown in fig. 38, viz. the surface-soil resting upon a
porous bed, and this upon one of a retentive character.
But the contour of the surface is such, that the main drains
must be formed in the middle of the section, which is the
lowest part in the case here illustrated. The substratum
of clay, rejecting the water, assists its accumulation at the
point marked P, and it will become necessary to provide a
drain of large dimensions in proportion to the extent of the
district to be served. If the general surface is favourable,
it will be better to arrange the principal number and extent
of drains at right angles to the section, as here shown,
rather than parallel with it. The main central drain will
be thus relieved of the violence of the rapid discharges
which the steep drains would send into it, and less danger
will accrue from floods descending from the higher lands.

116 POSITIONS OF STRATA.

130. An arrangement of strata, which is very apt to discover springs rising to the surface, is shown in fig. 41, in

Fig. 41.

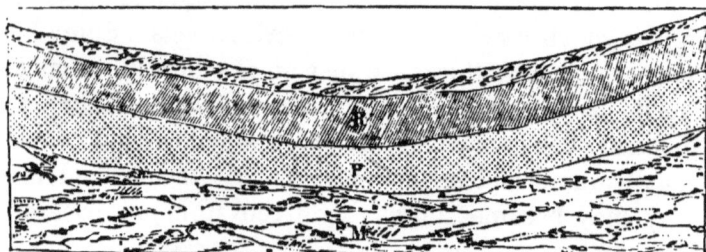

which the district appears to have a concave surface, with the porous stratum occupying the lower position. The water contained in the higher portions of this bed will burst forth at any outlets that may be formed through the clay; and indeed, if the latter be not of great depth, it will frequently force passages for itself, and thus augment the lower surface accumulations, which are collected at the middle of the section in consequence of its form, and the density of the upper clay. Efficient drainage will, in this case, require that the channels intersect the porous stratum, and, if the depth be not too great, the beds of the drains should reach that of the stratum. The position of the drains on the plan should also be determined, with a view to cut off the water from the gravelly bed at the higher parts of the section, and thus relieve the central main drain at R, which would otherwise become overloaded. Three main drains, therefore, or more, if the sides of the basin be of great extent, should be laid, viz. one at the middle of the section, and two at the higher part, on either side of, and parallel with it.

131. In fig. 42, the strata are represented in positions which produce swamps or morasses. Thus, at the point D, at the foot of a porous bed, lying upon one of clay, which rises from that point, the accumulation of water will require a main drain to be laid, bounding the base of the

Fig. 42.

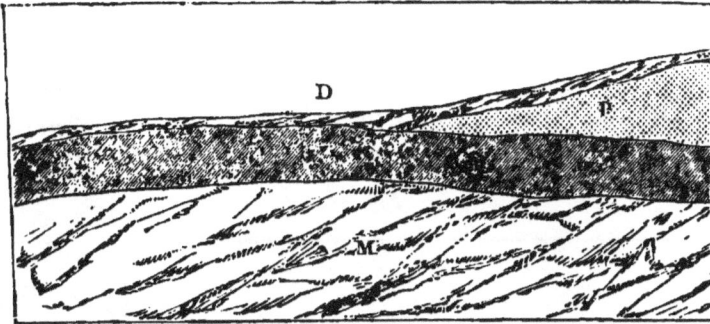

permeable stratum throughout the entire district, and to have a capacity in proportion to the extent of that stratum. If the clay be of inconsiderable thickness, the main drains should intersect it completely. In this arrangement it will ·be manifestly useless to cut channels *above* the point D, except as shallow feeders to the mains. This section illustrates one of the reasons of the failure of the methods formerly adopted of attempted drainage without consulting the structural condition of the soil.

132. Sometimes a tongue of gravel, or other pervious material, will be found to extend into and under the clay, as shown in fig. 43, in which a main drain at D, whatever

Fig. 43.

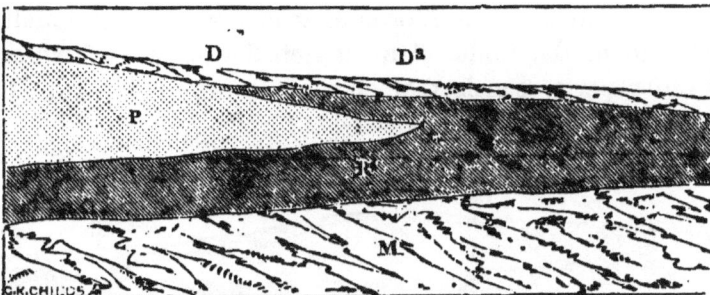

its dimensions may be, will not be sufficient to intercept the drainage water which passes through the bed, and will require another main at Dᵃ. In this case, indeed, the principal drain should be laid at this point; otherwise, that portion of the district lying between D and Dᵃ will remain in a moist and swampy state.

133. If, however, the position of the strata be reversed, and the clay runs into and beneath the porous material, as represented in fig. 44, the main drain at D should, if prac-

Fig. 44.

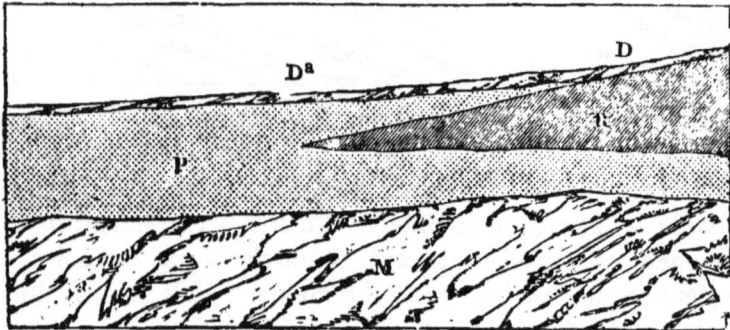

ticable, be cut through the clay, so that the water may be assisted in draining from it, and keeping the space from D to Dᵃ in a healthy condition. At the latter point, the depth of the main should be such as to reach the bed of the clay, and prevent the water running back towards the point R, in case the inclination has a tendency to produce that effect.

134. A patch of gravel or similar material is occasionally met with in the midst of a district, the surface of which, in other parts, consists of clay, as shown in fig. 45. In

Fig. 45.

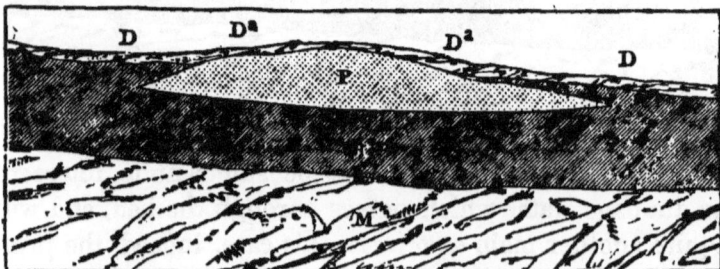

this case two sets of drains will be required, viz. at the points D D and Dᵃ Dᵃ; and the same remarks as to the relative depths of these mains will apply as already made in referring to fig. 43.

135. Fig. 46 shows a similar patch of clay running into

Fig. 46.

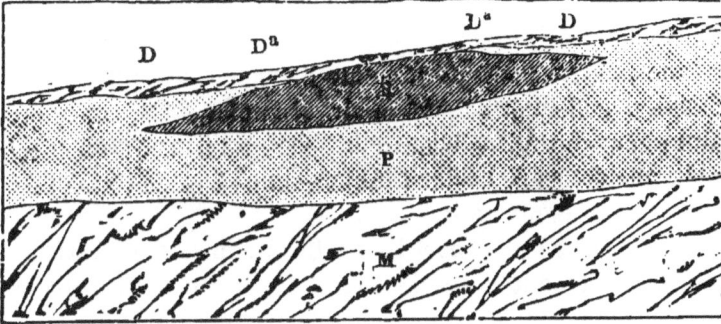

and under the gravel, requiring also two sets of main drains, which will be more effective in proportion to their depth, and the most so if they reach the bed of the clay, and thus prevent its injurious retention of the drainage water from the gravel or sand overlying its edges.

136. When the general surface of the district has a considerable inclination, as shown in figs. 47 and 48, the me-

Fig. 47.

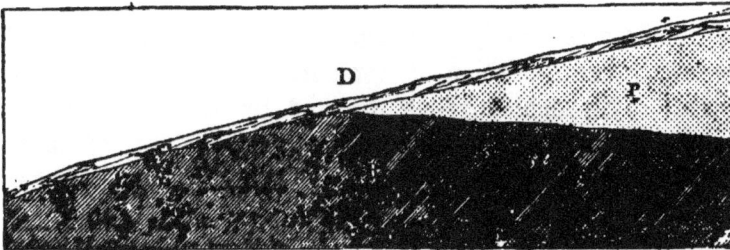

thods of drainage to be adopted will be varied according to the relative positions of the materials. Thus, if the porous material be above, as in fig. 47, the main drain should be

Fig. 48.

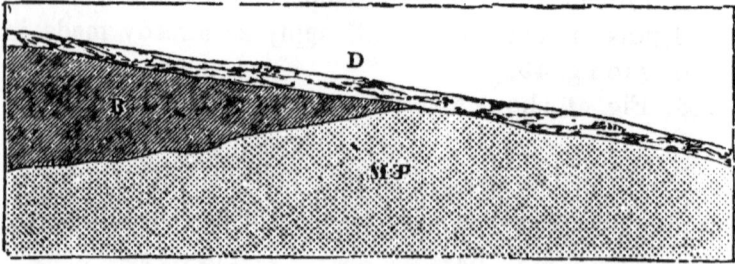

at the point D; but, if the clay lie upon a stratum of less density, as in fig. 48, the main should be laid at a lower situation, where it will naturally receive all the water which accumulates at D, besides that contained in so much of the lower bed as is above it.

137. If a bed of gravel lie in the hollow of a stratum of clay, as represented in fig. 49, the surface of the district

Fig. 49.

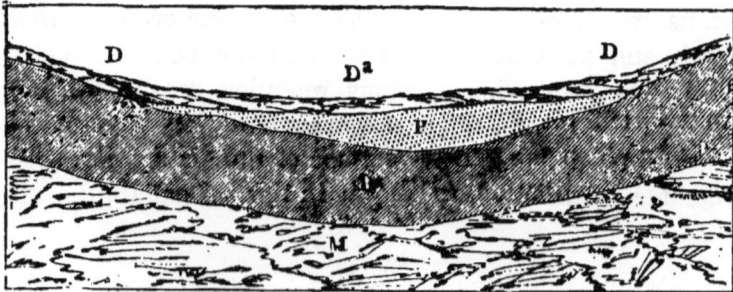

will remain tolerably dry except at the lowest point D^a, where the accumulation of water from the higher parts, resisted in its disposition to descend by the substratum of dense texture, will make a principal main drain of ample dimensions necessary. Auxiliary mains are also required at D D, to drain the clay surface above these points, and save the porous bed from the saturation which will naturally occur unless thus prevented.

138. In hilly districts, clays and gravel are often found in alternate layers, which outcrop on one side of the hill,

as sketched in fig. 50, and render a series of main drains necessary at the points marked D. By these drains, the

Fig. 50.

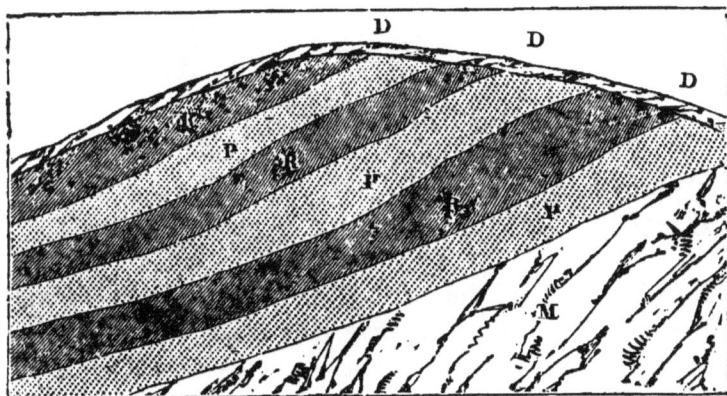

water which gathers in the retentive strata will be discharged at the lowest points on the surface, and prevent any mischievous excess on the soil. The intermediate portions of the porous materials which are exposed will readily get rid of their contents by percolation, and drains of comparatively small dimensions will be adequate to the efficient drainage of a section thus composed.

139. Sometimes the side of a hill displays a series of alternate and horizontal layers, as represented in fig. 51, in which case a small main drain should be laid at the exposed bed of each stratum, at the points marked D D, which will receive the contents of each porous layer, and prevent any injurious excess accumulating within the intermediate clays.

140. Having thus briefly noticed the several varieties of section which are likely to occur in the drainage of districts and lands, we have now to consider the form, size, and construction of drains which it will be advisable to adopt according to the circumstances of each case.

141. The rudest form of drain is that of an open cut or channel in the surface of the ground, for conveying the

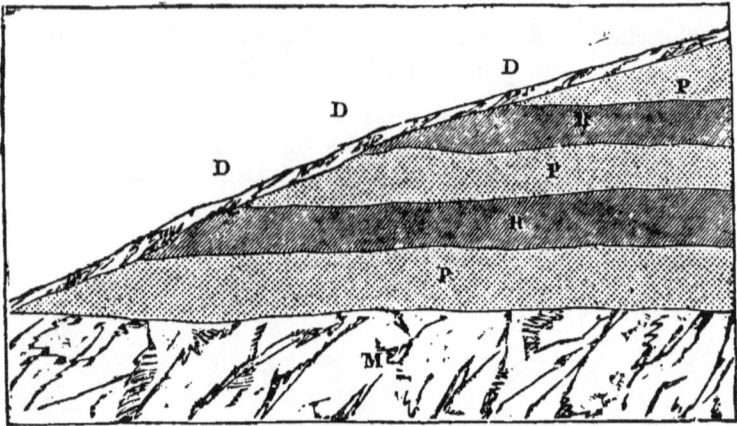

Fig. 51.

water which falls in the form of rain, or percolates through
the materials intersected, away into some lower position,
brook, or other receiver. These open drains are distinguished from the more complete form of underground or
covered drains formed by open channels, which are afterwards refilled, except at the lower part, along which a channel is preserved by one of several methods of construction.
Both of these methods appear to be of great antiquity,
having been certainly practised by the Romans, as recorded
by Palladius, Pliny, &c.

142. The arrangement and distances apart of open
drains have been usually determined by those of the ridges
and furrows. Previous to the introduction of "under-
draining," wet and strong lands were prepared for arable
culture by being ploughed up into the undulating shape
known as "ridge and furrow," the bottom of the furrow
forming a rude drain for the water from the adjoining
ridges. The wetness of the furrows or "thoroughs," as
sometimes called, and of the slips of land adjoining, however, occasioned the perishing of the crops, and led to the
adoption of shallow drains below the furrows, and commonly kept open with straw or brushwood. This was
termed "furrow" or "thorough" draining. In this manner

the ordinary width of the lands or ridges in each district indicates generally the distance at which the drains were placed, and the distances now most commonly observed in different districts, and on different soils, have reference to a width of ridge that was, or is, in use in those districts ; and it is "a fact worthy of remark, that throughout the country the statements of the number of feet from drain to drain is in almost every instance divisible (when reduced to inches) by *eighteen*, that being the space of ground in inches moved by a single turn of ordinary ploughing."* The long-established usages of each district may be regarded as indicating the requirements of that district, and the distance from furrow to furrow furnishes a kind of rude index of the comparative tenacity or porosity of the soil, or its capacity for retaining or transmitting water. The tabular statement, p. 124, (as prepared by Mr. Spooner,) illustrates the correspondence of the distances between ridges and drains with the character of the soil.

143. Open drains are applicable only as conductors of surface water, and for strong tenacious soils. To make them effective in draining from the body of the soil, the depth necessary renders open drains inadvisable; while, in loose soils, the inclination of the sides, which must be allowed in order to prevent their rapid destruction, occupies a most extravagant surface of the land. They are evidently inapplicable to land submitted to the plough, by which they are almost certain to be injured or destroyed, and thus have commonly been restricted to pasture land, whence they have been named *sheep-drains*. Even as thus limited, the use of open drains is of very doubtful advisability, inasmuch as they are always much exposed to injury, and to have their banks trodden down and destroyed. Admitting permanent utility as an object in drain-making, it is certain that covered drains should, in nearly all cases, both for arable and pasture districts, be preferred to open ones.

* Evidence of L. H. Spooner, Esq., of Balmacara House, Loch Alsh.

Width of Land or Ridge.		No. of turns of the Plough (18 in. wide) to the land.	Some of the Districts in which the respective widths of Ridge are in common use.	General character of the Soil.	Distance from Drain to Drain, in common use.
ft.	in.				
7	6	5	Common in the county of Essex.	Tenacious and uniform clay.	7 ft. 6 in., 15 ft., 21 ft., or every furrow, every other furrow, every third furrow, &c.
16	6	11	Parts of Surrey, Sussex, Kent, Middlesex, &c.	Same as above, fine and silthing clays, with beds of fine sand interspersed.	Drains 1 rod apart.
18	0	12	Parts of Yorkshire, Northumberland, South of Scotland, &c.	Clays, containing coarse sand and grit, interspersed with shale and slate fragments.	Drains 18 ft. or 1 rod (Scotch measure) apart.
21	0	14	Common in the above and the Midland Counties, &c.	Calcareous soils and clays, lighter than the above, with frequent intermixtures of sand and gravel.	Drains 21 ft. apart.
24	0	16	Very common in the Midland Counties and the Highlands.	Clays, similar to the above, with rotten sandstone rock and more frequent intermixtures of gravel, &c.	Drains 24 ft. apart.
30	0	20	Very generally adopted in the lighter clays throughout the country.	The lighter description of clays and clay gravels.	Drains 30 ft. apart.
33	0	22	Parts of Berkshire, Herts, Suffolk, Cambridgeshire, &c.	Chalk districts, stone, brush, gravelly, and sandy soils, and the lighter description of lands, usually springy soils.	Drains 33 ft. or rods apart.
36	0	24	Same as above, and very general.	......	Drains 36 ft. or rods (Scotch measure) apart.

The application of furrow drainage to the two last is comparatively of recent date.

144. For suburban and road drainage, the reasons for preferring covered to open drains have still greater force than those applicable to land drainage. These reasons are not only economical, but also sanitary. Open drains, presenting a commonly stagnant water surface to the atmosphere, produce an unwholesome evaporation. Decayed vegetable matter accumulates in these drains or ditches, and emits the most offensive. effluvia. Near the metropolis there are many large open watercourses, which serve to carry away flood waters, when such occur, but at other times the small quantity of water in these channels moves sluggishly over their rugged beds, or lodges in stagnant pools. These ditches sometimes serve as outfalls for the drainage of suburban houses, and the effluvium then becomes at times highly noxious and even fatal. The courses of these ditches were marked by excessive ravages of cholera among the adjoining population.

145. In carrying out land-drainage, the open roadside ditches are usually found to present most serious obstructions to the work; but if road-drainage were placed, as it should be, in proper subordination to the general system, covered tubular drains for the roads would of themselves effect considerable land-drainage, and in some districts closely intersected with by-ways and public footpaths, they would sometimes supersede the necessity for any other drainage. On a very stiff clay soil a road drain might, perhaps, not act more than from 12 to 15 ft. on either side of it, but in freer soils a single drain would frequently serve a width of from 1 to 2 chains. These road drains, properly constructed, would generally answer as excellent outfalls for the drainage of the land.

146. The extent of evaporating surface of stagnant moisture with decomposing vegetable and animal matter presented by the open ditches on both sides of a mile of road, equals from three quarters to one acre; that is, by the substitution of covered drains, three quarters to one acre would be gained as dry road, or cultivable land for each

mile of road, besides removing a frequent cause of accidents
with horses and vehicles.

147. Covered drains, being simply intersticial courses
formed beneath the surface, may be constructed in a great
variety of ways, which may be partly determined by the
proximity of the suitable materials. One of the simplest
forms, and most generally applicable, consists of a layer of
stones in the bed of the drain, which is afterwards filled up
with the soil taken out of it in order to deposit the stones,
as shown in fig. 52. In these drains there is a liability to

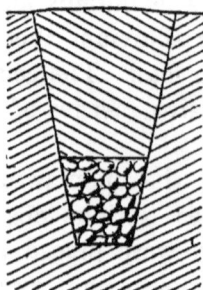 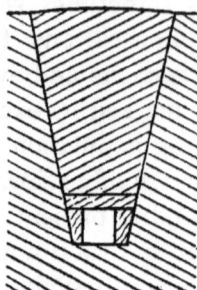 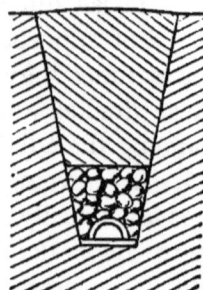

Fig. 52.   *Fig. 53.*   *Fig. 54.*

become less active, by particles of soil being forced down
or brought into the water, and clogging the spaces left for
its passage. If stratified stone is cheaply obtainable, the
better arrangement represented in fig. 53 should be
adopted, consisting of side stones, and one cover over them,
leaving an open space or duct through which the drainage
water passes of course more fluently than through the spaces
between the stones, as shown in fig. 52.

148. A compound drain, composed of a layer of loose
stones, and an artificial duct formed with a flat tile on the
bed of the drain, and covered with a semi-cylindrical tile, as
shown in fig. 54, combines the advantages of the two preceding drains. This form is commonly denominated the
sole and tile drain, and may, in most parts of the country,
be constructed at less cost than the stone duct shown in

fig. 53. It has also the advantage of greater permanency, being less liable to displacement of the parts. In the drain shown in fig. 53, the same arrangement of stones over the duct may of course be introduced; but unless the work is very carefully done, and the covering with the flat stones rendered perfect, the loose stones are liable to fall into the duct, and thus destroy its utility.

149. In clays and tenacious soils, drains such as that shown in fig. 55 are sometimes formed by cutting the lower

Fig. 55.   *Fig. 56.*   *Fig. 57.*

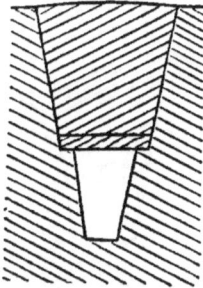 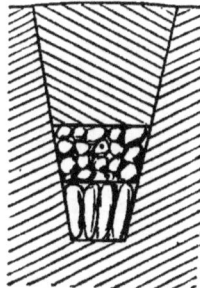 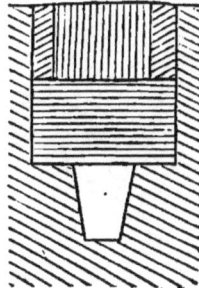

part narrower on each side, and thus leaving shoulders, on which a flat stone being supported, an open space is left below, forming a natural duct or open passage for the water. The permanence of this, *the shoulder drain*, is somewhat insecure, as it depends solely upon the shoulders being preserved, and the qualification of the material to resist all damage to the open parts of the drain. Another form of rough stone drain is represented in fig. 56, for which the larger stones are assorted, and placed in the bed with a layer of small stones upon them. It must be remarked of this, however, as of every rough stone drain, that its permanent action, depending upon the small spaces left between the stones, is very liable, in the course of time, to become much impaired or destroyed by the particles of soil and solid matters brought along with the drainage water; and on this account, especially, these drains are far inferior to those constructed with permanent open ducts.

150. Fig. 57 shows a drain suitable for bog and peaty soils, with which the drain is filled up, leaving an open space below for the passage of the water. The principal objection to this form of construction in peat is, that the effect of dry seasons is to contract the materials, which then get shifted by the superincumbent weight, and sometimes choke the watercourse below. For the purpose of protecting the water-way of drains, turf is occasionally placed over the stones, as shown in fig. 58, where the water-

Fig. 58.   Fig. 59.   Fig. 60.

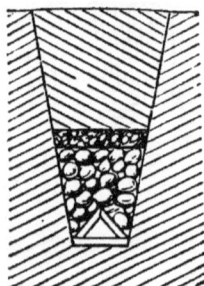 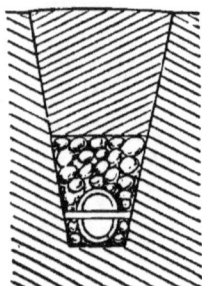 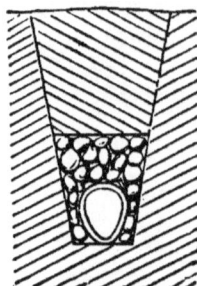

course is formed with three flat stones, or otherwise tiles, arranged as the sides of a triangle, and leaving an open duct between them. This duct is covered with a layer of loose stones, by which the stones forming the duct are kept in their places, and upon these stones a layer of turf is placed before filling the drain up with the soil.

151. Fig. 59 represents a compound drain, having two clear watercourses, and a layer of loose stones for intersticial drainage. The two courses are formed with two semi-cylindrical tiles, and a flat tile or sole between them. In all drains formed with soles and tiles, these are laid so that the joints in one break with those in the other; by which the joints are rendered less liable to be dislocated or disturbed than they would be if the joints in the soles and the tiles were laid coincident with each other.

152. The most complete and undoubtedly permanent form of drain is that which consists of an open channel

formed entirely of single pieces of tile-work or piping. These are now generally acknowledged to form the most superior drains; and, in nearly all places in this country, their cost will not much exceed that of the imperfect drains formed with loose stones. A drain of this construction is shown in fig. 60, where the earthen pipe is represented of an egg-shaped section, and a layer of loose stones placed above it. If drains be thus formed, the joints accurately laid, and the whole work carefully .done, the drainage will remain in a perfect and unimpaired condition for a very long period.

153. Drains are liable to injury by vermin, as well as vegetation, the roots of trees, &c., acting in a very injurious manner when their progress is interrupted by underground constructions for drainage. Drains should, therefore, be laid apart from trees, or these cleared away before con- structing the drains. The liability to injury by vermin is one feature in which pipe drains are superior to all others constructed of several parts, or depending partly upon the permanent position of the soil in which the drain is formed.

154. The form of construction being determined, the *size* of the drains is the next object of consideration. For- merly drains were commonly made of small depth. But deeper ones having been subsequently constructed, and con- siderable efficiency in the effect obtained, a great desire has arisen for *deep drains*. In arable districts one imperative condition as to the depth of drains is, that the lower and constructed part of the drain shall be below the action of the plough and other agricultural implements. The struc- ture, depth, and position of the strata are also circum- stances that will deserve regard in fixing the depth, as already explained at length in describing the several varie- ties of sections; and, besides these, another consideration, which must be kept in mind, is, the rate of fall which can be obtained, according to the levels of the surface of the district, to assist the discharge of the contents of the drain.

As a general principle, if it were impossible to allow all
these circumstances their due weight in arriving at a decision as to the depth of the drains, *deep* drains are doubtless more safe and likely to be efficient than *shallow* drains;
but while all the facts exist, and may be ascertained, by
which the depth should be regulated, it is mere blind prejudice which advocates *deep drains* in all cases and under
all circumstances.

155. On the *depth* of drains, the following observations
by the late Mr. Smith, of Deanston, are deserving of careful
consideration :—" Estimating the thorough drainage of
land by the cubic contents of the soil, reckoning from the
level of the bottom of the drainage to the surface of the
ground, can give no exposition of the agricultural effect,
because it has not yet been fully determined by experiment
or in practice how far it is beneficial to the growth of
plants to remove the *free* water from the lower regions of
the subsoil. One set of experiments over a course of three
years has been furnished by Mr. Hope, of Foreton Burn,
in East Lothian, from which it appears that the results
were in favour of moderate depths of drains; and the
practice in the Fens of Lincolnshire shows that the most
beneficial distance from the surface for the free water is
about 2 ft. In dry seasons, when the water in the level
ditches falls below 2 ft. from the surface, the crops are
found to suffer, and it is customary to dam up the water to
that level.* Water will rise some inches in soil by capillary
or molecular attraction; but in such cases the water never
fills the fissures or interstices of the soil to such an extent
as to exclude the atmospheric air, but merely attaches itself
to the surface of the particles of soil, and of the smaller
cells and channels in the soil, where it remains available
to the roots of plants, and without any of the bad effects
resulting from stagnant *free* water. Until the great point

* The expediency of this practice is questioned by J. A. Clarke, Esq., in
an elaborate report upon the farming of Lincolnshire. See Journal of the
Royal Agricultural Society, vol. xii. p. 326.

can be fully and practically determined as to the proper
distance for retaining a supply of water, the depth to which
land should be drained cannot be pronounced. The rule,
when ascertained, will probably be found to vary with the
nature and condition of the soil. In removing water falling on the surface, it has been found in practice, and which
agrees with a great theory, that having the artificial channels at near distances, and not over deep, is most effective
in the immediate and complete removal of the free surface
water. Distances of from 18 to 24 ft., with depths of from
2 ft. to 3 ft., have been found, over extensive tracts, and in
soils of various texture, to effect complete thorough-drain-
age for agricultural purposes."

156. As an advocate of deep draining, Mr. Elkington
must be named as connected with some very successful experiments in treating land, which astonished the good farmers of the last century, who had been accustomed to pay
very little attention to the improvement of their lands in
this manner, and had been satisfied to trust the aqueous
condition of their broad fields to the hedge-ditch and ridge-
furrow. Mr. Elkington, having a farm called Princethorp,
in the parish of Stretton-upon-Dunsmore, county of War-
wick, of which the soil was very poor, and so wet that the
sheep rotted by hundreds, turned his attention to the best
means of draining it. For this reason he began operations
in a field of wet clay soil, made nearly a swamp, and in
some parts a shaking bog by the springs of water which
issued from an adjoining bank of gravel and sand. Mr.
Johnstone, who published an account of Mr. Elkington's
"system," thus describes his proceedings:—"In order to
drain this field, he cut a trench about four or five feet deep
a little below the upper side of the bog, or where the wetness began to make its appearance; and after proceeding
with it so far in this direction, and at this depth, he found
it did not reach *the main body of subjacent water*, from
whence the evil proceeded. On discovering this, Mr.
Elkington was at a loss how to proceed. At this time,

while he was considering what was next to be done, one of his servants accidentally came to the field where the drain was making, with an iron crow or bar, which the farmers in that country use in making holes for fixing their sheep-hurdles. Mr. Elkington, having a suspicion that his drain was not deep enough, and a desire to know what kind of strata lay under the bottom of it, took the iron bar from the servant, and after having forced it down about four feet below the bottom of the trench, in pulling it out, to his astonishment, a great quantity of water burst up through the hole he had thus made, and ran down the drain. This at once led him to the knowledge of wetness being often pro-duced by water confined further below the surface of the ground than it was possible for the usual depth of drains to reach, and induced him to think of employing an auger, as a proper instrument in such cases."

157. These proceedings took place in the year 1764, and it is very evident, from Mr. Johnstone's account of Mr. Elkington's *discoveries*, and the principles upon which he conducted his draining operations, as distinguished from the methods then in common use, that these methods were adopted without any reference whatever to the leading cir-cumstances which properly regulate the steps to be taken. Thus the three leading points observed by Mr. Elkington, were, " 1st, finding out the *main spring*, or cause of the mischief;" "2nd, taking the level of that spring, and ascer-certaining *its subterraneous bearings;*" a measure never prac-tised by any, till Mr. Elkington discovered the advantage to be derived from it; "and 3rd, making use of the auger to reach or *tap* the spring, when the depth of the drain is not sufficient for that purpose."

158. The process of tapping is evidently available only when the spring is fed from a higher level, so that the pres-sure shall suffice to force the water upward through the auger-hole. Another method is sometimes adopted as a substitute for the auger-hole or vertical bore, namely, dig-ging a well of depth proportioned to the pressure, and

filling this well with loose stones, through which the water will rise, and thence pass away along the drain, with the bed of which the well communicates. Similar auger-holes, or wells, may be adopted to effect the precisely opposite object, viz. to make a downward passage of the drainage water from the drains which intersect an upper and clay stratum only, into a more porous bed beneath, in the body of which the water will become dispersed.

159. The cheapest method of forming open drains in grass land is by turning a furrow-slice over with the plough, and afterwards trimming it with the spade, the lines for the drains being previously marked with poles. The *cost* of these drains will not exceed one halfpenny per rood of six yards, which is the measure we shall adopt in all cases where the quantity is stated in *roods*. This mode of operating is inadvisable if the grass be rough and long, so that the plough is apt to become choked, or if swampy places occur. In such cases, it is far better to do all the work with the spade, by which the cost will be increased to 2*d*. per rood. if the drain be formed about 9 in. wide in the bed, 18 or 20 in. at top, and about 18 in. deep, which is a good size for the minor or sub-drains. Covered drains may be formed in grass-land 6 in. wide in the bed, 18 at top, and about 16 in. deep, at 4*d*. per rood, by cutting out the upper turf, the whole width across the cut, with the spade, casting out the lower portions subsequently, and then carefully replacing the turf, thus leaving an open space below, equal to the quantity cast out. The permanence of this drain is, however, very insecure; and, if cattle are admitted on the surface, they will certainly tread the turf to the bed of the drain, and thus destroy it. The first cost will, moreover, nearly equal that of a pipe drain, which needs a much narrower cut, and will remain permanently efficient.

160. In land to be planted with forest trees, open drains are always to be recommended, as covered ones are certain of destruction by the natural tendency of the roots of the trees to choke them in their search for moisture during dry

seasons. The main drains should be laid along the hollows in the surface, and made at least 3 ft. deep, with a flat bed 1 ft. wide, and the banks inclined at the rate of 1½ base to 1 perpendicular, except in firm clay soils, in which the banks may be formed much steeper. The minor drains should be for clays not less than 20 in. deep, and light soils 14 in., with a bed in both cases 9 in. in width, and the inclination of the banks regulated as for the mains. The cost of the former will be about 1½d. per rood, and the latter 3d. per rood. The cost of the mains will be in nearly the same proportion, according to the quantity of soil removed. The best distance at which to lay the minor drains from each other will vary in extreme cases from 5 to 40 yards, according to the levels of the site and character of the soil, the retentive clays requiring the drains closer than the lighter soils.

161. A general and *most important principle* as to the *capacity* of drains of all kinds whatsoever is, that it should *exceed* rather than *be deficient of* the dimensions ordinarily required to discharge the quantity of water for which provision is to be made. The principal use of a drain being to attract water towards it through the soil, besides passing the water thus collected away, its dimensions cannot be adequately estimated by simply considering the quantity to be conveyed within any given time. These dimensions should, therefore, be such as to present large surfaces of the soil intersected, and, other circumstances being the same, the efficiency of the drain will be in proportion to the extent of the surfaces, that is, to the depth of the drain. But, on the other hand, if the greater depth of the drain causes it to intersect porous strata overcharged with water from higher land, it will become injurious rather than beneficial, and this evil will be much aggravated if the greater depth be admitted as a reason for the proportionate infrequency of the drains. There can be no doubt that, in tenacious soils, shallow drains laid closely are, within certain limits, more useful than deep drains laid wide apart;

but, if contiguity can be observed, the deeper they are made the better, in ordinary cases.

162. Some reasons to guide the depth of drains may be derived from a consideration of the action of the soil upon the water which reaches it, as produced by its mechanical structure. Thus, in light and porous soils, the force or gravity is active in carrying the water to the bottom of the stratum; whereas, in the dense clays and soils, a certain capillary action is exercised upon the water introduced to them, which tends to raise it from the bed, and sustain it in general diffusion throughout the mass. Therefore, while porous soils evince little or no water on the surface, the lower part of the layer will be kept in a state of excessive wetness if it lies upon a clay bed; and, if its thickness be such that the roots of the vegetation reach the wet, the depth of the drains should at least equal that of the porous soil, so that the entire body may be relieved of the water. On soils of this nature, shallow drains are utterly useless, unless they happen to reach an impervious subsoil, and conduct the water into mains of greater depth.

163. In arable land, the minimum depth for covered drains may be estimated upon the depth to which the plough penetrates, and making such an allowance below this depth as will secure the materials of the drain from disturbance under any circumstances. Mr. Stephens calculates the depth of a furrow-slice with a two-horse plough at 7 in.; but in cross ploughing, 9 in. If four horses be used, the depth of the furrow will be 12 in.; and if the four-horse plough follow the common one, the depth will be increased to 16 in. Subsoil ploughing will penetrate 16 in. below the common furrow of 7 in. Allowing 3 in. between the lowest disturbed part of the soil and the surface of the materials in the drain, and restricting the effectiveness of the drain to that portion of it which is below the ploughed surface of 7 in. in depth, the minimum depth of drains should be such as to allow 19 in. below the furrow-slice, or 26 in. below the surface and above the constructed portion

of the drain, and so much more than this if subsoil ploughing be practised. Allowing 6 in. for the depth of the drain occupied by the pipes or tiles, Mr. Stephens estimates 33 in. as the minimum depth of drains in porous subsoils, and 50 in. in clay subsoils, with an additional 6 in. in each case if stones are employed as filling materials in the drain.

164. The size for the water-passage or duct of a drain should be determined by reference to a variety of circumstances, the combined influence of which may generally be estimated in practice, although not reducible to any very exact rules. Thus, the quantity of rain which falls upon the surface has to be considered, not as an annual or season quantity, but as a maximum per diem. Then, the nature of the soil and the state of the atmosphere, as affecting the ratio of evaporation, require attention. Beyond these considerations, the general level of the district in relation to the surrounding country, by which the tract to be drained may be made the recipient of foreign waters, on the one hand, or kept in a dry condition by the action of gravity, on the other, must be noticed. Again, the structure of the soil affects the quantity of the water which passes through it, and also the rapidity of its passage; and the amount of water to be met with will be modified by the part of the stratum at which the drain is situated. Thus, in porous materials, smaller ducts will suffice in the top of the layer than are required below; and the dimensions must be increased in proportion to the depth of soil above. As an auxiliary fact in enabling us to determine the capacity of the ducts of drains, the frequency of them upon the plan of the district will be greatly influential.

165. The best evidence on these points, viz. the dimensions and distance of drains, is to be gathered from the records of extended practice. In the weald-clay of Kent, which is commonly of a very tenacious character on the surface, but milder below, the body of the water naturally passes downwards until arrested by a more retentive stra-

tum, and, therefore, the deeper the drains the more efficiently they will act. In other parts of the weald, the soil is
compounded of the supersoil or cultivated earth and of a
strong clay, upon which it lies. This soil admits of percolation; but the tenacious clay beneath it does not, and, if
this clay be at a considerable depth from the surface, there
will be little utility in carrying the drains into it. In these
strong clays, not subject to springs, drains $2\frac{1}{2}$ ft. deep have
been found *more* efficacious than those made 4 ft. deep.
In the heavy lands of Norfolk, the drains which answer
best are $2\frac{1}{2}$ ft. deep, and laid at the distance of 22 ft. apart.
When they are made deeper, in clay in which flint and
chalk boulders are found dispersed about, the labour of
taking out the lower bed of 16 or 18 in. is very expensive,
costing in that county from 6 to 8 pence per rod of $5\frac{1}{2}$ yards.
In the clay-lands of Hampshire, the drains made from 30
to 36 in. in depth, and 18 to 24 ft. apart, have been found
most successful. We can readily understand that, as vegetation requires a certain amount of moisture, it is possible
to *drain* land so effectually that sufficient moisture is not
left to fulfil the purposes of cultivation, and the clay soils,
which are so reluctant both to receive and to discharge
water, will yet suffer a slow and sure deprivation through
the agency of deep drains, which will be injurious to the
health of vegetation; while drains of less depth would have
left the lower part of the stratum in a damp condition, and
capable, by the capillary action of the soil itself, of supplying the entire mass with a genial moisture. In Lincolnshire it is a known fact, that if the water in the ditches is
reduced to a level below 3 ft. in depth from the surface, the
grass-land is, in dry summers, most decidedly injured. In
the neighbourhood of Folkingham, a tract of clay land was,
several years ago, drained with tiles laid 3 ft. 6 in. deep,
and the surface, which was in broad bands with high ridges,
was levelled. After a short period, however, the texture of
the clay became so solid that the surface-water could not
get down to the drains, and it became necessary to alter the

method. On the same lands, drains now made 18 or 24 in.
deep are found entirely successful. In the neighbourhood
of Newcastle-on-Tyne, some clay lands have been drained
by drains laid $2\frac{1}{2}$ ft. deep and 20 ft. apart, with highly satisfactory results. In various parts of Scotland, the subsoils
of retentive clay have been more completely drained by
drains $2\frac{1}{2}$ ft. deep and 18 ft. apart, than by 4-ft. drains laid
36 ft. apart. In the counties of Worcester, Hereford, &c.,
the best drains in the clays are those laid from 2 to 3 ft. in
depth; those made 4 and 5 ft. deep being found far less
effective. Mr. Tebbet of Mansfield, near Nottingham, states
that the *best* way he has adopted on strong clay lands is
putting the drains 14 ft. apart and 2 ft. deep; while he
finds other clays that will draw at 18 to 24 ft. apart, and
2 to 3 ft. in depth for the drains.

166. A kind of average scale for the dimensions and distances of drains may be drawn from the experience we have
hitherto had in the draining of land. Classifying the varieties of soils into three divisions, as *Compact* or *Heavy*, *Medium*, and *Porous* or *Light*, each of which may be subdivided
into several degrees of retentiveness or porosity, the distance
of the drains apart may be graduated from 15 to 66 ft., and
their depth range from 2 ft. 6 in. to 4 ft. 6 in., as in the
Table, p. 139, which has been adopted by the General
Board of Health in their "Minutes of Information."

167. The *cost* of draining is necessarily a theme of deep
consideration in the execution of any plan which appears
likely to be most successful. The following records state
the size and distance of the drains, the nature of the soil,
and the total expense *per acre*.

The first eight of these cases are quoted from Mr. Smith's
(of Deanston) Pamphlet. They are instances of hard subsoils, with tiles and soles.

Nos. 9 to 16 are cited by Mr. Josiah Parkes, in the 6th
vol. of the "Journal of the Royal Agricultural Society," the
drain-pipes being supposed to be made upon the estate, and
costing 6s. per thousand.

A TABLE OF THE COST OF LAND-DRAINAGE PER ACRE.

THE differences in the quality of soils, that lead to differences in the depth and distance of the Drains, are also such as to affect the cost of digging the Drains. An increase of depth necessarily causes an increase of cost, from the mere circumstance of more earth having to be moved. But the same reason that causes Drains to be made closer, namely, the stiffness of the soil, renders them more difficult to dig, and hence increases the price of digging. This will explain how it happens, in the following Table, that the cost per rod is greater, not only as the depth increases, but as the distance of the Drains is less. Of two soils drained at the same depth, the expense of draining a rod will be least in that for which the Drains are furthest apart, which is where the soil is of the freest or least tenacious description.

Description of Soils.		Distance of Drains apart.	Depth of Drains.	Number of Yards of Drains per Acre.	Cost of cutting and filling per Chain.	Cost of cutting and filling per Acre.	Number of Drain Pipes of 12 in. long required per Acre.	Cost of Drain Tiles per Acre, at 30s. per 1000.	Total Cost per Acre.
		ft.	ft. in.		£ s. d.	£ s. d.		£ s. d.	£ s. d.
Compact or heavy Soils.	Compact tenacious gravelly Clay	15	2 6	968	0 1 8	3 13 4	2905	4 7 2	8 0 6
	Stiff adhesive Clay	16½	2 6	880	0 1 7	3 3 4	2640	3 19 2	7 2 6
	Friable Clay	18	2 9	807	0 1 6	2 15 1½	2420	3 12 7	6 7 8
	Free soft Clay	21	2 9	692	0 1 4	2 2 0	2076	3 2 3	5 4 3
Medium Soils.	Clayey Loam	22	3 0	660	0 1 8	2 10 0	1980	2 19 5	5 9 5
	Marly Loam	24	3 0	605	0 1 6	2 1 3	1814	2 14 5½	4 15 8
	Gravelly Loam	27	3 3	538	0 2 0	2 17 2	1613	2 8 4½	5 5 6½
	Friable Loam	30	3 3	484	0 2 0	2 4 0	1452	2 3 6½	4 7 6½
Porous or light Soils.	Light gravelly Loam	33	3 6	440	0 2 10	2 16 8	1320	1 19 7	4 16 3
	Light Marly Loam	36	3 9	403	0 2 8	2 9 4	1209	1 16 3	4 5 7
	Sandy Loam	39	4 0	373	0 2 6	1 19 8	1117	1 13 6	3 7 2
	Soft light Loam	42	4 0	346	0 2 4	1 16 9	1037	1 11 1½	3 7 10½
	Sandy Soil	45	4 3	325	0 2 4	1 16 5	974	1 9 2½	3 3 7¾
	Light gravelly Sand	49½	4 3	293	0 3 4	2 5 0	880	1 7 4½	3 12 10
	Deep do. do.	55	4 6	264	0 3 0	1 16 0	792	1 1 9	2 19 9
	Coarse gravelly do.	60	4 6	242	0 4 0	2 4 0	726	1 1 9	2 5 9
	Loose do.	66	4 6	220	0 3 4	1 13 4	660	0 19 9½	2 13 1½

The remaining cases are also given by Mr. Parkes, viz. in the "Gardener's Chronicle," the tiles being made upon the estate, and drawn by the tenants

No.	Soils.	Depth of Drains.	Distance between the Drains.	Cost of Labour per Acre.			Cost of Pipes or Tiles per Acre.			Total Cost per Acre.		
		Ft.	Ft.	£	s.	d.	£	s.	d.	£	s.	d.
1.	Clay	..	15	2	11	4	3	0	11½	5	12	3½
2.	Sandy clay	..	18	2	2	10½	2	10	9¾	4	13	8½
3.	Ditto	..	21	1	16	9	2	3	6½	4	0	3½
4.	Free stony subsoil	..	24	1	12	1	1	18	1½	3	10	2½
5.	Ditto	..	27	1	8	7	1	13	10¼	3	2	5¼
6.	Porous	..	30	1	5	8	1	10	6	2	16	2
7.	Ditto	..	33	1	3	4	1	7	8½	2	11	0½
8.	Sand or gravel	..	36	1	1	7	1	5	4	2	6	11
9.	Uniform clay	3	33	1	0	0	0	7	11	1	7	11
10.	Ditto	3	33	1	0	0	0	7	11	1	7	11
11.	Ditto	3 to 4	33	1	6	8	0	7	11	1	14	7
12.	Ditto	4½ to 4	40	1	2	0	0	6	6	1	8	6
13.	Clay, with some stones	4	50	1	6	6	0	5	3	1	11	9
14.	Clay. Hard gravelly subsoil	3 to 3½	49½	1	15	6	0	5	4	2	0	10
15.	Ditto	4	49½	1	15	6	0	5	4	2	0	10
16.	Various. Clay, gravel, sand	4	66	1	6	8	0	4	0	1	10	8
17.	Clay. Gravelly subsoil	3½ to 4	33	2	10	0	0	7	11	2	17	11
18.	Heavy clay	4	36		..			..		4	11	7
19.	Various clay	4	36		..			..		3	15	5
20.	Strong clay	4	30 to 33		..			..		4	4	2
21.	Strong land	4	39		..			..		4	15	7
22.	Weak blue clay	4	30		..			..		4	13	11
23.	Whitish stubborn clay	4	36		..			..		4	16	1
24.	Strong clay and gravel	4	33 to 36		..			..		5	3	4
25.	Whitish clay	4	36		..			..		4	4	8

168. The several items of cost of draining a rectangular field of 20 acres, with drains 3 ft. deep, and 22 ft. apart, may be averaged thus:—

	£	s.	d.
Main drain, 60 rods, at 8½*d.* per rod for cutting and filling	2	2	6
Drain-pipes, 990, at 40*s.* per thousand	1	19	7½
Minor drains, 2261½ rods, at 4½*d.* per rod	42	8	0
Drain-pipes, 37,320, at 30*s.* per thousand	55	19	7
	£102	9	8½

Equals 5*l.* 2*s.* 6*d.* per acre.

The main drains are supposed to be 3 ft. 6 in. deep, 20 in. wide at top, and 4 in. in the bed, with pipes 4 in. in diameter. The minor drains are supposed to be 3 ft. deep, 15 in. wide at top, and 3 in. in the bed, with pipes 3 in. in diameter.

169. The several items of cost of draining a similar rectangular field of similar soil, and prices for cutting and filling in proportion to the sectional area of the drains; the field being, as before, 20 acres in extent, with drains $4\frac{1}{2}$ ft. deep, and 45 ft. apart, may be estimated thus :—

	£	s.	d.
Main drain, 60 rods, at 1s. 2d. per rod for cutting and filling	3	10	0
Drain pipes, 990, at 40s. per thousand . .	1	19	$7\frac{1}{2}$
Minor drains, 1109 rods, at 10d. per rod for cutting and filling	46	4	2
Drain pipes, 18,300, at 30s. per thousand .	27	9	0
	£79	2	$9\frac{1}{2}$

Equals 3l. 19s. $1\frac{1}{2}d$. per acre.

The main drains are supposed to be 5 ft. deep, 24 in. wide at top, and 4 in. in the bed, with pipes 4 in. in diameter. The minor drains are supposed to be 4 ft. 6 in. deep, 21 in. wide at top, and 3 in. in the bed, with pipes 3 in. in diameter.

170. Mr. Smith gives estimates for drains constructed with reference to the nature of the soil, which may be arranged as in the Table given on p. 142.

Mr. Smith also mentions a district of 10,000 acres of stiff compact clay soil in Scotland, which has been satisfactorily drained with drains 2 ft. deep, and laid 20 ft. apart.

171. The principal circumstances which determine the cost of drainage works are—the labour of cutting and filling the drains; the material of which the drain itself is formed; and the outlets for the discharge of water. Of these, the last increases in proportion as the ground is steep and irregular, or unusually flat, and can only be included, in a general estimate, where the surface gently undulates; the material also varies greatly in cost, arising, in

Soils.	Depth of Drain.		Distance of Drains.	Cost per Acre.											
				Cutting and Filling.			Materials.—Tubes, at 20s. per thousand.			Chargeable for Mains, Outfalls, Superintendence, &c.			Total.		
	Ft.	In.	Ft.	£	s.	d.	£	s.	d.	£	s.	d.	£	s.	d.
Alluvial clay	3	0	21	2	1	10	2	2	0	0	18	0	5	1	10
Upland clay or till, full of stones.	2	9	21	3	2	9	2	2	0	1	2	0	6	6	9
Compact gravelly drift, with boulder stones.	2	9	24	3	4	0	1	16	$3\frac{1}{2}$	1	1	0	6	1	$3\frac{1}{2}$
Open sand and gravel, with moorish bottom.	4	0	40	3	6	0	1	1	$9\frac{1}{4}$	1	2	0	5	9	$9\frac{1}{4}$
Peat moss, forming its o channel.	3	6	18	1	4	5		..		0	6	0	1	10	5

the case of tiles, in the supply being near at hand, and equal to the demand, or otherwise; and, in the case of stones, in the distance of carriage.* The following table supposes

Description of Drains.	Depth of each Drain.		Width at top.	Width at bottom.	Average Width.	Running yards of Drain to the cubic yard.	Sandy Soils, light Loams, and light Clays; easy digging. At 4d. per cubic yard.				Stiffer Clay and Gravel, requiring some pickwork. At 6d. per cubic yard.				Hard Clay and close Soils, requiring pickwork. At 8d. per cubic yard.			
							Per yard.		Per rod.		Per yard.		Per rod.		Per yard.		Per rod.	
	ft.	in.	in.	in.	in.		s.	d.	s.	d.	s.	d.	s.	d.	s.	d.	s.	d.
Stone Drains, as in Fig. 56.	4	0	18	8	13	2+†	0	2	0	11	0	3	1	$4\frac{1}{2}$	0	4	1	10
	3	6	16	8	12	$2\frac{1}{2}$−	0	$1\frac{3}{5}$	0	9	0	$2\frac{2}{5}$	1	$1\frac{1}{4}$	0	$3\frac{1}{5}$	1	$5\frac{1}{2}$
	3	0	12	8	10	$3\frac{1}{2}$+	0	$1\frac{1}{7}$	0	$6\frac{1}{4}$	0	$1\frac{5}{7}$	0	$8\frac{1}{2}$	0	$2\frac{1}{3}$	1	$0\frac{1}{2}$
Pipe Tile Drains.	4	0	18	3	$10\frac{1}{2}$	$2\frac{1}{2}$+	0	$1\frac{3}{5}$	0	9	0	$2\frac{2}{5}$	1	$1\frac{1}{4}$	0	$3\frac{1}{5}$	1	$5\frac{1}{2}$
	3	6	16	3	$9\frac{1}{2}$	$3\frac{1}{4}$	0	$1\frac{3}{13}$	0	7	0	$1\frac{11}{13}$	0	$10\frac{1}{4}$	0	$2\frac{6}{13}$	1	2
	3	0	12	3	$7\frac{1}{2}$	$5\frac{1}{15}$	0	$0\frac{3}{4}$	0	$4\frac{1}{4}$	0	$1\frac{1}{5}$	0	$6\frac{1}{2}$	0	$1\frac{1}{2}$	0	$8\frac{1}{4}$

* See Mr. Spooner's evidence—"Minutes of Information," collected by the General Board of Health, 1852.

† The signs + and − imply a small fraction greater or less than the number stated.

In the price per rod, the fractional parts are reduced to the farthings nearest to them.

two sets of drains, the one opened for stones (as illustrated in fig. 56), the other for pipe-tiles, and at depths of 3 ft., 3½ ft., and 4 ft. respectively. The table shows the average width of cutting for each size and sort, and the number of lineal yards required to equal a solid yard.

172. Stones, as used for the filling of drains, are of two kinds, viz. the pebbly, or round stones, obtained from the sea-coast, or channels of inland streams, and the fragments produced by breaking up stratified or other rocks, and procured from the quarry. Of these, the former are much superior as the materials for drains, as they preserve the interstitial channels more permanently than the angular scraps from the quarry, the several projections of which are liable both to block up the spaces, and to be broken off by ramming, and thus interfere very mischievously with the passages for the water. As to the size of the stones, the standard commonly prescribed, namely, the "size of a goose's egg," is as good as any. At any rate, none should exceed 4 in. in diameter, or be less than 2 in. In all cases the stones should be assorted according to size, and used separately. Carelessness, in this respect, often leads to the complete choking of the drain, by the smaller stones filling up the spaces between the larger ones, and forming an impermeable dam across the drain.

173. Mr. Roberton, of Roxburghshire, who has paid much attention to the construction of rough stone drains, adopts these dimensions for them, viz. 33 in. deep, 7 in. wide at bottom, and 9 in. wide at the height of 15 in. from the bed of the drain, which is the space filled with stones in the manner shown in fig. 52, p. 126. Fifteen cubic feet of stones will fill this space in a rood of 6 running yards of such a drain. Mr. Stirling makes his drains of this description: 30 in. deep in the furrows, 5 in. wide in the bed, and 8 in. wide at a height of 15 in. from the bed. A rood of this drain will be filled to a depth of 15 in. by 12·3 cubic feet of stones. Inasmuch as the durability and efficiency

of these drains will be nearly in proportion to the space allotted to the stones, it is desirable, if the means will allow such an expense, to make the bed of the drain somewhat wider than here stated. Mr. Stephens prefers 9 in. width of bed, and 18 in. depth of stones, in a drain 36 in. deep.

174. The *cost* of Mr. Roberton's drains is thus stated by Mr. Stephens:—The drains being laid from 30 to 36 ft. apart, and the subsoil favourable to drainage. The averages of these distances gives 70 roods, of 6 yards each, of drains to the imperial acre.

	£	s.	d.
Opening drains 33 in. deep, and 7 in. wide at bottom, at 5½d. per rood of 6 yards, for 70 roods	1	12	1
Preparing stones 4 in. diameter, at 4d. per rood	1	3	4
Carriage of stones, at 4½d. per rood . .	1	6	3½
Unloading carts, and moving screen barrow, at ¾d. per rood	0	4	4½
Filling in earth, at ¼d. per rood . . .	0	1	5¼
Extra expense in the main drains . . .	0	10	0
Per acre of 70 roods . . .	4	17	6½
Or per rood of 6 yards . .	0	1	4¾

In another instance, the expenses were as follows:—

	£	s.	d.
Opening drains 28 in. deep, and 7 in. wide at bottom, at 4d. per rood of 6 yards, for 70 roods	1	3	4
Preparing stones, at 2½d. per rood . .	0	14	7
Carriage of stones, at 2¾d. per rood . .	0	16	0½
Carried forward	2	13	11½

	£	s.	d.
Brought forward	2	13	11½
Unloading carts, and moving screen barrow, at ½d. per rood			
	0	2	11
Filling in earth, at ¼d. per rood	0	1	5½
Extra expense in the main drains	0	10	0
Per acre of 70 roods	3	8	4
Or per rood of 6 yards	0	0	11¼

From these two instances, Mr. Stephens deduces 1s. 1d. as the average cost per rood, the average depth being 30½ in., and he has calculated a Table, which we give here, omitting the smaller fractions :—

Subsoils to which the Distances are applicable.	Distances between the Drains, in Feet.	Roods of Drains per Acre.	Cost per Acre.
Hard Till . . . . {	10	242	£13 2 2
	11	220	11 18 4
	12	202	10 18 6
Stiff Clay . . . . {	13	186	10 1 10
	14	173	9 6 10
Sandy Clay . . . . {	15	161	8 14 9
	16	151	8 3 10
	17	142	7 14 2
	18	135	7 5 11
	19	127	6 18 0
	20	121	6 11 1
Free and Stony . . . {	21	115	6 4 10
	22	110	5 19 2
	23	105	5 14 7
	24	101	5 9 5
	25	97	5 4 3
	26	93	5 0 10
	27	90	4 17 1
	28	86	4 13 7
	29	83	4 10 4

H

Subsoils to which the Distances are applicable.	Distances between the Drains, in Feet.	Roods of Drains per Acre.	Cost per Acre.
Open	30	81	£4 7 5
	31	78	4 4 7
	32	76	4 2 0
	33	73	3 19 5
	34	71	3 17 1
	35	68	3 14 11
	36	67	3 13 0
	37	65	3 10 10
Irregular Beds of Gravel or Sand, and irregular open Rooky Strata . . .	38	64	3 9 0
	39	63	3 7 2
	40.	61	3 5 6

175. In using tiles and soles, as shown in fig. 54, p. 126, the width of the drain in the bed will be determined by the breadth of the soles. For tiles the internal diameter of which is from 3 to 4 in., the soles are commonly 7 in. in breadth. The length of the tile and sole is of course the same; but this length varies in different localities. The Ainslie machine tiles are 15 in. long when burnt; those made by the Marquis of Tweeddale's machine are 14 in.; but the more common length is 12 in. Machine-made tiles are in all cases much superior to those made by hand, being more thoroughly compressed, and consequently more dense. Containing *more* clay, they are still *thinner* than hand-made tiles. The 15-in. tiles are less subject to displacement in the drain, and less handling is, of course, wanted to make up any given length. The angular junctions are best formed by semicircular notches in the sides of the tiles into which the ends of the others are fitted. Tiles for main drains are commonly 4 in. wide, and 5 in. high in the clear, and the soles 9 or 10 in. in width, the thickness of both being about ¾ths of an inch. The ordinary price of tiles may be taken as about 20s. per thousand, and of the soles, half that of the tiles, or 10s. per thousand. To

the cost of the tiles an addition has to be made, averaging from 30s. to 40s. per acre, (the drains being 15 ft apart,) for the expenses of carriage, loading and unloading, &c., which expenses do not appertain to the common loose stone drains as last described. With the addition of loose stones *over* the tiles, as shown in fig. 54, the cost will, of course, be considerably enhanced, as compared with that of common loose stone drains, nearly to the extent of the cost of the tiles, and the extra incidental expenses of from 30s. to 40s. per acre, the drains being 15 ft. apart. This will, however, form one of the most perfect, durable, and capacious drains that can be applied, and will command attention wherever these objects are sought with minor reference to cost.

176. Tiles in the form of complete tubes or pipes are, however, more perfect instruments of drainage than the separate soles and tiles, being less liable to derangement, and requiring only half the handling. Pipe-tiles, 2 in. in diameter, may be bought at the works at 18s. per thousand, and 15 in. in length; those 1½ in. in diameter at 15s., and 1 in. in diameter at 10s. per thousand. Mr. Stephens gives the following comparative summary of the cost, per acre, of the three kinds of drain, viz. stones, soles and tiles, and pipe tiles, the drains being 30 in. deep, and 15 ft. apart:—

	£	s.	d.
Loose stone drains	8	14	9
Sole and tile drains	7	10	8¾
Pipe-tile drains	5	8	9
Showing a saving by using			
Soles and tiles instead of stones, of . .	1	4	0¼
Pipe tiles instead of stones	3	6	0
Pipe tiles instead of soles and tiles . . .	2	1	11¾

Mr. Mechi states his expense of draining an acre of " strong clay land" (in Essex), the drains being 40 ft. apart, and average depth 4 ft., amounts to 2l. 9s. 6d. To this should be added about 1l. for the extra incidental expenses which his estimate does not include, and the total cost per acre

will be 3*l.* 9*s.* 6*d.* Without questioning the skill and experience of this gentleman, whose advocacy of thorough
draining merits due appreciation, we are satisfied that he
would find greater ultimate advantage by expending a little
more money in making his drains more frequent and less
in depth.

177. In comparing these several methods of forming
drains, it is perhaps scarcely necessary to state the truth,
that *first cost* does not afford an adequate test of efficiency
and permanent economy. To obtain this, subsequent current expenses, and the resulting condition of the soil for
agricultural purposes, must also be brought into the account,
and will show the benefits to be derived from the use of
pipe tiles still more forcibly.

178. The flat stone drains shown in figs. 53 and 55, are
economical forms of construction in some localities, viz.
where the required slabs are cheap and abundant. The
shoulders, shown in fig. 55 as being left in the soil, may be
in some cases better substituted by clearing the drain out
level, and introducing two slabs of stone, meeting at the
bottom, and having a stone wedged in between them at the
top, the upper portion being filled in with loose stones. If
the flat stones at the sides can be obtained 6 or 8 in. broad,
and 1½ in. thick, at about 4*d.* per ton, an acre of land may
be drained with drains 32 in. deep below the crowns of
the gathered-up ridges, and laid at distances of 15 ft., at a
cost of about 2*l.* 12*s.*, exclusive of carriage of materials,
and the ploughing by which the upper ridges are formed.
Constructed as in fig. 55, if the covering flags can be
procured 12 in. wide, and 2 in. thick, at the same rate
per ton, the acre may be similarly drained for 8*s.* or 9*s.*
less.

179. For the draining of bogs, the arrangement represented in fig. 57 is peculiarly applicable, all the materials
being on the spot, as the whole drain is refilled with the
peat itself, which is well known to resist the action of
water with impunity. Provided the cutting of the drains

be done in summer, when the material quickly becomes dried, and sufficient time is allowed between the successive operations of cutting and refilling to effect the required consolidation, no better method is yet devised for effecting this kind of work. The liability of the moss in the bed of the drain to subside in detached parts, and with great irregularity, is certain to destroy any arrangement of tiles, . pipes, or other artificial materials.

180. Another mode of forming drains with the natural materials, called *plug-draining*, has yet to be mentioned, rather to make our list complete than for any extended applicability of which it is susceptible. Plug drains are practicable only in subsoils of very firm clay entirely free from stones, and which never become thoroughly dry except by evaporation. Their use is further limited to lands, the occupation of which is that of permanent pasture. The usual form of open channel being formed in the clay, wooden blocks are introduced, which fit the lower part of the channel to a height of 8 or 10 in., and are convex or arched on the top. Upon these blocks, or *suters*, or *plugs*, the clay is returned and rammed down in a very careful and thorough manner. The plugs are then withdrawn, or drawn forward, for the formation of another length, and leave an open space or duct below for the passage of the water. In Gloucestershire, this kind of drain, 2 ft. deep, has been executed at from 4*d.* to 7½*d.* the rood of 6 yards, but the whole of the operations have to be conducted with extreme care, when the soil is free from frost or much wet, the ramming, moreover, requiring stout labourers, while much of pipe-draining may be performed by women and children, and the finished work is, after all, peculiarly liable, like all other drains formed with natural materials, to the destructive attacks of moles and other underground vermin.

181. These being the principal forms and constructions of land-drains, some notice of the successive operations to be carried on, and the implements to be employed, is re-

quired to complete our account of the draining of districts and lands.

182. The preliminary survey of the district to be drained having been made, and such indications of the nature of the soil as this survey affords noted, the next desideratum is, precise information as to the relative level or altitude, from a fixed datum, of every portion of the surface to be drained. These levels are obtained by the spirit-level in the usual manner, but they should be so complete as will enable us to lay down a *plan* of the district, with the *levels* of *equal altitude* marked upon it. These levels will appear as *lines* upon the map. Thus the highest point will be represented by a dot. The space around it, at one degree of altitude less, will appear as a continuous line encircling the dot. The space which has one degree of altitude less than this is represented by another encircling line around the preceding, and so on, down to the lowest altitude. These encircling lines, called *contour lines of equal altitude*, will, of course, be more or less irregular, according to the superficial character of the district. They were used in the French Survey, in 1818, having been suggested in an Essay read before the French Academy of Sciences so early as 1742, and since adopted for the Irish Survey in the year 1838. A precise idea of these lines may be formed, by supposing a block of stone, of an irregular conical form, to be immersed in a vessel of water. If the water be drawn off so as to lower its level equally, say one inch, at each successive discharge, the lines on which the water meets the surface of the cone will be the true contour lines of equal altitude. The *degree* of altitude to be adopted for each contour line will, of course, vary in amount, according to the prominence of the hills and the exactness required, being less in proportion as the district is flat or slightly undulating. From a map thus plotted, combining, as it does, a true plan and infinite sections of the district, any vertical section can at any time be accurately prepared without further reference to the ground,

while an immense advantage is obtained in the true indi-
cation of all brooks, and outcroppings of the substrata of
the soil.

183. The levels being ascertained and recorded, the observations necessary to complete the information upon
which the general arrangement of the drains can be determined have next to be made, by means of boring into and
examining the structure of the subsoil at various points of
the district. The tools, or *boring-irons*, employed for this
purpose, consist principally of the *auger*, by which cylindrical holes are made in the soil, and their contents brought
up to the surface ; the *punch*, with which compact gravel
and soft rock are perforated ready for the action of the
auger ; and the *chisel*, or *jumper*, with which, by the aid of
sledge-hammers, the necessary holes are made in hard
rocks which resist the auger and chisel. The auger is
from 2 to 4 in. in diameter, and about 15 in. long in the
shell. Its form is that of a cylinder, with a longitudinal
opening throughout its length, and a sharp cutting edge or
nose of steel secured to the entering end of it. The form
of the punch is that of a pyramid of four sides, from 8 to
12 in. in length, and from 2 to 4 in. square on the base,
which is the upper end of the tool when in use. The
chisel is a flat tool, with a broad and cutting end, and is
made of various sizes, according to the size of hole required.
Each of these tools has a screw formed at the upper end,
which will fit rods of iron made about 3 ft. long and 1 or
1¼ in. square, used to lengthen the apparatus when the
hole becomes deeper than the length of the cutting tool.
A cross handle of wood, fitted to a tapped socket, that fits
upon the tools and rods, is used for the purpose of turning
the augers.

184. When the nature of the subsoil has been ascertained by boring, and the arrangement of main and minor
drains determined, the work of forming the drains is commenced by marking the lines upon the surface with poles,
and driving stakes to indicate the intended position and

direction of the cuttings. A common garden line is then used to mark off the sides of the cuttings according to the width they are intended to have. The ground is then rutted with the spade along the line, and a rut is made at each side of the drain. From 20 to 30 yards are thus lined out at each stage. The different tools to be then employed will vary according to the structure of the subsoil to be removed.

185. For ordinary soils, such as clays, loams, and small gravel, and the usual combinations of these materials, the common spade, the ditcher's shovel, and foot and hand-picks, are the tools employed. The foot-pick has a cross handle of wood, and a tramp fixed at about 15 in. from the point of the tool, on which the foot of the workman is placed to drive it into the soil, and which forms a fulcrum on which stones or lumps of the hard dry clay are raised. The ditcher's shovel resembles a common spade, but is somewhat stronger, and rounded off on each side from the haft, forming a rounded point at the lower end. If the sides of the drain have a tendency to fall in before the work can be completed, they must be sustained by planks and short struts of timber wedged in between them. In order to test the correct dimensions of the drain, a gauge is useful, which consists of an upright stem, on which two or three cross-pieces are fitted to move up and down. The stem being graduated, these cross-pieces may be shifted and fixed so as to correspond with the intended limits of the cutting, and applied wherever thought necessary.

186. The uniformity of the fall of drains is tested with three staves, each consisting of an upright stem, and a cross-piece on the top fixed at right angles, so that the form of the instrument resembles that of the letter T. Two of these staves are made about 2 ft. in height, and the third one equal to 2 ft. added to the depth of the drain. The two shorter ones are held perpendicularly, one at each end of the drain, upon the surface of the ground, while the long one is shifted along the bed of the drain, and the

prominent or depressed points marked for correction. Besides these staves, a common mason's level, with a plumb line, is a very useful instrument for testing the inclination of the drains. Mr. Denton has designed a still more complete apparatus for this purpose, which he names the A level, from its resemblance to that letter in form, having a cross-shifting limb, by-which the inclination of one of the legs is adjusted, so that a plumb line from the apex indicates the intended fall, one of the legs having a sliding vernier, and the other having cross hairs for ranging the levels to the end of the drain, with a graduated staff.

187. In the formation of loose stone drains, after the · stones are assorted as to size, and deposited in the drains (for which purposes Mr. Roberton, of Roxburghshire, has introduced a very complete and portable form of harp or screen, provided with a movable tailboard), they are evenly distributed in the drain with a strong rake, and beaten down with a heavy beater, having a cross-handle for raising it. The best form of shovel for the stones is that known as the frying-pan, or lime shovel, having a raised rim around the hinder part of it, and being made of a capacious size.

188. The cutting of tile drains, requiring very exact work, is best performed with tools fitted peculiarly for it. Thus the narrow drain spade produces more perfect work in throwing out the loosened soil, and also trimming the sides of the drain, than the common spade. Pipe drains, being much narrower in the bed than any others, are best finished with small tools designed for the purpose. These are the narrowest drain spades, made only $3\frac{1}{2}$ or 4 in. wide at the mouth, and having a stud or tramp in front for the heel of the workman, to assist in pressing the tool into the soil. Narrow hoes are also useful in removing small stones, &c., from the bed; and for removing any wet and loose earth that may accumulate in it after the cutting of the drain, a scoop formed like a long narrow hoe with raised sides is very effective, being. in use, drawn forward by the workman.

The bed of these drains may also be very nicely finished and adjusted with a trowel, formed with the edges parallel, and rounded only near the front.

189. The objects to be attained in draining implements are strength, combined with lightness, and handiness; width, sufficient to remove at one cut all the soil required to be taken out of a drain at any one part of it; such a shape of the blade, that it shall, when lifted, raise the earth cut, leaving the least possible quantity of broken earth to be removed subsequently. Spades of various widths are needed for cutting the parts of a drain, so that it shall be finished 'with a regular tapering section with a top width sufficient to enable the drainer to work in it, and a bottom width only equal to receive the material to be deposited. Many implements have been introduced of various shapes, and some of them possessing considerable merit; but we have space here to notice only the two principal sorts of drain spades, which have been found in practice superior to all others for working in common soils, and of increased superiority in proportion to the difficulty which the material to be excavated presents to ordinary operations. Of these, the " Markly spade," introduced by Mr. Darby, represented

Fig. 61. *Fig.* 62. *Fig.* 63.

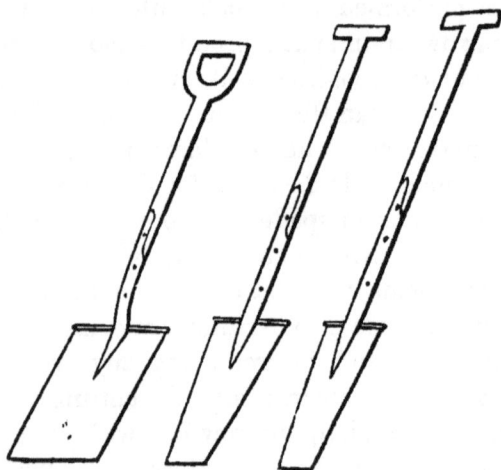

of different sizes in figs. 61, 62, and 63, is admirably formed for removing the lowest 16 or 18 in. of a drain, and has proved its superiority in " shape and make" as adopted for cutting through and lifting out a hard gravel or rocky bottom. Figs. 64, 65, and 66, represent a kind of spade ori-

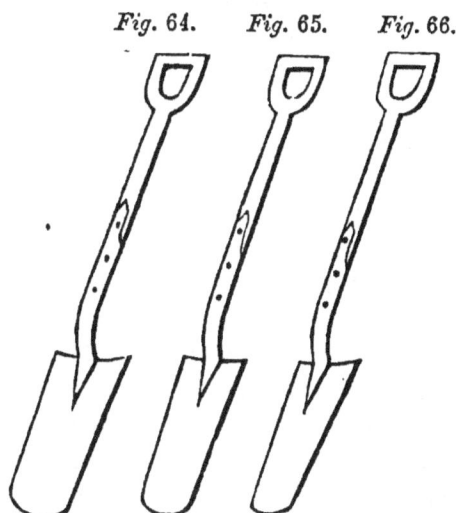

Fig. 64.   *Fig. 65.*   *Fig. 66.*

ginally used in Kent, and since introduced into Rosshire, and other parts of Scotland, where it has been found to improve the method of cutting the drains, and also to cheapen their formation to a material extent. With such implements as these, a drain 10 to 12 in. wide at top, 3 in. wide at bottom, and 36 to 40 in. deep, may be opened by a single cut from each of these sizes.* Drains 40 to 48 in. deep require a double cut in width and depth with the largest of the three spades.

190. Bog drains are formed with some tools of peculiar form, which may be noticed. Thus the surface turf is cut along the lines of the intended drain with an edging iron of a crescent-like form, and sharpened round its outer edge. For cutting the turf out, a broad-mouthed shovel is used.

* Mr. Spooner's evidence—" Minutes of Information," collected by the General Board of Health, 1852.

The moss, being cut into square peats with this shovel, is lifted from the drain with a three-pronged fork made for the purpose, or if footing be found for the workmen, the peats may be neatly cut from the bed, and thrown out with spades fitted with the handle at a large angle to the spade, which may thus be conveniently used in a horizontal position.

191. In soils where peat is plentiful, this material is sometimes cut or compressed into form, and baked so as to constitute open ducts when laid together in the drain. Mr. Calderwood, of Ayrshire, introduced, some years ago, a tool fitted to cut peats into a massive semi-cylindrical form by one cut, and without any waste of material, the hollow in one peat fitting the exterior of another. It is said that one man accustomed to the work can cut from 2000 to 3000 of these peats in a day. They are afterwards dried in the sun, and stacked till required. Several highly-ingenious machines have been invented for forming peat tiles, and also for moulding the pipe tiles of clay of various forms, to secure ready and accurate joints, by which improvements the cost of production has been within late years 'most materially reduced.

192. Some part of the work of cutting drains has, at various times, been attempted with ploughs of different forms and construction, fitted, wherever they are applicable, to effect some economy in the cost of labour for the work. Ploughs to facilitate the cutting of drains have been in use for many years in districts where alluvial clays prevail, and when used, they have been found to economise the cost of drainage considerably; but owing to the great number of horses (from 8 to 12) required to work them, and the difficulty first experienced by ordinary farm-servants in trying to manage so many horses and such large implements, the use of them has been heretofore much restricted.

193. An implement has, however, been introduced, which, as a draining plough, has far surpassed all previous efforts and which accomplishes the entire work of opening the

ground, depositing the pipes, and making good again in an admirable automatic style. This implement is the "draining plough" of Messrs. Fowler, Harris, and Taylor, of Temple Gate, Bristol, and will be found described and illustrated in the Second Part of the Rudimentary Treatise on Draining

INDEX.

---

www.ingramcontent.com/pod-product-compliance
Lightning Source LLC
Chambersburg PA
CBHW021348210326
41599CB00011B/793